Power Ultrasound in Electrochemistry

Power Ultrasound in Electrochemistry

From Versatile Laboratory Tool to Engineering Solution

Edited by

BRUNO G. POLLET

The University of Birmingham
Edgbaston, United Kingdom

WILEY

A John Wiley & Sons, Ltd., Publication

This edition first published 2012

Registered office
John Wiley & Sons Ltd, The Atrium, Southern Gate, Chichester, West Sussex, PO19 8SQ, United Kingdom

For details of our global editorial offices, for customer services and for information about how to apply for permission to reuse the copyright material in this book please see our website at www.wiley.com.

Library of Congress Cataloging-in-Publication Data

Power ultrasound in electrochemistry : from versatile laboratory tool to engineering solution sonoelectrochemistry / edited by Bruno G. Pollet.
p. cm.
Includes bibliographical references and index.
ISBN 978-0-470-97424-7 (cloth)
1. Sonochemistry. 2. Sonic waves–Industrial applications. 3. Electrochemistry. I. Pollet, Bruno G.
QD801.P69 2012
541′.370284–dc23

2011043940

A catalogue record for this book is available from the British Library.

HB ISBN: 9780470974247

Set in 10/12pt Times by Aptara Inc., New Delhi, India.
Printed and bound in Malaysia by Vivar Printing Sdn Bhd

1 2012

MIX
Paper from responsible sources
FSC® C013604

I dedicate this book to:

A wonderful Man and a brilliant Scientist –
Professor John Phil Lorimer

and

Mes parents et ma petite soeur que j'aime énormément

Contents

Foreword

When I think back to my first excursions into the world of ultrasound and its effects on chemical reactions it takes me back to 1975 when I obtained my first permanent academic post as an organic chemist at an institution that was then called Lanchester Polytechnic but later became Coventry University. The department I joined was Chemistry and Metallurgy, reflecting the applied nature of science courses at that time. One day I was walking through a metallurgy laboratory and saw an ultrasonic bath being used to clean metal samples. The process intrigued me for I could see that the ultrasonic bath was producing a large amount of energy as evidenced by the disturbance of the water with which it was filled. It occurred to me that this was perhaps a form of energy which might be employed to influence chemical reactivity using as an example a simple solvolysis reaction. However, the initial results were puzzling but I was sharing an office with a physical chemist, the late Phil Lorimer, but neither of us had heard of using ultrasound as a source of energy to promote chemical reactivity. Together we pursued this new subject and met many problems in convincing the UK science fraternity that we were 'on to something big'. We produced our first paper in 1980 as a *Chemical Communication*, in which we reported a small (twofold) enhancement in the hydrolysis rate of 2-chloro-2-methylpropane. By 1986 the idea of using ultrasound to influence reactions had greatly expanded worldwide and we were involved in organising the first ever international conference on sonochemistry at Warwick University.

So where does electrochemistry fit into the development of sonochemistry? Phil Lorimer was originally an electrochemist and so had an interest in all things that might influence electrochemical processes. Together with another colleague, David Walton, we began to apply ultrasound to electrochemistry in the late 1980s and discovered that it could, for example, modify the electrochemical oxidation mechanism of cyclohexanecarboxylate. In 1990 we published a review using the term 'Sonoelectrochemistry' for the first time in a peer-reviewed journal. This review forced us to look at the literature surrounding the uses of ultrasound in electrochemistry and brought to light a number of research publications that had not previously been drawn together. Other sources have been unearthed since then, including the pioneering work of Young and Kersten in 1936 on the effects of ultrasonic radiation on electrodeposits. This was perhaps the first observation of improvements in hardness and brightness induced by ultrasound. Walker reinvestigated and advanced the work in the 1970s and in 1993 wrote a comprehensive review of his and other work entitled 'Ultrasonic Agitation in Metal Finishing'. It is surprising to me that the very early work on sonoelectrochemistry has not been cited extensively. This is the case with the 1953 paper of Yeager and Hovorka entitled 'Ultrasonic Waves and Electrochemistry'. It provided a survey of the electrochemical applications of ultrasonic waves that were discussed in terms of (i) the effects of ultrasonic waves on electrode processes, (ii) electrokinetic phenomena involving ultrasonic waves, and (iii) the use of ultrasonic waves as a tool in the study of the structure of electrolytic solutions.

Our own work in sonoelectrochemistry progressed at a pace and in 1995 we took on one of our own young and bright chemistry graduates to study for a PhD. He obtained his doctorate three years later with a thesis entitled 'The Effect of Ultrasound upon Electrochemical Processes' and his name was Bruno G. Pollet – the editor of this book. Bruno became more and more involved in the work on sonoelectrochemistry and looked at a range of topics, including the effect of ultrasonic frequency and power upon electrochemical systems, from the theory and modelling to 'real' industrial applications.

The group of authors that Bruno has assembled for this book have been able to cover many of the main areas of sonoelectrochemistry with contributions on fundamentals, analysis, organic synthesis, nanoparticles, polymerisation and much more. I recommend this book to you as a compendium of current thoughts and approaches to sonoelectrochemistry written by experts in the field.

Tim Mason
May 2011

About the Editor

Bruno recently joined The University of Birmingham from Industry. He is an expert in the area of Proton Exchange Membrane Fuel Cell (PEMFC), Electrochemical Engineering and Sonoelectrochemistry (www.sonoelectrochemistry.com). He is currently responsible for the Hydrogen Fuel Cell Vehicle projects and PEMFC and Membrane Electrode Assembly (MEA) activities at the university. He is CEO of West Midlands Fuel Cells Ltd, CTO of H2Tech (Hawaii), Head of the Birmingham PEM Fuel Cell Research Group, Associate Director of The Centre for Hydrogen and Fuel Cell Research, Visiting Professor Fuel Cell Nanomaterials Center at The University of Yamanashi (Japan), Operations and Delivery Director of the UKRC Doctoral Training Centre in Hydrogen, Fuel Cells and their Applications and Programme Director of the MRes and PhD with Integrated Studies in Hydrogen, Fuel Cells and their Applications. He has worked for Samuel Banner Ltd (Banner Chemicals Group) and Albion Chemicals Ltd (now Brenntag (UK) Ltd) in Sales and Marketing, Johnson Matthey Fuel Cells Ltd (Johnson Matthey Plc) as Test Facility Scientist, Membrane Electrode Assembly (MEA) Design Scientist and Programme Leader, SmartWater Europe Ltd as Research Manager, and Coventry University as Head of Sonoelectrochemistry, Project Development Manager and Lecturer in Environmental and Physical Sciences. He has also worked as an EPSRC and EU Research Fellow in the field of Fuel Cells and Electrochemiluminescence at the Liverpool Electrochemistry Group. Bruno was awarded an Engineering Diploma in Chemistry and Material Sciences from the Université Joseph Fourier (Grenoble, France), a BSc (Hons) in Applied Chemistry from Coventry University and an MSc in Analytical Chemistry from The University of Aberdeen. He also gained his PhD in Physical Chemistry in the field of Electrochemistry (Sonochemistry and Sonoelectrochemistry) under the supervision of Professor J. Phil Lorimer at Coventry University. Bruno has published several publications and chapters in the field of Fuel Cells, Sonoelectochemistry and Sonochemistry.

He is a Fellow of The Royal Society of Chemistry (FRSC) and Member of the:

- International Society of Electrochemistry (ISE)
- Electrochemical Society (ECS)
- International Journal of Hydrogen Energy (IJHE) Editorial Board
- Journal of Nano Energy and Power Research (JNEPR) Editorial Board
- Member of The Society of Chemical Industry (SCI) – Electrochemistry Technology Group
- Electrocatalysis Editorial Board

List of Contributors

Mahito Atobe Equipe Sonochimie et Réactivité des Surfaces, Institut UTINAM UMR CNRS 6213, Université de Franche-Comté, 25009 Besançon, France

Dr Craig E. Banks Faculty of Science and Engineering, Manchester Metropolitan University, Chester Street, Manchester, M1 5GD, UK

Verónica Sáez Bernal Departamento de Química-Fisica, Universidad de Alicante, Apdo 99. 03080, Alicante, Spain

Professor Peter R. Birkin School of Chemistry, University of Southampton, Highfield, Southampton, SO17 1BJ, UK

Pedro L. Bonete Ferrández Departamento de Química-Fisica, Universidad de Alicante, Apdo 99. 03080, Alicante, Spain

Cédric Costa PEM Fuel Cell Research Group, School of Chemical Engineering, University of Brimingham, Birmingham, B15 2TT, UK

Oliver J. Curnick PEM Fuel Cell Research Group, School of Chemical Engineering, University of Brimingham, Birmingham, B15 2TT, UK

Marie-Laure Doche Equipe Sonochimie et Réactivité des Surfaces, Institut UTINAM UMR CNRS 6213, Université de Franche-Comté, 25009 Besançon, France

María Deseada Esclapez Departamento de Química-Fisica, Universidad de Alicante, Apdo 99. 03080, Alicante, Spain

José González-García Departamento de Química-Fisica, Universidad de Alicante, Apdo 99. 03080, Alicante, Spain

Professor Jean-Yves Hihn Equipe Sonochimie et Réactivité des Surfaces, Institut UTINAM UMR CNRS 6213, Université de Franche-Comté, 25009 Besançon, France

Jaanus Kruusma Faculty of Science and Engineering, Manchester Metropolitan University, Chester Street, Manchester, M1 5GD, UK

Fabrice Lallemand Equipe Sonochimie et Réactivité des Surfaces, Institut UTINAM UMR CNRS 6213, Université de Franche-Comté, 25009 Besançon, France

Professor Timothy J. Mason Sonochemistry Centre, Coventry University, Priory Street, Coventry CV1 5FB (UK)

Jonathan P. Metters Faculty of Science and Engineering, Manchester Metropolitan University, Chester Street, Manchester, M1 5GD, UK

Dr Bruno G. Pollet PEM Fuel Cell Research Group, School of Chemical Engineering, University of Brimingham, Birmingham, B15 2TT, UK

Abdeslam Et Taouil Equipe Sonochimie et Réactivité des Surfaces, Institut UTINAM UMR CNRS 6213, Université de Franche-Comté, 25009 Besançon, France

Francis Touyeras Equipe Sonochimie et Réactivité des Surfaces, Institut UTINAM UMR CNRS 6213, Université de Franche-Comté, 25009 Besançon, France

Professor David J. Walton Emeritus Professor, Coventry University, Priory Street, Coventry, CV1 5FB, UK

Acknowledgements

The editor would like to thank all involved in the production of this book.

Introduction to Electrochemistry

Bruno G. Pollet and Oliver J. Curnick

I.1 Introduction

This summary introduces some fundamental aspects of electrochemistry, explaining the various electrochemical phenomena occurring at the electrode surface when a potential is applied across it. For this purpose, electrode kinetic and mass-transport parameters will be defined in detail.

I.2 Principles of Electrochemistry

All chemical interactions involve the interaction of electrons at the atomic or molecular level, so that, in a sense, all chemistry is electrochemistry. The fundamental process in electrochemistry is the transfer of electrons between the surface of the electrode and the molecules of a chemical species in the region adjacent to this surface. The nature of this region has a significant effect on the current response, thus it is very important to have some idea about its structure.

Several models have been proposed for the interfacial region. In the simplest model, the charge on the electrode is balanced by a layer of solvated ions of opposite charge held at the electrode surface by coulombic attraction. This region is called the 'electric double layer' [1]. A consequence of this arrangement is that the potential drop between the electrode and the solution occurs across an interfacial region which is a few nanometers thick, leading to a high electric field.

Other models have also shown that the interfacial region can be viewed as two layers of equal and opposite charge separated by a dielectric material (Figure I.1). The first region consists of adsorbed solvent molecules and anions and is defined as the inner Helmholtz plane (IHP). The next layer is defined as the outer Helmholtz plane (OHP) and consists of solvated cations held in this plane by coulombic attraction extending into the diffusion layer where there is competition between the ordering effect of coulombic attraction and disordering of thermal motion. In other words, any electrode immersed in an electrolyte can be modelled as a resistor and capacitor connected in series (an rC system) (by analogy with electrical circuits) where the double layer act as a capacitor C and the ionic medium as a

Power Ultrasound in Electrochemistry: From Versatile Laboratory Tool to Engineering Solution, First Edition. Edited by Bruno G. Pollet.

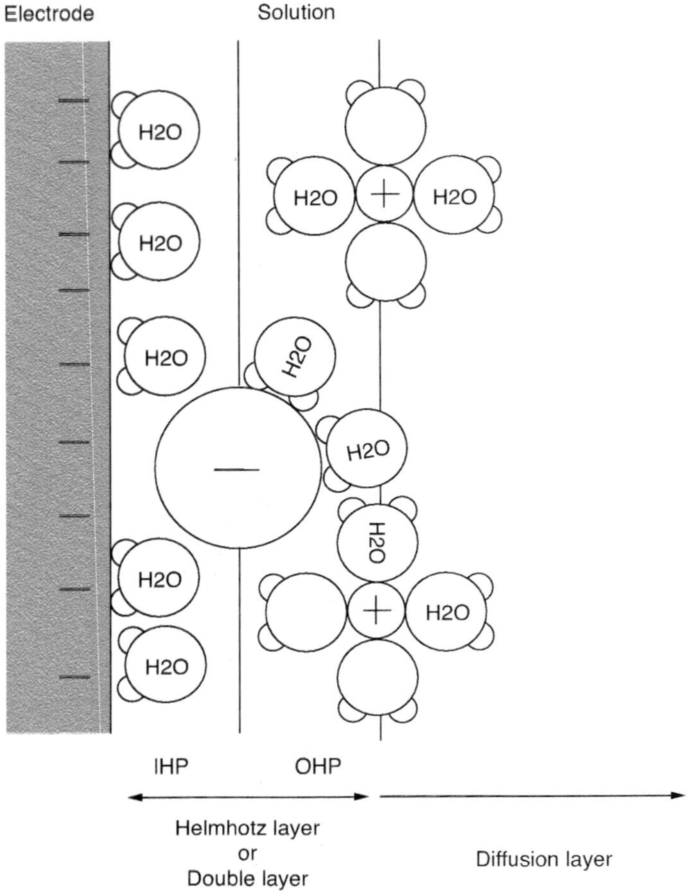

Figure I.1 Relative positions of the inner and outer Helmholtz planes of electrode double layer.

resistor r. The product rC is very important in electrochemistry since it determines the rate at which the current flowing to or from the electrode responds to a change in the applied potential, in the absence of any electrochemical reactions taking place at the surface [1].

The next section discusses electrochemical reactions occurring in the interfacial region, that is, the electron-transfer kinetics and the mass-transport of the electroactive species within the double layer.

I.3 Electron-Transfer Kinetics

When a metal (R) is dipped into a solution of its ions (O) an equilibrium such as

$$M^{n+} + ne^{-} \leftrightarrow M \tag{I.1}$$

or

$$\mathrm{O} + ne^- <=== \overset{k_f}{\underset{k_r}{}} ===> \mathrm{R}$$

is established at its surface. Such an electrode will adopt a potential difference with respect to the solution, whose value is a measure of the position of the equilibrium, which in turn depends on the concentrations of the O and R species at the electrode surface.

Ideally, a redox process is governed by the Nernst equation (Equation I.2) [2], which describes the relationship between the electrode potential, $E_{\mathrm{O/R}}$, and the concentrations at the electrode surface of the electroactive species O and R (assuming that the activity coefficients of O and R are unity). The Nernst equation is then [2]

$$E_{\mathrm{O/R}} = E^{\mathrm{o}}_{\mathrm{O/R}} + \frac{RT}{nF} \ln \frac{C^{\mathrm{S}}_{\mathrm{O}}}{C^{\mathrm{S}}_{\mathrm{R}}} \tag{I.2}$$

where

R is the gas constant in J $\mathrm{K^{-1}}$ $\mathrm{mol^{-1}}$ ($R = 8.3184$ J $\mathrm{K^{-1}}$ $\mathrm{mol^{-1}}$ at 298 K),
T is the temperature in K,
F is the Faraday constant in C $\mathrm{mol^{-1}}$ ($F = 96\,484.6$ C $\mathrm{mol^{-1}}$),
$E_{\mathrm{O/R}}$ is the working electrode potential in V,
$E^{\mathrm{o}}_{\mathrm{O/R}}$ is the *formal* redox couple (or standard reduction potential – SRP) in V,
n is the number of electrons transferred per ion or molecule,
$C^{\mathrm{S}}_{\mathrm{O}}$ is the electrode surface concentration of O, M (electrode) in mol $\mathrm{cm^{-3}}$, and
$C^{\mathrm{S}}_{\mathrm{R}}$ is the electrode surface concentration of R, M (electrode) in mol $\mathrm{cm^{-3}}$.

[*Formal* implies that the activity coefficients are assumed to be unity.]

Experimentally, the electrode potential ($E_{\mathrm{O/R}}$) cannot be measured directly. However, it can be inferred from a measurement of the potential difference between the electrode and some second electrode placed into the same solution (cell potential, E_{cell}), provided the potential on the second electrode is well known. This requires a reference electrode, for example, a saturated calomel electrode (SCE) or a standard hydrogen electrode (SHE) [2–4].

Thus, by convention, one may write that the cell potential is

$$E_{\mathrm{cell}} = E_{\mathrm{O/R}} - E_{\mathrm{reference}} \tag{I.3}$$

If $E_{\mathrm{reference}} = 0$ V (as for the SHE), then,

$$E_{\mathrm{cell}} = E_{\mathrm{O/R}} \tag{I.4}$$

If no current flows through the electrochemical cell, that is, no electrochemical changes have occurred, the electrochemical cell is said to be at equilibrium. In other words, the electrode will adopt an equilibrium or reversible potential (E_{rev}).

Thus one may write that

$$E_{\mathrm{O/R}} = E_{\mathrm{rev}} \tag{I.5}$$

or

$$E_{\mathrm{rev}} = E^{\mathrm{o}}_{\mathrm{O/R}} + \frac{RT}{nF} \ln \frac{C^{\mathrm{S}}_{\mathrm{O}}}{C^{\mathrm{S}}_{\mathrm{R}}} \tag{I.6}$$

Equation (I.6) implies that there is a dynamic equilibrium at the electrode surface, that is, the oxidation of R and the reduction of O occur at the same rate. Experimentally, it is observed that, at the reversible potential, no net current flows in the cell, that is, the forward and the reverse currents are equal in magnitude [3]. Thus one may write

$$I_f = I_r = I_o \tag{I.7}$$

where I_f and I_r are the partial currents for the forward and reverse electrochemical reactions respectively and I_o is an important kinetic parameter of an electron-transfer reaction known as the exchange current at E_{rev}. I_o is a measure of the intrinsic ability of O and R to take part in electron-transfer reactions at the electrode surface; for example, large values indicate that electron transfer is facile.

In 'real' situations, information on the electron-transfer processes cannot be obtained using a two-electrode system at equilibrium. Electrode kinetic parameters such as I_o can only be determined if the equilibrium $O + ne^- \leftrightarrow R$ is 'disturbed', that is, a potential difference is applied to the electrochemical cell. In order to quantify relationships between current and potential, it is necessary to employ a three-electrode system in which the potential difference is varied between a working electrode (W.E.) [on which the electrode reaction occurs] and a reference electrode (R.E.) and the current, developed by one or several electrode reactions at the working electrode, is measured between a counter electrode (C.E.) and the working electrode. These three electrodes are linked to a potentiostat (Figure I.2).

The steady potential resulting from the rapid establishment of the equilibrium in Equation (I.6) can be explained as follows: no net current is flowing when the forward and reverse rates of the reaction are equal. The further such an equilibrium lies to the right, the more negative is the electrode potential. If the working electrode potential is made more negative than the equilibrium potential, the equilibrium may re-assert itself in order to satisfy the Nernst equation, that is, the surface concentration of O and R have to take up new values [see Equation (I.6)]. In this case, C_R^S increases and C_o^S decreases. Thus, a decrease in the ratio C_o^S/C_R^S is observed and a cathodic current will be noted.

It should be emphasized that these predictions are based on thermodynamic parameters. It is important to note that the partial currents flowing in the electrochemical cell at any

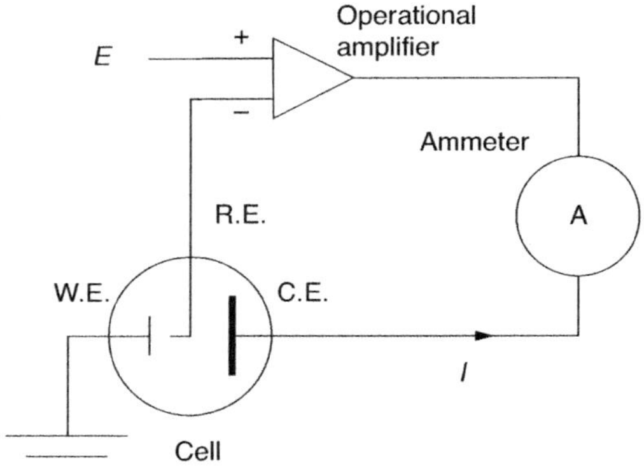

Figure I.2 *Potentiostatic assembly for electrode potential measurements.*

potential depend on the electron-transfer kinetics. Thus, at any potential the measured current, I_{net}, is related to the forward and reverse partial currents and is given by

$$I_{net} = I_r - I_f \tag{I.8}$$

For simplicity, it is assumed in the following discussion that the rate of heterogeneous electron transfer is the rate-limiting step, that is, other factors such as mass-transfer effect are not considered.

Since I represents the number of electrons reacting with O per second, or the number of coulombs of electric charge flowing per second, the question 'What is I?' is essentially the same as 'What is the rate, v, of the reaction $O + ne^- \leftrightarrow R$?' The following equations demonstrate the direct proportionality between the Faradaic current and electrolysis rate (Faraday's law) [1]:

$$I = \frac{dQ}{dt} \tag{I.9}$$

and

$$\frac{Q}{nF} = \frac{w}{M_r} = N \tag{I.10}$$

where

I is the current flowing in A,
Q is the quantity of electricity passed in C,
t is the time in s,
n is the valence of the element deposited,
F is the Faraday constant in C mol^{-1} ($F = 96\,484.6$ C mol^{-1}),
w is the mass of substance deposited in g,
M_r is the relative atomic mass of the element deposited in g mol^{-1}, and
N is the number of moles electrolysed in mol.

Thus if the reaction rate, v, is given by

$$v = \frac{dN}{dt} \tag{I.11}$$

then, from Equations (I.9) and (I.10)

$$v = \frac{I}{nF} \tag{I.12}$$

Since electrode reactions are heterogeneous, their reaction rates are described in units of moles per unit area per second; that is,

$$v = \frac{I}{nFA} \tag{I.13}$$

or

$$v = \frac{i}{nF} \tag{I.13a}$$

where i is the current density ($i = I/A$) in A cm^{-2}, A is the electrode surface area (cm^2) and v is in mol cm^{-2} s^{-1}.

If one assumes, for dilute solutions, that the activity approximates to concentration, one may write

$$v = kC^{S} \tag{I.14}$$

Employing Equations (I.13) and (I.14), the current is given by

$$I = nF\mathrm{A}kC^{S} \tag{I.15}$$

Thus one may write that the partial currents for the forward and reverse reactions as:

$$I_{f} = nF\mathrm{A}k_{f}C_{o}^{S} \tag{I.16}$$

and,

$$I_{r} = nF\mathrm{A}k_{r}C_{R}^{S} \tag{I.17}$$

where n is the number of electrons transferred, A is the electrode surface area (cm^2), C^{S} is the surface concentration (mol cm^{-3}) and k_f and k_r are the rate constants (cm s^{-1}) for the forward and reverse heterogeneous reaction.

Unlike conventional reaction rate constants, which vary only with temperature, these rate constants are dependent both on temperature and the applied potential, E_{app}. It is found experimentally that the forward and reverse rate constants vary exponentially as given below [3]

$$k_{f} = k_{o}\exp[(-\alpha nF/RT)(E_{app} - E^{o})] \tag{I.18}$$

and

$$k_{r} = k_{o}\exp[((1-\alpha)nF/RT)(E_{app} - E^{o})] \tag{I.19}$$

where k_o is the *formal* or the apparent heterogeneous rate constant and α is known as the electron-transfer coefficient.

Thus, the measured current, I_{net}, becomes

$$I_{net} = I_{r} - I_{f} = nF\mathrm{A}\left(k_{r}C_{R}^{S} - k_{f}C_{o}^{S}\right) \tag{I.20}$$

or substituting k_f and k_r from Equations (I.18) and (I.19)

$$I_{net} = nF\mathrm{A}k_{o}\left\{C_{R}^{S}\exp[(1-\alpha)(nF/RT)(E_{app} - E^{o})] - C_{o}^{S}\exp[(-\alpha nF/RT)(E_{app} - E^{o})]\right\} \tag{I.21}$$

Equation (I.21) is known as the Eyring equation and relates the surface concentrations to the net current flow, I_{net}. The equation is ideal and cannot be tested experimentally. However, one such equation which can be tested is the Butler-Volmer equation, which relates the bulk concentrations (measurable) with the net current. Discussion below will show how the Eyring equation can be transformed to the Butler-Volmer equation provided that the Nernst equation is considered.

If it is assumed that the surface concentrations and the bulk concentrations of O and R are only equal when the electrode is at equilibrium, that is, $C_{o}^{S} = C_{o}^{*}$ and $C_{R}^{S} = C_{R}^{*}$ when $E_{cell} = E_{rev}$, then rearranging the *Nernst* equation [Equation (I.6)] gives

$$\left(C_{R}^{*}/C_{o}^{*}\right)\exp[(nF/RT)(E_{rev} - E^{o})] = 1 \tag{I.22}$$

If both sides of Equation (I.22) are raised to the power α, one obtains

$$\left(C_{\mathrm{R}}^{*}/C_{\mathrm{o}}^{*}\right)^{\alpha} \exp[(\alpha nF/RT)(E_{\mathrm{rev}} - E^{\mathrm{o}})] = 1 \tag{I.23}$$

(where $1^{\alpha} = 1$) and if both sides of Equation (I.22) are raised to the power $(\alpha - 1)$, one also obtains

$$\left(C_{\mathrm{R}}^{*}/C_{\mathrm{o}}^{*}\right)^{(\alpha-1)} \exp[(\alpha - 1)(nF/RT)(E_{\mathrm{rev}} - E^{\mathrm{o}})] = 1 \tag{I.24}$$

If both sides of Equation (I.23) are multiplied by $C_{\mathrm{o}}^{\mathrm{S}}$, one may write

$$C_{\mathrm{o}}^{\mathrm{S}} \left(C_{\mathrm{R}}^{*}/C_{\mathrm{o}}^{*}\right)^{\alpha} \exp[(\alpha nF/RT)(E_{\mathrm{rev}} - E^{\mathrm{o}})] = C_{\mathrm{o}}^{\mathrm{S}} \tag{I.25}$$

similarly if both sides of Equation (I.24) are multiplied by $C_{\mathrm{R}}^{\mathrm{S}}$, one may write

$$C_{\mathrm{R}}^{\mathrm{S}} \left(C_{\mathrm{R}}^{*}/C_{\mathrm{o}}^{*}\right)^{(\alpha-1)} \exp[(\alpha - 1)(n/RT)(E_{\mathrm{rev}} - E^{\mathrm{o}})] = C_{\mathrm{R}}^{\mathrm{S}} \tag{I.26}$$

Inserting Equations (I.25) and (I.26) into Equation (I.21) leads to Equation (I.27)

$$\begin{aligned} I_{\mathrm{net}} = nF\mathrm{A}k_{\mathrm{o}} \left(C_{\mathrm{R}}^{*}\right)^{\alpha} \left(C_{\mathrm{o}}^{*}\right)^{(1-a)} \{&\left(C_{\mathrm{R}}^{\mathrm{S}}/C_{\mathrm{R}}^{*}\right) \exp[(1 - \alpha)(nF/RT)(E_{\mathrm{app}} - E_{\mathrm{rev}})] \\ &\times \left(C_{\mathrm{o}}^{\mathrm{S}}/C_{\mathrm{o}}^{*}\right) \exp[(-\alpha nF/RT)(E_{\mathrm{app}} - E_{\mathrm{rev}})]\} \end{aligned} \tag{I.27}$$

It is also found experimentally that there is a deviation of the applied potential, E_{app}, from the reversible potential, E_{rev}. This 'perturbation' is termed overpotential, η (see later in this section) and is given by:

$$\eta = E_{\mathrm{app}} - E_{\mathrm{rev}} \tag{I.28}$$

Inserting Equation (I.28) into Equation (I.27) leads to Equation (I.29)

$$\begin{aligned} I_{\mathrm{net}} = nF\mathrm{A}k_{\mathrm{o}} \left(C_{\mathrm{R}}^{*}\right)^{\alpha} \left(C_{\mathrm{o}}^{*}\right)^{(1-\alpha)} \{&\left(C_{\mathrm{R}}^{\mathrm{S}}/C_{\mathrm{R}}^{*}\right) \exp[(1 - \alpha)(nF/RT)\eta] \\ &\times \left(C_{\mathrm{o}}^{\mathrm{S}}/C_{\mathrm{o}}^{*}\right) \exp[(-\alpha nF/RT)\eta]\} \end{aligned} \tag{I.29}$$

or

$$I_{\mathrm{net}} = I_{\mathrm{o}} \left\{\left(C_{\mathrm{R}}^{\mathrm{S}}/C_{\mathrm{R}}^{*}\right) \exp[(1 - \alpha)(nF/RT)\eta] - \left(C_{\mathrm{o}}^{\mathrm{S}}/C_{\mathrm{o}}^{*}\right) \exp[(-\alpha nF/RT)\eta]\right\} \tag{I.30}$$

with $I_{\mathrm{o}} = nF\mathrm{A}k_{\mathrm{o}}\,(C_{\mathrm{R}}^{*})^{\alpha}\,(C_{\mathrm{o}}^{*})^{(1-\alpha)}$ being the exchange current. Often the exchange current is normalised to unit area to provide the exchange current density, $i_{\mathrm{o}} = I_{\mathrm{o}}/A$.

If the solution is well stirred or currents are kept so low that the surface concentrations do not differ appreciably from the bulk values, that is, $C_{\mathrm{o}}^{\mathrm{S}} = C_{\mathrm{o}}^{*}$ and $C_{\mathrm{R}}^{\mathrm{S}} = C_{\mathrm{R}}^{*}$, then Equation (I.30) becomes

$$i_{\mathrm{net}} = i_{\mathrm{o}}\{\exp[((1 - \alpha)nF/RT)\eta] - \exp[(-\alpha nF/RT)\eta]\} \tag{I.31}$$

which is known as the Butler-Volmer equation.

Equation (I.31) must be regarded as the fundamental equation of electrode kinetics and it shows how the net current varies with the exchange current density, the overpotential and the electron-transfer coefficient. For practical purposes, it is convenient to consider the limiting behaviour of Equation (I.31) for small and large values of the arguments of the exponential terms. Experimentally, two limiting forms of Equation (I.31) are used:

(a) At small overpotentials.

For small values of overpotential, that is, when $\eta \rightarrow 0$, the exponential terms can be written as Taylor expansions, that is, $\exp(x) \approx (1 + x)$. Thus, Equation (I.31) becomes

$$i_{net} = \frac{i_o nF}{RT}\eta \tag{I.32}$$

that is, the current density is directly proportional to the overpotential. Thus a plot of i_{net} versus η is linear. By analogy with Ohm's law ($V = Ir$), here (RT/nFi_o) can be regarded as an impedance and is often referred to as a Faradaic (or charge transfer or ohmic) resistance. This is particularly important in AC impedance measurements. In practice, the linear approximation can be used for $|\eta| > 10/n$ mV.

(b) At large overpotentials.

At large positive overpotentials, that is, when $\eta \rightarrow +\infty$ that is, $\exp[(-\alpha nF/RT)\eta] \rightarrow 0$, then Equation (I.31) can be approximated to

$$i_{net} \approx i_o \exp[(1 - \alpha)(nF/RT)\eta] \tag{I.33}$$

or

$$\log i_{net} \approx \log i_o + (1 - \alpha)(nF/(2.3RT))\eta \tag{I.34}$$

[where $\ln(x) = 2.3 \log(x)$] or

$$\eta \approx -2.3RT \log i_o/((1 - \alpha)nF) + 2.3RT \log i_{net}/((1 - \alpha)nF) \tag{I.35}$$

At large negative overpotentials, that is, when $\eta \rightarrow -\infty$, Equation (I.31) can be approximated to

$$i_{net} \approx -i_o \exp[(-\alpha nF/RT)\eta] \tag{I.36}$$

or

$$\log |i_{net}| \approx \log i_o - \alpha nF/(2.3RT)\eta \tag{I.37}$$

or

$$\eta \approx 2.3RT \log i_o/(\alpha nF) - 2.3RT \log |i_{net}|/(\alpha nF) \tag{I.38}$$

The logarithmic relationships in Equations (I.35) and (I.38) are known as the Tafel equations in the form of $\eta = \mathrm{a} + \mathrm{b} \log i_{net}$. In practice, the Tafel approximation is generally used for $|\eta| < 120/n$ mV.

A plot of versus η *vs.* $\log i_{net}$ will be linear in this high overpotential region, and $\log i_o$ and α can be found from the intercept (a) and slope (b) respectively, as shown in Figure I.3.

Having discussed the important parameters in electrode kinetics, the next section considers the effect of overpotential on the electrode potential measurements.

All galvanic cells are said to operate reversibly when they draw zero current, that is, operate at the reversible potential, E_{rev}. However, if the electrode potential is deliberately altered to a value more anodic or cathodic to its equilibrium value, then current will immediately flow in such a direction so as restore the equilibrium, that is, normal battery

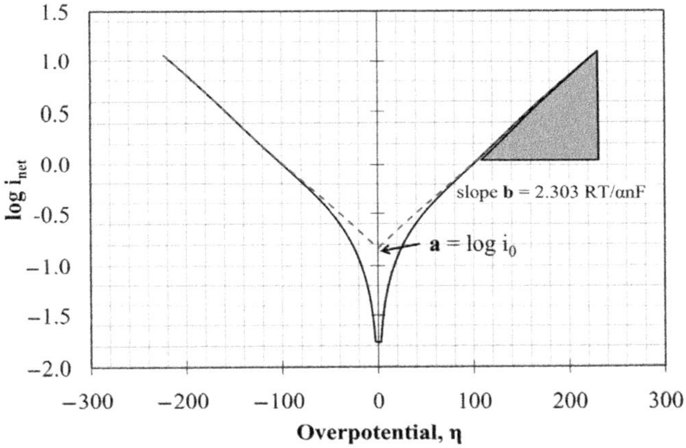

Figure I.3 *A typical Tafel plot for a reversible reaction.*

discharge or recharge. As described earlier in this section, this perturbation of the electrode potential (E_{app}) is known as the overpotential, η, and is described by Equation (I.28).

The kinetic steps found in all electrode processes are:

(i) transport of ions from the bulk,
(ii) ionic discharge, and
(iii) conversion of discharged atom to a more stable form.

The first step gives rise to (a) concentration overpotential (η_C) while the latter two give rise to (b) activation overpotential (η_A). In any system there may be a third overpotential called (c) ohmic overpotential (η_R), which arises due to the finite resistance of the electrolyte solution to the passage of charge.

Thus

$$\eta = \eta_C + \eta_A + \eta_R \tag{I.39}$$

In general, an overpotential leads to a fall in current and the galvanic cell ceasing to operate.

(a) Concentration Overpotential (η_C)

This is an important effect, which arises due to changes in concentration in the vicinity of the electrode surface created by the electrochemical reactions which occur there. Consider two electrodes of a metal M placed in a solution of M^{n+} ions. If no potential is applied across the two electrodes, no current will flow since the potential of both electrodes is the same. If a potential difference is applied between the electrodes, one electrode becomes a cathode and the other an anode. At the cathode, M^{n+} ions are discharged at a faster rate than they dissolve and at the anode M passes into solution more rapidly than M^{n+} ions are discharged. Unless the replenishment is immediate and complete, a discrepancy develops between the surface and the bulk concentrations. Thus, η_C grows and decays slowly on application or interruption of the current flow at a rate characteristic of the diffusion coefficients of the species involved. η_C is the only form of overpotential which is affected by stirring [6].

(b) Activation Overpotential (η_A)

If a reaction is to proceed at a reasonable rate and produce an efficient quantity of product, a significant increase in the applied potential over the equilibrium value is necessary. This excess potential is known as the activation overpotential (η_A). η_A increases rapidly and exponentially after a polarising current is caused to flow. It decreases when the current flow is stopped.

(c) Ohmic Overpotential (η_R)

Ohmic overpotential arises from the passage of an electric current through an electrolyte solution surrounding the electrodes. An ohmic drop (Ir) in potential occurs between the electrodes due to the poor conductivity of the electrolyte. This effect may be reduced by separating the reference electrode from the working solution using a glass capillary. This ohmic overpotential may also be caused by the formation, on the electrode surface, of an adherent layer (e.g. oxide films) of reaction product, which is a relatively poor conductor of electricity.

Thus, electrode reactions depend on the rate constants of the overall electron-transfer process, the concentrations of reactants and products at the electrode surface, their rate of diffusion and the physical and chemical nature of the electrode.

I.4 Determination of Overpotentials

There are two methods of determining the overpotential of an electrolytic cell. Namely by: (i) the decomposition voltage and (ii) the discharge potential methods.

I.4.1 Decomposition Voltages

A graph of current versus cell voltage gives a decomposition curve and allows decomposition voltages to be determined. The decomposition voltage (E_D) is defined as the minimum potential difference that must be applied between a pair of electrodes before decomposition occurs and a current flows [5]. An experimental value of the decomposition voltage can be obtained by extrapolating the second branch of the curve back to zero current.

The overpotential of the system may be obtained using Equation (I.40)

$$\eta = E_D - E_{rev}^{cell} \tag{I.40}$$

where

$$E_{rev}^{cell} = |E_{rev,c} - E_{rev,a}| \tag{I.41}$$

with $E_{rev,a}$ and $E_{rev,c}$ being the reversible potential of the anode and cathode respectively.

I.4.2 Discharge Potentials

This method requires the study of the electrodes reactions separately potentiostatically. Curves can be plotted for the anode and the cathode separately and extrapolated to give the respective anodic discharge potential, $E_{d,a}$, and cathodic discharge potential, $E_{d,c}$. The amount by which the applied electrode potential exceeds the reversible potential, E_{rev}, for the electrode concerned is the sum of the anode or cathode overpotential η_a and η_c respectively that is,

$$\eta_a = E_{d,a} - E_{rev,a} \tag{I.42}$$

$$\eta_c = E_{d,c} - E_{rev,c} \tag{I.43}$$

Thus, the overpotential of the system may be obtained using the following equation:

$$\eta = \eta_{anode} + \eta_{cathode} \tag{I.44}$$

where

$$\eta_{anode} = \eta_{A,a} + \eta_{C,a} + \eta_{R,a} \tag{I.45}$$

and

$$\eta_{cathode} = \eta_{A,c} + \eta_{C,c} + \eta_{R,c} \tag{I.46}$$

Having discussed the electron-transfer kinetic processes occurring at the surface of an electrode, the next section will describe how electroanalytical techniques may be used to obtain both hydrodynamic and electrode kinetics information for an electrode process.

I.5 Electroanalytical Techniques

The electroanalytical techniques employed in electrochemistry may be divided into two distinct groups, namely: (i) voltammetry and (ii) amperometry [6].

I.5.1 Voltammetry

Voltammetry is the measurement of current response to an applied potential. The magnitude of this current is affected by both the rate of electron transfer between the working electrode and the solution and by the rate of mass transport from the bulk solution to the electrode surface. There are two classes of voltammetry, namely: (i) cyclic voltammetry (CV) conducted on a stationary electrode and (ii) hydrodynamic voltammetry, that is, voltammetry based on forced controlled convection.

I.5.1.1 Cyclic Voltammetry

Potential sweep techniques have been applied to a wide range of systems, enabling kinetic parameters to be determined for a large variety of mechanisms. An 'electrochemical spectrum' or a cyclic voltammogram indicates, at a given scan rate, at which potential an electrochemical process or series of processes occurs. In CV, the potential is swept between two programmed potentials at a programmed scan rate and on reaching the final potential the scan is reversed at the same scan rate. The current is recorded against the potential applied to the working electrode. In order to obtain quantitative information on the electrode reaction (e.g. kinetic parameters) it is necessary first to perform qualitative experiments. This is achieved by observing how the peaks appear and disappear as the potential limits and scan rate are varied. There are three types of reactions for which the shape of the cyclic voltammogram differs from each other, namely: (a) reversible (b) quasi-reversible and (c) irreversible reactions.

(a) Reversible Reactions

If one considers the following reaction, $O + ne^- \leftrightarrow R$, at a very low scan rate, the voltammogram recorded will appear as a steady-state I vs. E curve. When the scan rate increases the peak heights increase. In reversible reactions, the electron-transfer rate is, at all potentials, greater than the rate of mass transport. It has been shown that the cathodic peak potential ($E_{p,c}$) is independent of the scan rate (f).

The diagnostic tests for reversible systems (at 298 K) are that:

- The peak separation of the anodic to the cathodic peak potential is equal to $59/n$ mV,
- The ratio between the anodic and the cathodic peak current is equal to 1, and
- The peak potentials are independent of the scan rate f.

Examples of typical reversible redox systems are: the Cp_2Fe^{2+} (ferrocene)/Cp_2Fe^+ (ferricinium) or the MV^{2+} (methyl viologen)/MV^+ or the $[Ru(NH_3)_6]^{3+}$ (hexaammine ruthenium(III))/$[Ru(NH_3)_6]^{4+}$ (hexaammine ruthenium(IV)) redox couples. It should be noted that the reversibility of these redox couples depends strongly on the nature of the electrolyte and the electrode. For example, the reduction of $[Ru(NH_3)_6]^{3-}$ on a platinum electrode leads to a reversible cyclic voltammogram with a peak-to-peak separation of 58.9 mV whereas on a glassy carbon electrode the same reaction gives a quasi-reversible cyclic voltammogram (see next section) with a peak-to-peak separation of approximately 61 mV.

(b) *Quasi*-Reversible Reactions

Electrochemical reactions are said to be *quasi*-reversible if the rate of the electron transfer with respect to that of the mass transport is insufficient to maintain Nernstian equilibrium at the electrode surface. In other words, the electron-transfer rate becomes 'comparable' to the mass-transport rate.

Equation (I.47) gives the dependence of the peak current (I_p) on the scan rate (f) and the concentration at 298 K.

$$I_p = \pm(2.69 \times 10^5) n^{3/2} A\, C_i^* D_i^{0.5} f^{0.5} \tag{I.47}$$

where A is the area of the working electrode (cm^2), n is the number of electrons transferred, $C_i{}^*$ is the concentration of the species in the bulk solution (mol cm^{-3}), D_i is the diffusion coefficient of that species (cm^2 s^{-1}) and f is the scan rate (V s^{-1}). Equation (I.47) is known as the Randles-Sevick equation.

The cyclic voltammetric diagnostic tests for *quasi*-reversible processes (at 298 K) are that:

- the peak potential increases with $f^{1/2}$ but is not proportional to it,
- the ratio of the anodic and cathodic peak current is equal to 1 provided that $\alpha_A = \alpha_C = 1$,
- the peak separation of the anodic to the cathodic peak potential is greater than $59/n$ mV and increases with scan rate, and
- the cathodic peak potential shifts negatively with increasing scan rate.

Examples of typical quasi-reversible redox systems are: the $Fe(CN)_6{}^{3-}$ (ferricyanide)/$Fe(CN)_6{}^{4-}$ (ferrocyanide) and the $Fe(EDTA)^-$ (ferric EDTA)/ $Fe(EDTA)^{2-}$ (ferrous EDTA) redox couples.

Two important kinetic parameters determined from *quasi*-reversible cyclic voltammograms are: (i) the half-wave potential ($E_{1/2}$) and (ii) the apparent heterogeneous or formal rate constant (k_o). These may be obtained when a careful study of a *quasi*-reversible cyclic voltammogram is carried out.

(i) Half-wave potentials ($E_{1/2}$) are obtained by recording the anodic and cathodic peak potentials E_{pa} and E_{pc} respectively and using the following equation

$$E_{1/2} = \frac{(E_{pa} + E_{pc})}{2} \tag{I.48}$$

Whilst half-wave potentials can be used to identify the relative amounts of several components in a mixture of species, they are sensitive to the presence of different complexing species, including supporting electrolyte, and therefore they are mainly used for 'fingerprinting' with extreme caution.

(ii) Apparent heterogeneous rate constant (k_o) can be determined using cyclic voltammetry from *quasi*-reversible systems. Following the method described by Nicholson [Equation (I.49)], the increase in ΔE_p ($E_{pc} - E_{pa}$) may be used to determine the rate of heterogeneous electron transfer. Working curves, which relate ΔE_p to a kinetic parameter (ψ) has been published [1,2]. The formal rate constant for an electrochemical reaction is determined by the following equation:

$$k_o = \psi \left(D_o^{-1} nF/RT\right)^{1/2} (D_R/D_o)^{\alpha/2} \tag{I.49}$$

The ratio of the diffusion coefficients in many cases is approximately 1 and the transfer coefficient (α) is generally 0.5.

(c) Irreversible Reactions

Electrochemical reactions are said to be irreversible if the electron-transfer rate becomes larger than the mass transport when the scan rate is increased. The two most marked features for a cyclic voltammogram of a totally irreversible system are either (i) a peak is observed on the forward scan but no peak is observed on the reverse scan or (ii) the two peaks obtained on the forward and reverse scan are asymmetrical to each other.

Equation (I.50) shows the dependence of the peak current (I_p) on the forward scan rate (f) and the concentration ($C_i{}^*$) (at 298 K).

$$I_p = \pm 0.4463\, nF\mathrm{A}(\alpha n_\alpha)^{0.5}(F/RT)^{0.5} C_i^* D_i^{0.5} f^{0.5} \tag{I.50}$$

Here A is the area of the working electrode (cm^2), n is the number of electrons transferred, α is the transfer coefficient, n_α is the number of electrons transferred in the rate-determining step, C_i^* is the concentration of the species i in the bulk solution (mol cm^{-3}), D_i is the diffusion coefficient of that species (cm^2 s^{-1}) and f is the scan rate (V s^{-1}).

The diagnostic tests for cyclic voltammetry of an irreversible process (at 298 K) are that:

- there is no reverse peak,
- the cathodic peak current is proportional to $f^{1/2}$,
- the cathodic peak current shifts by $-30/(\alpha_C n_\alpha)$ mV for each decade increase in f, and
- the separation between the peak potential and the half-peak potential is equal to $48/(\alpha_C n_\alpha)$ mV.

A typical irreversible system is the $S_2O_3^-/S_2O_3^{2-}$ (thiosulfate) redox couple.

In summary, the reversibility of a redox couple will depend both on the kinetics of electron transfer (i.e. the formal rate constant, k_o) and the mass transport conditions.

The next section shows how the kinetic parameters can be determined when the electrochemical system is subjected to forced convection, that is, agitation of the electrolyte solution.

I.5.1.2 Voltammetry Based on Forced Controlled Convection: Hydrodynamic Voltammetry

Before proceeding much further, the nature of various mass-transfer processes should be considered. In electrochemistry, there are three types of mass transfer.

Diffusion – the movement of an electroactive species down a concentration gradient. This occurs whenever there is a chemical change at a surface. At the electrode surface there is always a boundary layer (up to 10^{-2} cm thick) in which the concentrations of O and R are a function of distance from the electrode surface.

Migration – the mechanism by which charged electroactive species move through the electrolyte due to a potential gradient. The forces leading to migration are purely electrostatic and the charge can be carried by any ionic species in the electrolyte solution.

Convection – the movement of electroactive species due to mechanical forces, for example, rotation of the electrode, vibrations, stirring of the electrolyte, flowing the electrolyte through the cell and so on.

To extract quantitative data from electrochemical experiments, the mode of mass transport must be controlled. The problem with the use of stationary electrodes for other than thermodynamics measurements in electrochemistry is one of transport. As soon as any net current passes at the electrode the composition of the solution at the electrode surface changes. Steady-state measurements are irreproducible due to the effects of stray convection produced by thermal gradients and mechanical shock. The answer is to use a system in which there is reproducible mass transport. There are a number of ways of achieving this using forced convection electrode geometries. These include using dropping mercury, tube, wall jet and rotating electrodes. This group of electrochemical experiments in which forced convection is employed is called hydrodynamic techniques. In these techniques the mass transport is a combination of convection in the bulk solution and diffusion, which occurs in the diffusion layer. By analogy with polymer extrusion, the maximum liquid flow occurs away from the walls. Thus, away from the electrode surface, the solution flow is turbulent but this turbulence decreases nearer the working electrode surface (Figure I.4). There is transition from turbulent flow to laminar flow, in which the solution layers slide past each other parallel to the electrode. The solution layer closest to the electrode surface is stationary and is called the Nernst diffusion layer of thickness δ.

Hydrodynamic voltammetry can be used to determine the thickness of the diffusion layer and the rates of heterogeneous charge transfer of the electroactive species. The main feature of a rotating voltammogram, and in fact of all hydrodynamic voltammograms, is the appearance of a plateau at large potentials. This plateau corresponds to a limiting current, the magnitude of which is determined by the thickness of the diffusion layer. The limiting

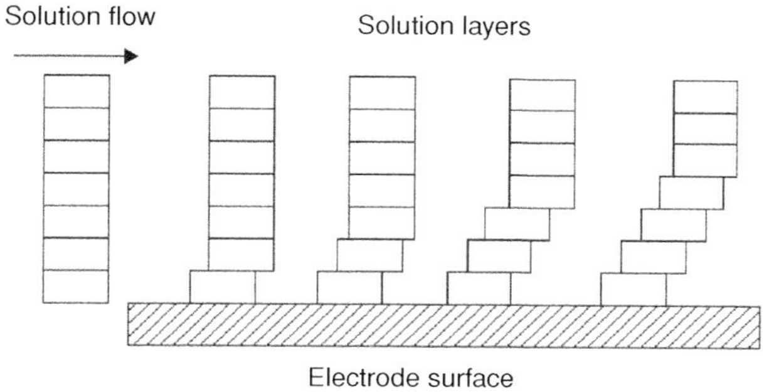

Figure I.4 *Representation of the development of a boundary layer over an electrode.*

current I_{lim} and the limiting current density i_{lim} for an electrochemical process is given by Equations (I.51a) and (I.51b) respectively

$$I_{\text{lim}} = \frac{nFAD_{\text{o}}C^*}{\delta} \quad \text{(I.51a)}$$

or

$$i_{\text{lim}} = \frac{nFD_{\text{o}}C^*}{\delta} \quad \text{(I.51b)}$$

where n is the number of electrons transferred during the electrochemical process, F is the Faraday constant, D_{o} is the diffusion coefficient, C^* is the concentration of the species in the bulk solution and d is the diffusion layer thickness.

For a rotating disc electrode (RDE), the dependence of the limiting current on the rotation speed (at a given temperature) is given by the Levich equation

$$I_{\text{lim}} = 0.620nFAD_{\text{o}}^{2/3}\upsilon^{-1/6}C^*\omega^{1/2} \quad \text{(I.52)}$$

where υ is the kinematic viscosity of the electrolyte solution ($\text{cm}^2\ \text{s}^{-1}$), and ω is the angular rotation rate in rad s^{-1}. The Levich equation can be reduced to the experimental form of

$$\text{d}I_{\text{lim}}/\text{d}C^* = \text{B}\,\omega^{1/2} \quad \text{(I.52a)}$$

where B ($= 0.620nFAD_{\text{o}}^{2/3}\upsilon^{-1/6}$) is a constant.

In a typical RDE experiment, potential sweeps are performed at several rotation rates, usually at a slow potential scan rate of [1–30 mV s^{-1}] in order to obtain *pseudo* steady-state conditions and to minimise contributions to current from double-layer charge/discharge (Figure I.5(a)). A Levich plot of I_{lim} *vs.* $\omega^{1/2}$ is linear with a slope equal to $0.620nFAD_{\text{o}}^{2/3}\upsilon^{-1/6}C^*$, from which the number of electrons n transferred in the reaction can be determined provided the other parameters are known.

For a rotating cylinder electrode (RCE), the dependence of the limiting current is given by the Eisenberg equation,

$$I_{\text{lim}} = 0.0791nFAD_{\text{o}}^{0.644}\upsilon^{-0.344}C^*\omega^{0.7} \quad \text{(I.53)}$$

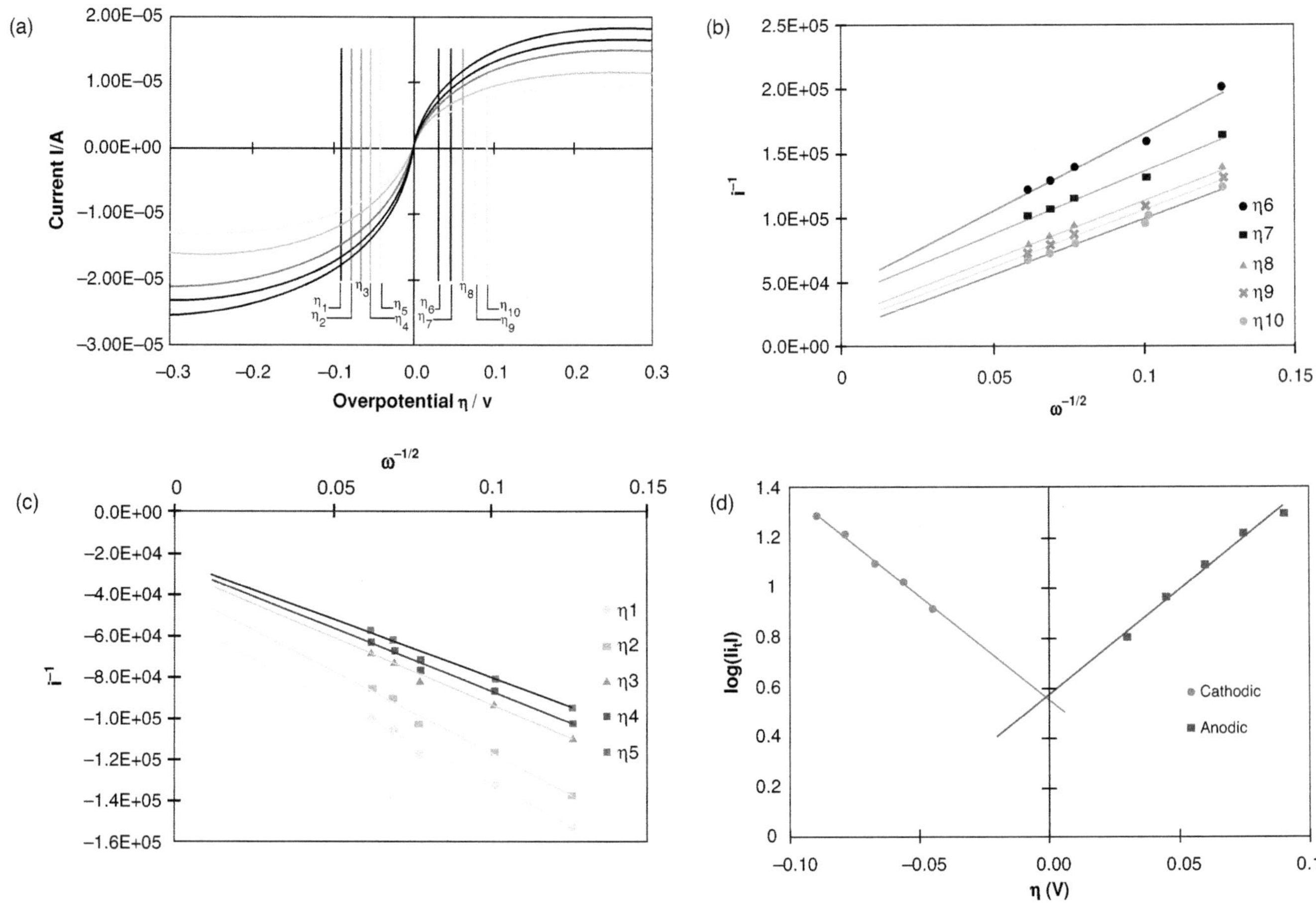

Figure I.5 *(a) Rotating disc voltammograms recorded at 25 mV s^{-1} in a 10 mM solution of potassium ferricyanide and potassium ferrocyanide in a background electrolyte of 0.5 M KCl, using a 5 mm Pt disc working electrode. Rotation rates ω = 600, 933, 1600, 2000, 2500 rpm. (b) Koutecky–Levich plots for the anodic branches and (c) cathodic branches of the voltammograms. (d) Tafel plot using I_k values determined from the intercepts of Koutecky–Levich plots.*

(Here ω is the rotation speed of the electrode and υ is the kinematic viscosity of the solution) The Eisenberg equation can be reduced to the experimental form of:

$$I_{\text{lim}} = B'\omega^{0.7} \tag{I.53a}$$

where B' $(= 0.0791nFAD_o{}^{0.644}\upsilon^{-0.344}C^*)$

Hydrodynamic voltammograms can also be used to obtain information on the kinetics of the electrode reaction. The current measured in a hydrodynamic experiment has both kinetic and diffusion components, as described by the Koutecky-Levich equation:

$$1/I = 1/I_k + 1/I_{\text{lim}} \tag{I.54}$$

Substituting equation (I.52) into (I.54) gives

$$1/I = 1/I_k + 1/(B\omega^{1/2}) \tag{I.55}$$

To determine the kinetic current, I_k, a series of potentials (typically five or six) within the mixed kinetic/diffusion controlled portion of the voltammogram are chosen, as shown in Figure I.5, and a Koutecky-Levich plot of $1/I$ *vs.* $1/\omega^{1/2}$ then yields a series of straight lines (Figure I.5(b,c)) [1–6]. The intercepts of these lines at $1/(B\omega^{1/2}) = 0$ (corresponding to an infinite rotation rate) yield values for $1/I_k$ $(= I_k{}^{-1})$ at each potential.

The exchange current (I_o) and the Tafel slope (see Equation (I.38)) for the electrode reaction can be found from a Tafel plot of these kinetic currents *vs.* their corresponding overpotentials (as shown in Figure I.5(d). For further information, the reader is invited to refer to the excellent book: *Electrochemical Methods: Fundamentals and Applications* by Allen J. Bard and Larry R. Faulkner (see References section).

I.5.2 Amperometry

I.5.2.1 Chronocoulometry

As its name implies, chronocoulometry is the measurement of charge (coulombs) with time (chrono). It involves the measurement of the charge-time response to a potential step excitation. The *i-t* response is integrated to give a monitored response of charge (Q) versus time. The resulting *Q-t* curve for the forward step is described by Equation (I.56)

$$\int_0^t i\,dt = Q = \frac{2nFAC^*D_O^{1/2}t^{1/2}}{\pi^{1/2}} = Jt^{1/2} \tag{I.56}$$

Chronocoulometry is also useful for measuring electrode surface area A, diffusion coefficient D_o, electroactive species concentration C^*, the number of electrons transferred n, the time required for complete electrolysis, adsorption of electroactive species, and kinetics and mechanisms of coupled chemical reactions. A plot of Q versus $t^{1/2}$ (Anson plot) can be used to determine the electrode area (A) provided D_o, C^*, n and the slope $(J = 2nFAC^*D_o{}^{1/2}t^{-1/2})$ values are known.

I.5.2.2 Electrolysis

Electrolysis is a chemical process by which chemical reactions are produced electrically either in solutions or molten salts. The most important industrial applications of electrolysis fall into four broad categories: metal recovery and electroextraction, electrochemical organic synthesis, electro-concentration of solids, cleaning and disinfections of liquids. The quantity of electrolysis products, their rate of production and often their nature depend on electrolysis conditions.

I.5.2.2.1 Parameters Important in Electrolysis The main parameters important in electrolysis are the following [5]:

- Faraday's Law
  There is a simple relationship between the amount of electricity passed through an electrolytic cell and the amount of substances produced at the electrodes. The quantity of electrical charge associated with one mole of electron can be expressed in terms of coulombs (C). In electrolysis, 96 484.6 C mol^{-1} is often referred as the Faraday constant. Faraday's law is described by Equation (I.9).
- Electrode Potential
  As discussed above, the electrode potential determines which electron-transfer reactions can occur. The potential is a major factor controlling the current efficiency, the space-time yield and the product quality.
- Electrode Material
  The ideal electrode material for most processes should be totally inert or stable in the electrolysis medium and permit the desired reaction with a high current efficiency at low overpotential. The better materials are expensive and it is more common for the active material to be either a coating or an inert substrate (titanium or carbon for anodes, steels for cathodes). The shape of the electrode is also important for particular processes. For example, electrodes are often constructed from meshes in order to maximise surface area, reduce cost and weight and enhance the release of gaseous products.
- Electrolysis Medium
  The properties of the electrolysis medium will be determined by the choice of the solvent, electrolytes and complexing agents, pH and the concentration of each constituent.
- Temperature
  Temperature is a parameter frequently used. Temperatures above ambient are employed because of their beneficial effects on the electrode kinetics and mass transport. The diffusion coefficient and the exchange-current density are all increased.
- Mass-Transport Regimes
  As discussed previously, the effectiveness of an electrolytic cell depends on the nature of the mass transport employed, that is, the type of forced agitation used. Thus, the flow of electroanalyte from the bulk solution to the surface of the electrode has an important effect on the quality and the amount of electrodeposit obtained.

It is known that when an electrolyte flows over an electrode, a boundary layer of solution develops. The formation of such boundary layer has particular importance since the electrode reaction takes place within it. It has been shown that the development and scale of the boundary layer depends on the relative importance of the inertial and viscous forces. The ratio of inertial:viscous forces is known as the Reynolds number, *Re*, and it is calculated using Equation (I.57)

$$Re = \frac{\rho l U}{\mu} = \frac{l U}{\upsilon} \qquad (I.57)$$

Here ρ is the density of the solution, μ its dynamic viscosity, υ its kinematic viscosity, U is the limiting flow velocity and l is the length of the electrode.

The boundary layer develops only below a critical value of Reynolds number (Re_c) where the viscous damping is sufficient to suppress any perturbations which arise – the flow is known to be laminar. However, above this critical Reynolds number, the viscous damping is no longer predominant and turbulence commences.

Experimentally, the Reynolds number may be determined provided that the rate of mass transport is known, that is, if the limiting current density is known. It was shown that it is possible to derive an expression for the magnitude of the rate of mass transport known as the Sherwood number, Sh, given by the following equation

$$Sh = H\ Re^{a} Sc^{b} = \frac{i_{\text{lim}} l}{nFDC^{*}} \tag{I.58}$$

where H, a and b are constants which may be obtained from experimental measurements of the limiting current density (i_{lim}) and Sc, the Schmidt number, is defined as

$$Sc = \frac{\mu}{\rho D} = \frac{\upsilon}{D} \tag{I.59}$$

D is the diffusion coefficient of the electroactive species.

N.B.1 For a disc electrode of radius r_e, the Reynolds number is $Re = U\ r_e/\upsilon$ and the Sherwood number is $Sh = i_{\text{lim}} r_e/(nFDC^{*})$

N.B.2 For a rotating disc electrode, $H = 0.620$, $a = 0.5$ and $b = 0.33$. For a rotating cylinder electrode, $H = 0.079$, $a = 0.7$ and $b = 0.36$.

Thus according to Equation (I.59), the increase in limiting current density and the increase in uniformity of concentration of the electroanalyte within the diffusion layer at the electrode surface will depend on the *Reynolds* number.

- Current Efficiency

  The current efficiency, $C.E.$, is the yield based on the electrical charge passed during the electrolysis,

$$C.E. = \frac{W_{\text{exp}}}{W_{\text{theor}}} \times 100\% \tag{I.60}$$

where W_{expt} and W_{theor} are the experimental and theoretical (Faraday's law) masses of electrodeposit respectively.

It should be noted that the current efficiency may also be calculated using Equation (I.61)

$$C.E. = \frac{(C_{\text{i}}^{*} - C_{\text{f}}^{*})}{C_{\text{f}}^{*}} \times 100\% \tag{I.61}$$

where C_{i}^{*} and C_{f}^{*} are the initial and final bulk concentrations of the electroactive species respectively, that is, bulk concentrations before and after complete electrolysis.

Values of $C.E.$ below 100% indicate that there is either a back-reaction occurring in the cell or that by-products are being formed. These latter materials are often due to the electrolysis of the background electrolyte rather than the starting material.

I.5.2.2.2 Electrolysis at Constant Potential Electrolysis with constant applied potential can provide more selectivity than constant current electrolysis since the potential of the working electrode cannot shift to a value which permits some undesirable reactions to occur, for example, the evolution of hydrogen in the deposition of metals.

It has been shown that the current resulting from a simple electrolysis process proceeding with 100% current efficiency under controlled potential conditions decreases in a manner which may be approximately described by a first-order decay expression, that is

$$I_{\mathrm{t}} = I_{\mathrm{o}} \exp(-kt) \tag{I.62}$$

where I_{t} is the current at any time t, I_{o} is the initial current and k is the rate constant for the disappearance of the electroactive species which depends on the cell geometry, the mass-transfer conditions and the nature of the electrode reaction.

As the electrolysis current is proportional to the concentration of electroactive material, Equation (I.62) becomes:

$$C_{\mathrm{t}}^{*} = C_{\mathrm{o}}^{*} \exp(-kt) \tag{I.63}$$

or

$$\ln(C_{\mathrm{t}}^{*}/C_{\mathrm{o}}^{*}) = -kt \tag{I.64}$$

where C_{t}^{*} is the concentration at any time t and C_{o}^{*} is the initial concentration.

A plot of $-\ln(C_{\mathrm{t}}^{*}/C_{\mathrm{o}}^{*})$ *vs.* t may be used to determine the rate constant for the disappearance of the electroactive species, k.

References

(1) Bard, A.J. and Faulkner, L.R. (eds) (2004) *Electrochemical Methods: Fundamentals and Applications*, Wiley, India.
(2) Bard, A.J. (ed.) (1985) *Standard Potentials in Aqueous Solutions*, Marcel Dekker, New York.
(3) Holze, R. (ed.) (2009) *Experimental Electrochemistry: A Laboratory Textbook*, Wiley-VCH, Weinheim.
(4) Hamann, C.H., Hamnett, A. and Vielstich, W. (eds) (2007) *Electrochemistry*, 2nd completely revised and updated edn Wiley-VCH, Weinheim.
(5) Pletcher, D. and Walsh, F.C. (eds) (1993) *Industrial Electrochemistry*, 2nd edn, Blackie Academic & Professional, London.
(6) Wang, J. (ed.) (2006) *Analytical Electrochemistry*, 3rd edn, John Wiley & Sons, New Jersey.

1

An Introduction to Sonoelectrochemistry

Timothy J. Mason and Verónica Sáez Bernal

1.1 Introduction to Ultrasound and Sonochemistry

Ultrasound is defined as sound with a frequency that is beyond the range of human hearing and this is generally considered to be above 20 kHz (20 000 hertz, hertz = cycle per second). The ultrasound frequency range can be broadly subdivided into two categories (Figure 1.1):

(i) **Diagnostic ultrasound** operates at very high frequencies (above 5 MHz); it is used in foetal scanning and in general for the non-destructive testing of materials. The important feature of diagnostic ultrasound is that the energy involved is too low to produce cavitation in tissues and so is safe to use in medical scanning.
(ii) **Power ultrasound** is mainly in the frequency range between 20 and 100 kHz (but can be extended up to 2 MHz). The energies involved are sufficiently high to produce cavitation in liquids and thus can be used to influence chemistry and processing (see later).

The origins of ultrasound can be traced back to the late nineteenth century with the discovery of the piezoelectric effect by Curie and the ultrasonic whistle by Galton in 1893. However, the first commercial application of ultrasonics did not appear until 1917 with Langevin's echo sounding technique. Langevin's discovery was the direct result of a competition organised in 1912 to find a method of detecting icebergs and so to avoid any repetition of the disaster which befell the Titanic. Essentially all imaging, from medical ultrasound to non-destructive testing (flaw detection), relies upon the same pulse echo type

Power Ultrasound in Electrochemistry: From Versatile Laboratory Tool to Engineering Solution, First Edition. Edited by Bruno G. Pollet.

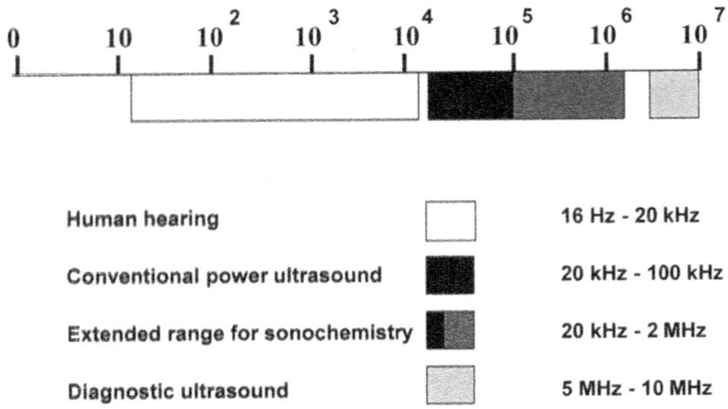

Figure 1.1 *Sound range.*

of approach, that is, a pulse of sound is transmitted through a material and is reflected from a boundary. The time of flight from emission to reception of the echo is then used to compute the distance to the boundary using the speed of sound in that medium. High frequency ultrasound (>1 MHz) is therefore employed for modern medical imaging and non-destructive testing.

The potential of using acoustic cavitation to create changes in materials was identified in the USA in the late 1920s [1–3]. Over the succeeding years some more research went into the applications of ultrasound in chemistry and the first review of this type of work was published in the 1940s [4].

By the 1950s, industry had adopted power ultrasound (which at that time was known as 'Macrosonics') for a range of applications that were gathered together in a series of five publications [5]. At that time there were six general industrial applications and these are listed in order of commercial turnover in Table 1.1.

The majority of the applications shown in Table 1.1 can be considered to be 'engineering' and the whole area of ultrasonic applications in the chemistry laboratory, which had been identified in the 1940s, did not really develop until the 1980s. This renaissance is generally considered to be due to chemists becoming aware of potential alternative uses of ultrasonic equipment which was then appearing in the laboratory in the form of cleaning baths and ultrasonic probes (the latter more commonly in biological laboratories in the form of biological cell disruptors).

From Table 1.1, it can be seen that power ultrasound can be used in two basic ways to bring about changes in a material and these are:

(i) **Direct** mechanical transmission of vibration from the transducer onto a solid to induce vibrations in the surface of the material.
(ii) **Indirect** transmission of energy via cavitation induced in a fluid by the transmission of acoustic vibrations.

Each of these has a relevance to electrochemistry since (i) can be taken to represent induced vibrations at a solid electrode surface and (ii) can be taken to represent sonication of an electrolyte. In many cases both will occur simultaneously and influence the overall electrochemical process.

Table 1.1 Industrial applications of macrosonics: the major established areas. Reprinted with permission from Ultrasonics, Macrosonics in Industry 1. Introduction by E. A. Neppiras, 10, 1, 9–13 Copyright (1972) Elsevier Ltd.

1	**Cleaning**	Dispersing loosely held non-soluble particles and coatings Dissolving solid surface coatings Degreasing Descaling
2	**Plastic Welding**	Remote welding of rigid thermo-plastics Insertion Staking Splicing Continuous welding of sheet material
3	**Metal Welding**	Microbonding Continuous welding of sheet material Bonding metals to non-metals
4	**Chemical Processing**	Extracting: – perfume from flowers essential oils from hops juices from fruits chemicals from plants biomacromolecules from cells
5	**Metal Forming**	Wire-drawing Tube-drawing Plastic-forming of sheet metal
6	**Machining**	Ultrasonic machining with abrasive Vibration-assisted ECM Vibration-assisted milling, turning, drilling, grinding, polishing Teeth descaling Vibration-assisted rotary drills

1.2 Applications of Power Ultrasound through Direct Vibrations

The application of power ultrasound through the introduction of direct vibrations onto a solid surface is the basis of many engineering applications including welding, drilling, cutting and tube drawing. The drilling and cutting applications have now also become an important development in surgery.

1.2.1 Welding

A large proportion of ultrasonic equipment currently in industry is devoted to welding or riveting plastic mouldings for the consumer market. The equipment generally operates at around 20 kHz and a shaped tool or horn transmits (and amplifies) the vibrating motion to a shaped die pressing together the two pieces of material to be welded. The vibrational amplitude is typically 50–100 microns (μm).

Ultrasonic welding is generally used for the more rigid amorphous types of thermoplastic. It is particularly important that the vibrational energy is primarily transmitted to the joint rather than be absorbed by the body of the material, producing heat. This is because any warming of the bulk material can lead to a release of internal moulding stresses and produce distortion. Thermoplastics have two properties which make them particularly suited to ultrasonic welding: (a) low thermal conductivity and (b) melting or softening temperatures of between 100 and 200 °C. As soon as the ultrasonic power is switched off the substrate or bulk material becomes a heat-sink, giving rapid cooling of the welded joint. When the more traditional conductive heating is used for welding, however, the thermal gradient has to be reversed before cooling occurs, leading to long heating/cooling process cycles. Another major advantage of the use of ultrasound is the high joint strength of the weld, reaching 90–98% of the material strength. Indeed, test samples usually break in the body of the material and not at the weld itself.

1.2.1.1 Machining (Drilling and Cutting)

Ultrasonic drilling and cutting were originally introduced and developed for the accurate profiling of brittle materials such as ceramics and glass. It has been used in the aerospace industry since the 1970s for cutting glass and carbon fibre composites and has found many applications in the food industry. In dentistry it is used for removing plaque from teeth and in medicine it is used to improve the efficiency of scalpels and for bone cutting.

Ultrasonic cutting uses a knife-type blade attached through a shaft to an ultrasonic source. Essentially the shaft with its blade behaves as an ultrasonic horn and is normally operated at a frequency of 20 kHz. The cutting action is a combination of the pressure applied to the sharp cutting edge surface and the mechanical longitudinal vibration of the blade. Typically the tip movement will be in the range 50 to 100 microns peak to peak. Several advantages arise from this technology:

(i) The ultrasonic vibration of 20 kHz applies an intermittent force to the material to be cut and generates a crack (cut) at the tip, controlling its propagation or growth, and thereby minimising the stress on the bulk material.
(ii) The repeated application of the cutting tip to the product applies a local fatiguing effect which reduces significantly the overall force required to break the bonds of the bulk material.
(iii) In conventional cutting the blade has to compress the bulk material to allow a gap the width of the blade to pass through and this applies a tensile rupturing force at the crack tip. With ultrasonic cutting the whole blade moves or vibrates continuously as it stretches and contracts. This very high frequency movement effectively reduces the coefficient of friction to a very low level, enabling the blade to slide more easily through the bulk material.

1.2.1.2 Metal Forming

This is a rather specialised application of ultrasonics in which the circular die used for tube drawing (or extrusion) is radially vibrated. The vibrations provide an easier passage of the material through the die thus permitting a greater diameter reduction with each pass. It is also claimed that the surface finish produced using this technique is better than that obtained using conventional cold drawing techniques.

1.3 Applications of Power Ultrasound through Cavitation

There are very few references to ultrasound in chemistry from about 1955 to 1970; then in 1970 a major renaissance in the subject began to occur, which then accelerated. This revival of interest is undoubtedly due to the more general availability of commercial ultrasonic equipment. In the 1960s the ultrasonic cleaning bath began to make its appearance in metallurgy and chemical laboratories and in this period biology and biochemistry laboratories also began using ultrasonic cell disruptors on a regular basis. The next major reviews came in the late 1980s [6, 7].

Almost all of the technologies associated with cleaning and chemical processing depend upon the phenomenon of acoustic cavitation. Like any sound wave, ultrasound is propagated via a series of compression and rarefaction waves induced in the molecules of the medium through which it passes. At sufficiently high power the rarefaction cycle may exceed the attractive forces of the molecules of the liquid and cavitation bubbles will form. It is the fate of these cavities when they collapse in succeeding compression cycles which generates the energy for chemical and mechanical effects (Figure 1.2). This is a remarkable phenomenon induced throughout the liquid at ultrasonic frequencies that is, 20 kHz or above. In aqueous systems each cavitation bubble acts as a localised 'hotspot' generating temperatures of about 4000 K and pressures in excess of 100 MPa. The implosion happens with a collision density of 1.5 kg cm^{-2} and pressure gradients of 2 TPa cm^{-1}, with lifetimes shorter than 0.1 μs and cooling rates above 10^{10} K s^{-1} are involved.

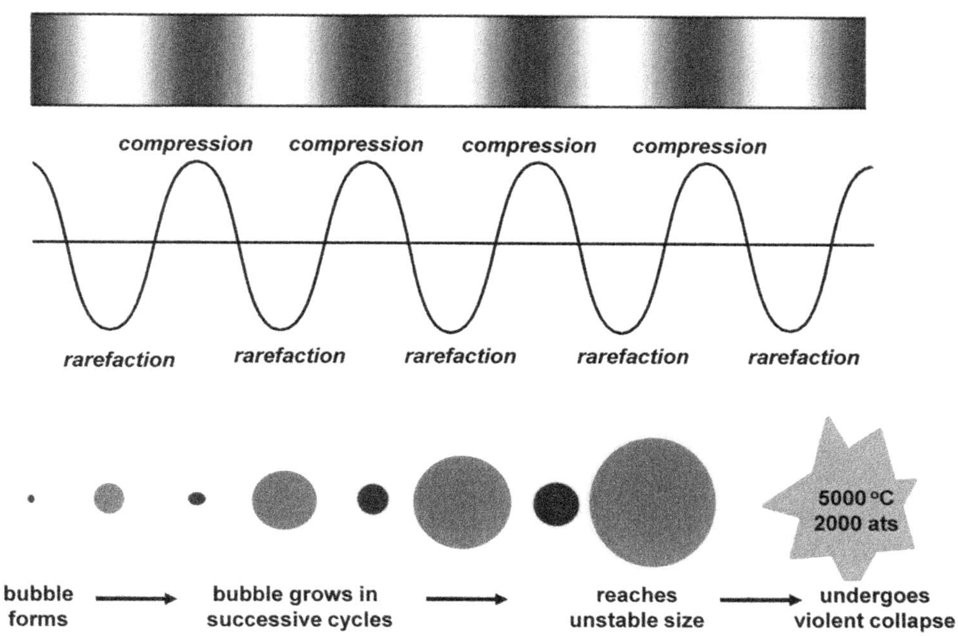

Figure 1.2 *Sinusoidal wave form and bubble collapse.*

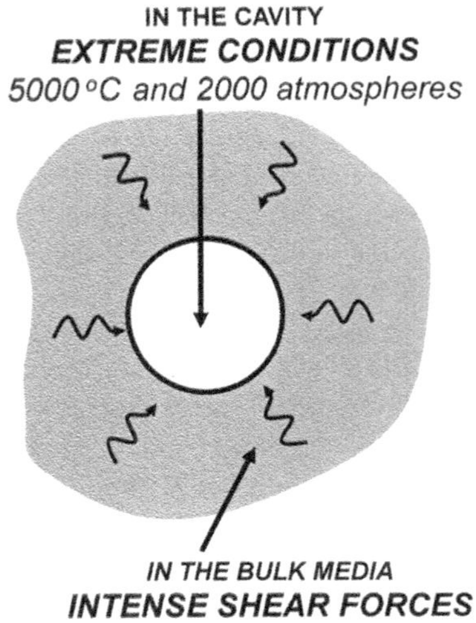

Figure 1.3 Cavitation collapse in a homogeneous medium.

1.3.1 Homogeneous Reactions

The mechanical and chemical effects of the collapsing bubble will be felt in two distinct regions: (a) within the bubble itself, which can be thought of as a high pressure and high temperature microreactor, and (b) in the immediate vicinity of the bubble where the shockwave produced on collapse will create enormous shear forces (Figure 1.3).

The cavitation bubble does not enclose a vacuum – it contains vapour from the solvent or any volatile reagent in the solution. Under the extreme conditions generated on collapse of the bubble the molecules in these vapours suffer fragmentation to generate reactive species of the radical or carbene type. The sonochemical decomposition of water into radicals is important in a number of applications. In environmental remediation the hydroxyl radical is a key species in the decomposition of chemical contaminants because it is an extremely strong oxidant and can attack aromatic compounds. Radicals can also attack and kill bacterial cells in sonochemical sterilisation. The generation of radicals can also be used to initiate emulsion polymerisation in aqueous systems, thus avoiding the use of chemical initiators.

1.3.2 Heterogeneous Reactions Involving a Solid/Liquid Interface

When the acoustic cavitation bubble collapses close to or on a solid surface it can only do so asymmetrically due to the surface, which leads to a microjet of liquid being directed towards the surface of the material at speeds of up to 200 m s^{-1} (Figure 1.4). This effect is equivalent to high pressure jetting and is the reason why ultrasound is used for cleaning. The jets can be sufficiently powerful to cause pitting of the surface (erosion). It can also serve to activate solid catalysts and increase mass and heat transfer to the surface by disruption of the interfacial boundary layers. It is primarily as a result of this jet that almost any

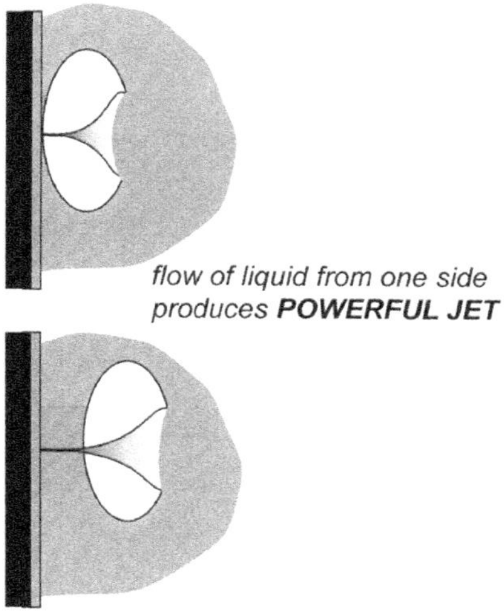

Figure 1.4 Cavitation collapse near to a surface.

electrochemical process can be improved by the application of power ultrasound, a field of research which has now become known as **sonoelectrochemistry**.

Cavitation in powder dispersions can cause deaggregation and fragmentation, which can increase the surface area available for reaction and provide efficient mixing.

1.3.3 Heterogeneous Liquid/Liquid Reactions

In any system involving immiscible liquids, cavitation bubble collapse at or near the interface between the two liquids will cause disruption and mixing – resulting in the formation of very fine emulsions. In chemistry such emulsions provide enormous interfacial contact areas between immiscible liquids. For this reason it has been used in conjunction with phase transfer catalysis. Ultrasonic emulsification has also been used for many years in the food industry for the production of tomato sauce, mayonnaise and other similar blended items. Emulsification also has an important role in the plastics industry because a significant proportion of synthetic polymers (over 50% of elastomers) are produced industrially by emulsion polymerisation.

1.4 Electrochemistry

In its simplest form electrochemistry is a study of reactions that take place as the result of applying an electrochemical potential to a chemical system. Electrochemical reactions take place at the interface between an electrode and the electrolyte solution. They involve the transfer of electrons between the electrode surface and the electrolyte or species in solution together with associated chemicals reaction(s) (see Figure 1.5).

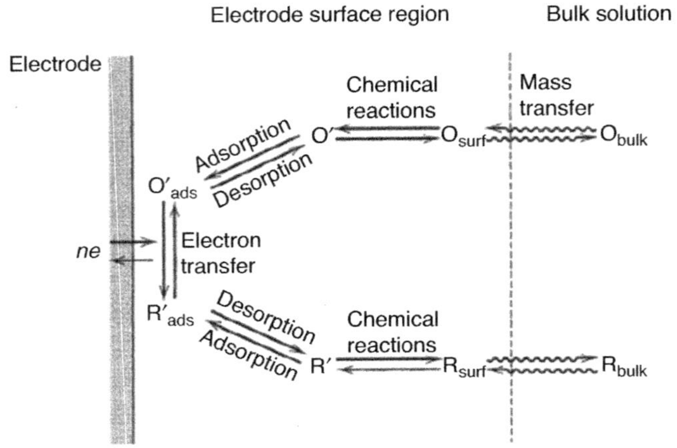

Figure 1.5 Scheme of the electrochemical processes on the electrode surface.

1.5 Sonoelectrochemistry – The Application of Ultrasound in Electrochemistry

Sonoelectrochemistry is the coupling of ultrasonic radiation to an electrochemical system. The first application of ultrasound in an electrochemical process dates back to the 1930s and the work of Moriguchi [8] who showed that under sonication water electrolysis using a platinum electrode took place at lower voltages and faster rates than under normal (silent) conditions. In succeeding years research in this field was carried out mainly by metallurgists [9] who were concerned with improving the efficiency of electroplating and the quality of the deposits. However, the term 'sonoelectrochemistry' did not appear in print until 1990 [10].

It might be thought that the effects of ultrasonic irradiation upon an electrochemical process would be limited solely to considerations of the heterogeneous system involving the electrode surface and the electrolyte layer close to this surface, but this is not entirely true. Homogeneous processes that take place in the bulk solution must also be taken into account since the electroactive species in the bulk solution can suffer reactions under the extreme conditions generated during the cavitational collapse and new mechanisms of reaction can arise [11]. In addition, if the electrolyte consists of immiscible liquids then sonication will generate and maintain a very fine emulsion.

The benefits of sonoelectrochemistry have been reported for a number of processes, including electrosynthesis, electroanalysis, bioelectrochemistry, the synthesis of conducting polymers, electroplating, nanomaterial preparation and electrocatalysis [10, 11].

The particular advantages which accrue from the use of ultrasound in electrochemistry include:

(i) degassing at the electrode surface,
(ii) disruption of the diffusion layer, which reduces depletion of electroactive species,
(iii) improved mass transport of ions across the double layer,
(iv) continuous cleaning and activation of the electrode surfaces.

All of these effects combine to provide enhanced yield and improved electrical efficiency.

Given the remarkable effects recorded in the early years of the study of sonoelectrochemistry, it is perhaps surprising that the subject took so long to become accepted. This is almost certainly due to the complexity of the processes induced by ultrasound and the high sensitivity of the results to the experimental conditions such as cell geometry, electrode position and electrode–ultrasound emitter distance as well as the effects of the ultrasonic parameters themselves, such as acoustic power and frequency [12]. However, in recent years, the development of new electrochemical and sonochemical equipment has encouraged a resurgence of this technology and new applications are being developed [13, 14].

There is particular interest in the applications of ultrasound to electrochemistry as a means of making such processes more environmentally friendly. This includes the minimisation of waste and the replacement of toxic components, for example, alternatives to mercury electrodes, and the reduction or replacement of chemical additives. Sonoelectrochemistry can also be used for pollutant degradation and remediation, and to provide more efficient sensors for the detection of various species in water and in the atmosphere.

1.5.1 Ultrasonic Factors that Influence Sonoelectrochemistry

When the ultrasound field intensity is above the cavitation threshold, the collapse of the bubbles is also associated with the formation of shock waves [15] and microflows [16] among other phenomena [17]. In this section the mechanisms that impinge directly on electrochemistry as a result of cavitation will be considered.

1.5.1.1 Acoustic Streaming

When an acoustic wave travels through a medium, part of its energy will be absorbed inducing circular flow in the electrolyte. This phenomena is known as acoustic streaming [17]. Acoustic streaming can occur in three different regions [18]:

(i) in the bulk solution the pattern is as shown in Figure 1.6(a).
(ii) on the surface of solids for example, the reactor walls as shown in Figure 1.6(b);
(iii) at the boundary layer as shown in Figure 1.6(c).

The strength of acoustic streaming is directly proportional to the ultrasonic intensity, the attenuation coefficient of the medium and the surface area of ultrasonic emitter, and is inversely proportional to the bulk viscosity of the electrolyte and the speed of sound in the medium [18].

Acoustic streaming can reach parts of the electrolyte away from the direct influence of the ultrasonic emitter because the liquid driven by the radiation pressure can move away from this region [19]. Thus an increase in mass transfer can be observed even when the working electrode is located outside the volume directly affected by high intensity ultrasound [20].

The main effect of the acoustic streaming is to promote the movement of the solution, enhancing mass transfer of electroactive compounds to the electrode surface, and to reduce the diffusion boundary layer [11, 20].

1.5.1.2 Turbulent Flow

The movement of the cavitational bubbles under the influence of the ultrasound field can cause a chaotic and disorderly movement of the solution, giving rise to a turbulent flow. The levels of turbulence are high near to the emitting surface and decrease rapidly with

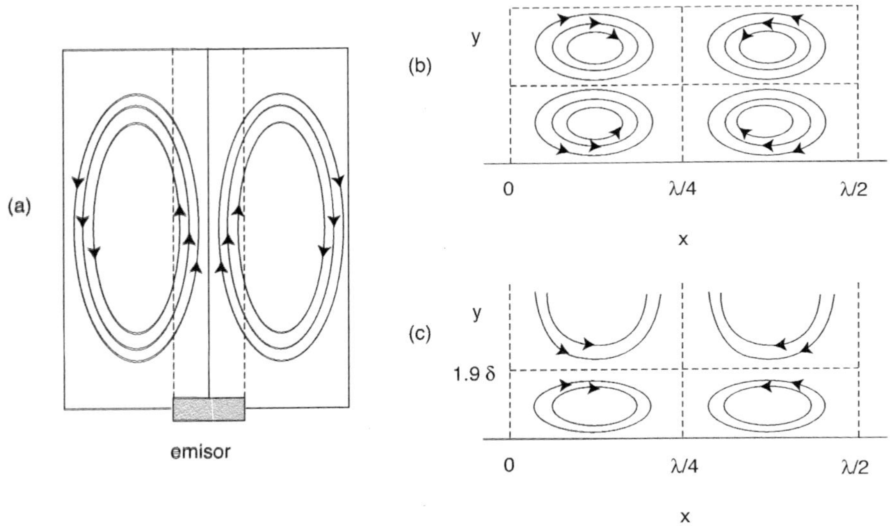

Figure 1.6 *Typical flow patterns of acoustic streaming. Reprinted with permission from Journal of Sound and Vibration, Acoustic Streaming by Sir James Lighthill, 61, 3, 391–418 Copyright (1978) Elsevier Ltd.*

increasing distance from it [21]. Turbulent flow is effective in improving mass transport processes within the solution and at the electrode surface and thus it can cause similar effects to acoustic streaming.

1.5.1.3 Microjets

Microjets are formed at any surface dipped in a liquid when the liquid is subjected to acoustic cavitation (see Section 1.3.2) [22]. If the surface is an electrode and collapse takes place close to the surface the liquid jet will directly impinge upon it, as shown in Figure 1.7 [23].

The formation of the jet and its shape was predicted theoretically by Plesset and Chapman [24] and subsequently confirmed through experimental observation by Lauterborn and Bolle [25]. These studies showed that bubbles that collapsed close to a surface do not remain spherical and so the final temperatures and pressures reached after collapse will be lower in comparison with the values reached during the symmetrical collapse of a bubble in the bulk solution.

Bubbles close to the electrode surface will also give rise to microstreaming (see Section 1.5.1.1) [26]. The combined effects of microjets and microstreaming contribute to the enhanced mass transport of electroactive compounds to the electrode. However, there are two other effects – electrode cleaning and surface activation. These effects will prevent fouling of the electrode surface and can also improve any electrodeposition processes [27].

1.5.1.4 Emission of Shock Waves

There is one more mechanical effect arising from cavitation in addition to shear forces and jetting in the bulk liquid and that is a shock wave generated in the last stages of the violent

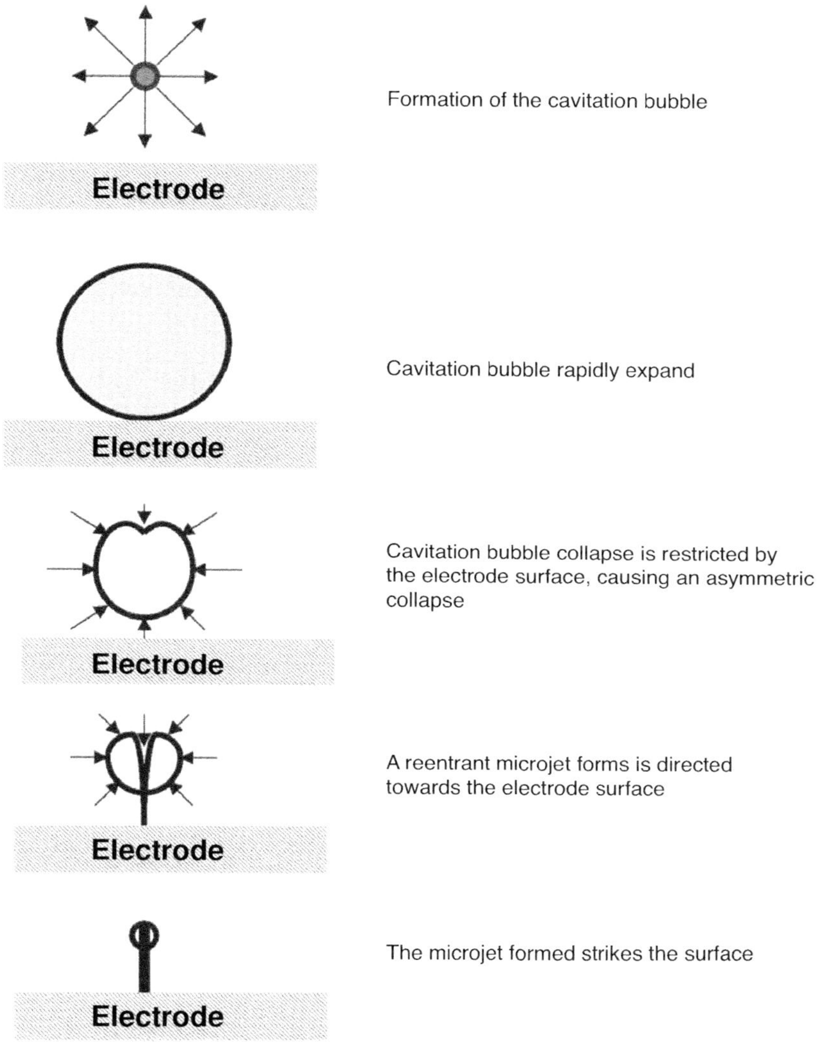

Figure 1.7 *Schematic diagram of the cavitating bubble implosion sequence at or near the electrode surface. The arrows indicate fluid pressure.*

collapse [28]. The attenuation of this pressure wave is rapid and so effects on the electrode surface will only take place when the collapse occurs close to it.

1.5.1.5 Chemical Effects

In addition to the acoustic effects above, which are mechanical in nature, there will also be sonochemical effects (see Section 1.3). In any liquid medium highly reactive radicals can be formed by the homolytic cleavage of any vapour contained in the cavitation bubble during the bubble's collapse. The nature of these radicals will depend on the solvent used and the presence of any dissolved gases. In aqueous media and in the presence of oxygen, radicals such as $HO^•$, $HO_2{}^•$ and $O^•$ have been detected [29].

Sonochemical effects [30–32] associated with the generation of radicals from the sonolysis of the electrolyte have been observed in several electrochemical process, for example, the electrodeposition of lead dioxide on glassy carbon [33] and the oxidation of phenol [34]. However, the influence of chemical effects in sonoelectrochemistry has been relatively little studied in comparison with the physical effects referred to above.

1.6 Examples of the Effect of Ultrasound on Electrochemical Processes under Mass Transport Conditions

The behaviour of simple electrochemistry redox agents such as Fe^{2+}/Fe^{3+}, $Fe(CN)_6{}^{4-}/Fe(CN)_6{}^{3-}$ [30] and Cr^{3+}/Cr^{4+} in aqueous media and ferrocene/ferrocinium in acetonitrile [11, 35] has been studied under the influence of an ultrasonic field. All of these processes exhibit a fast quasi-reversible electron transfer so that the electrode kinetics is considered to be under mass transport control.

Figure 1.8(a) shows the cyclic voltammogram for the electrochemical behaviour of ferrocene on a platinum electrode [35]. This curve presents the typical characteristic behaviour for a reversible one-electron transfer process. Figure 1.8(b) shows the cyclic voltammogram obtained under ultrasonic radiation at 20 kHz with 50 W cm^{-2} directed at the electrode surface with a horn electrode at a distance of 40 mm. Significant changes can be observed as a result of sonication. The mass transport is increased considerably, as evidenced by the greater current density in comparison with the peak current seen under silent conditions,

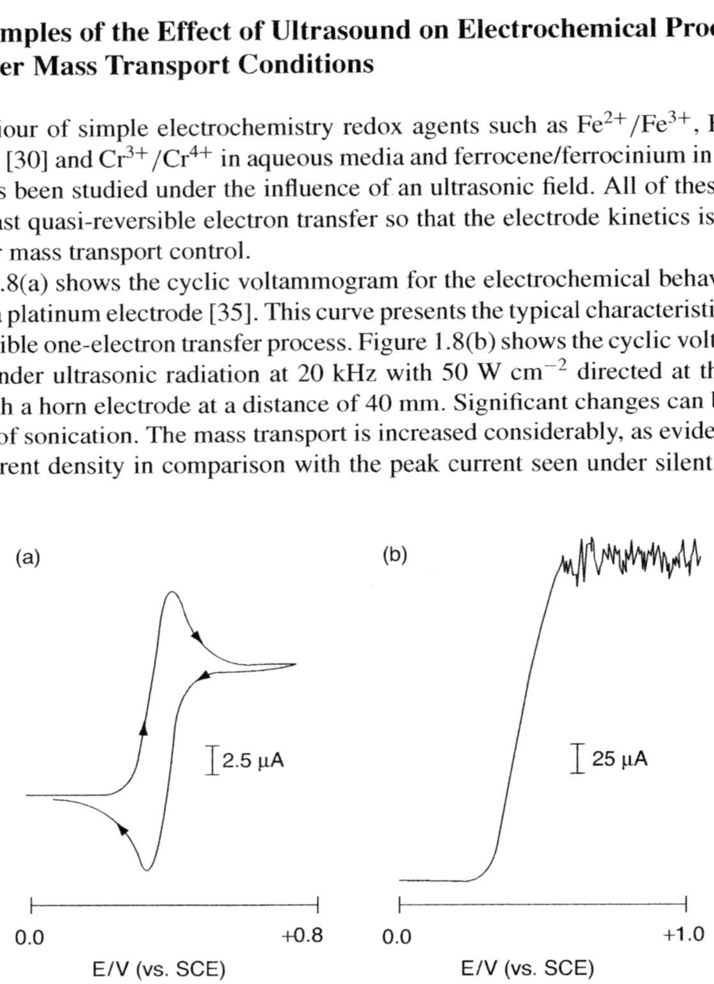

__Figure 1.8__ Cyclic voltammograms for the oxidation of 2 mM ferrocene in acetonitrile + 0.1 M NBu_4ClO_4, platinum electrode (2 mm diameter), 20 mV s^{-1} (a) silent conditions and (b) under 20 kHz, 50 W cm^{-2}, 40 mm electrode–horn distance. Reprinted with permission from Electrochimica Acta, Dual activation: coupling ultrasound to electrochemistry—an overview by R. G. Compton et al., 42, 19, 2919–2927 Copyright (1997) Elsevier Ltd.

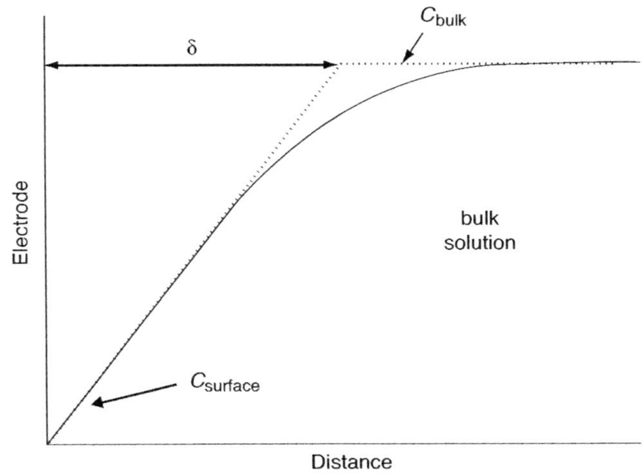

Figure 1.9 *Schematic diagram of the electrode–solution interface according to the Nernst diffusion layer model.*

and also the voltammetric response is qualitatively changed [11, 12]. The typical reversible cyclic voltammogram observed under silent conditions is transformed into a sigmoidal shape indicative of a constant rate of transport of the electroactive species to the electrode surface resulting in a sustained steady current.

The voltametric curve obtained under ultrasonic radiation can be further analysed using the Nernst diffusion layer model [11]. According to this model, the electrolyte can be divided into two zones (Figure 1.9). The closest region to the electrode surface with thickness δ is named the Nernst diffusion layer where it is assumed that there is a totally stagnant layer and thereby diffusion is the only mode of mass transport. In the region outside the Nernst diffusion layer, strong convection occurs and the concentration of all species can be considered to be constant. This model gives an approximate description of the mass transport processes at the electrode–solution boundary assuming a laminar sublayer close to the surface and an approximately linear concentration gradient across a thin layer close to the electrode surface.

The limiting current density (j_{lim}) based on this model is given by Equation (1.1):

$$j_{\mathrm{lim}} = \frac{nFD(C_{\mathrm{bulk}} - C_{\mathrm{surface}})}{\delta} \tag{1.1}$$

where j_{lim} is the limiting current density of the electrochemical process, n is the number of transferred electrons, F is the Faraday constant, D is the diffusion coefficient, C_{bulk} and C_{surface} are the concentration in the bulk solution and in the surface respectively, and δ is the diffusion layer thickness. According to Equation (1.1) the limiting current density increases when diffusion layer thickness decreases.

Marken *et al.* [36] reported that under extreme conditions the minimum diffusion layer achievable in aqueous solution is as thin as 0.7 μm. The limiting currents for the oxidation of ferrocene in the presence of an ultrasonic field was found to increase with increasing

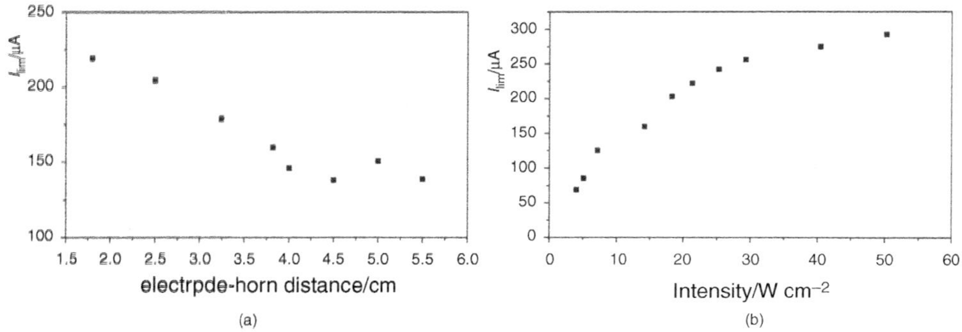

Figure 1.10 *Limiting currents for the oxidation of 2 mM ferrocene in acetonitrile + 0.1 M NBu_4ClO_4, platinum electrode (2 mm diameter), 20 kHz, (a) effect of the electrode surface–ultrasound horn distance and (b) effect of the ultrasonic intensity. Reprinted with permission from Electrochimica Acta, Dual activation: coupling ultrasound to electrochemistry—an overview by R. G. Compton et al., 42, 19, 2919–2927 Copyright (1997) Elsevier Ltd.*

ultrasonic intensity and with decreasing distance between the ultrasonic horn electrode [37], as shown in Figure 1.10. Subsequent studies using other redox couples and a range of ultrasonic configurations produced similar results [11, 38].

In summary, it has been shown that ultrasonic radiation enhances mass transport of electroactive materials to the electrode surface. It has been demonstrated that this enhancement is dependent upon the geometry of the ultrasonic horn relative to the electrode surface.

1.7 Experimental Methods for Sonoelectrochemistry

1.7.1 Cell Construction

A number of experimental arrangements have been used for the introduction of ultrasound into an electrochemical system. The most simple of these is to immerse a conventional electrochemistry cell directly in the water of an ultrasonic bath [11]. This is shown in Figure 1.11 and, in order to maximise the energy transfer of energy from the bath into the electrolyte, the cell should have a flat and thin bottom. Spherical containers are not as good since they will reduce energy transfer through some refection of the acoustic waves [22]. Other workers [39, 40] have used the whole ultrasonic bath as the electrochemical cell by placing the electrodes directly into the bath containing the electrolyte.

Using the simple experimental set-up of an immersed electrochemistry cell, the power transmitted inside is only between 1 and 5 W cm^{-2} and the results obtained depend on the position of the cell within the bath. In addition, the electrodes will interfere with and deflect the acoustic wave and cause an uneven distribution of the ultrasound field [41]. As a general rule it is essential, when using a bath as the source of ultrasound, that all of the operational parameters must be fixed in the experiments otherwise the reproducibility of results will be poor [42]. However, despite these disadvantages, a number of studies have been carried out using this configuration [35] including polymerisation reactions and metal electroplating.

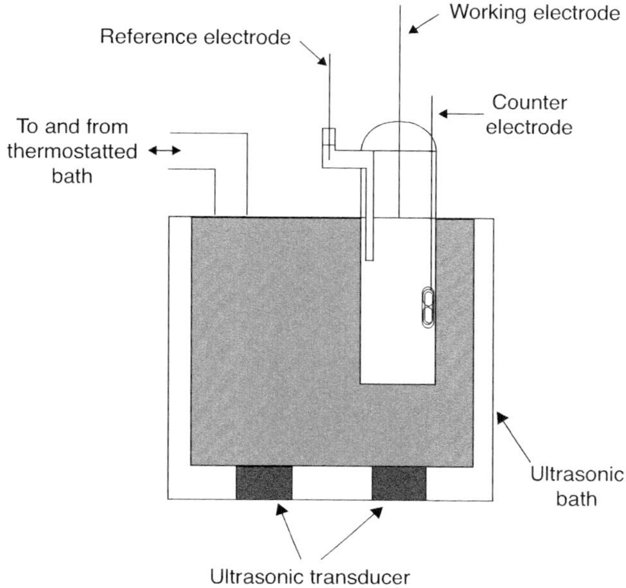

Figure 1.11 *Electrochemical cell dipped in an ultrasonic bath. Reprinted with permission from [33] Copyright (1998) Royal Society of Chemistry.*

A method by which ultrasound can be introduced directly into an electrochemical system is by using an ultrasonic probe dipped into it. An ultrasonic probe consists of a transducer driving an acoustic horn which transmits the vibrational energy to its tip and, in the most used configuration, where both the electrodes and the probe are immersed in the electrolyte solution with the horn tip facing the working electrode at a known distance [43]. This arrangement allows the ultrasonic waves to be directed onto the electrode surface and provides more efficient power control in comparison with the ultrasonic bath. An example of a thermostatted cell with the ultrasonic horn inserted centrally, from the top and opposite the working electrode, is shown in Figure 1.12. This sonoelectrochemistry arrangement allows the ultrasonic intensity and the distance between horn and electrode to be varied [11].

The ultrasonic probe system has several advantages over an ultrasonic bath for sonoelectrochemistry, in particular the fact that the acoustic power applied can be several hundred times higher. This can be applied directly to the electrochemical system and can be controlled by both the amplitude of vibration of the tip and the electrode to probe tip distance. This arrangement does, however, have a few limitations, such as:

(i) Erosion of the horn tip, releasing metal particles (usually titanium alloy) into the system.
(ii) Temperature control due to the heating effect of the horn.
(iii) The need for bipotentiostat control due to the presence of the metal tip of the ultrasound system in direct contact with the electrochemical system [11].

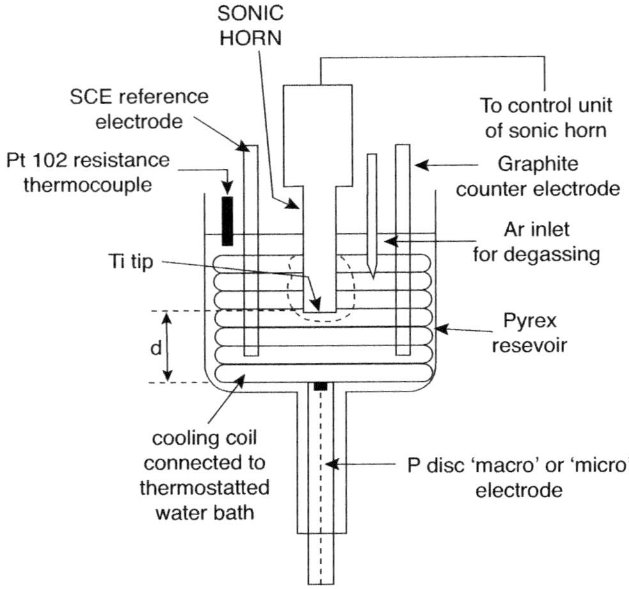

Figure 1.12 *Diagram showing an experimental set-up using an ultrasonic horn in sonoelectrochemistry experiments. Reprinted with permission from Electrochimica Acta, Electrode processes at the surfaces of sonotrodes by R. G. Compton et al., 41, 2, 315–320 Copyright (1996) Elsevier Ltd.*

Another configuration for a sonoelectrochemical cell is to use the horn itself as the working electrode and was first introduced by Reisse *et al.* [44]. Such an electrode is referred to as a sonotrode (or sonoelectrode) [11]. Subsequently the sonotrode system involving consecutive pulses of electrolysis for deposition and ultrasound to release the deposit has been used in the preparation of nanopowders [13]. Figure 1.13 shows the experimental set-up used when a titanium probe (20 kHz) acts both as a cathode and an ultrasound emitter. The electroactive part of the sonoelectrode is the planar circular surface at the bottom of the horn and the immersed cylindrical part into the electrolyte is covered by an isolating plastic jacket. The ultrasound probe is connected to a generator and a potentiostat using a pulse driver.

Some sonoelectrochemistry arrangements using flow system have also been reported [34,45].

1.7.2 Stability of the Electrodes Under Sonication

The action of microjets from the collapse of cavitational bubbles on an electrode surface can have both beneficial and deleterious effects. Certainly it can activate the surface and improve the performance of the electrodes by cleaning [10] but it can also damage the electrode surface through erosion [46]. The erosion process can change the surface area of the electrode and so the hardness of the electrode material becomes an important consideration.

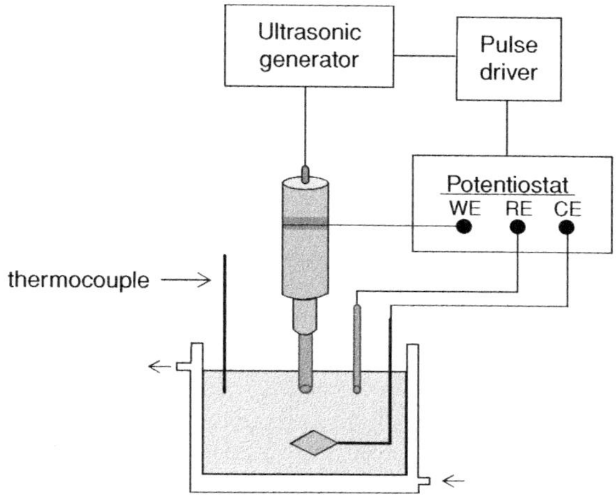

Figure 1.13 *Sonoelectrochemistry set-up used in the production of nanopowders (WE = working electrode, RE = reference electrode and CE = auxiliary electrode).*

Madigan *et al.* [48] reported a study relating to the effects of sonication on common electrodes surfaces at low frequencies (20–30 kHz), including both metals and non-metals [47], analysing the surface after ultrasonic treatment sonication by scanning electron microscopy. Applying an ultrasonic irradiation of 475 W at 20 kHz for 120 s and using 1 mm electrode surface–ultrasound horn distance, non-metals such as glassy carbon and Ebonex were severely pitted in aqueous media. By contrast, metals such as Pt, Au, W and Pd remained largely undamaged and no appreciable surface roughening was observed under the same experimental conditions.

Compton *et al.* [46] investigated the influence of ultrasonic irradiation on the erosion and roughening of Pt and Al, by means of *in situ* AC impedance spectroscopy together with subsequent atomic force microscopy characterisation of the sonicated surfaces. This study was carried out applying 60 W cm^{-2} of ultrasonic intensity at 20 kHz and the electrode was placed 35 mm away. The surface of these electrodic materials was roughed after sonication in 0.1 M TBAP/acetonitrile solution even after an exposure time as short as 120 s. This work corroborated the suggestion that electrode surface roughness increased with increased exposure time and reduced horn–electrode distance.

Several analyses about the effects of an ultrasound field on the surface of a glassy carbon electrode have been reported in the literature [33, 47]. When glassy carbon electrodes were sonicated in water using high intensity (475 W) and very short distance (2 mm) a dramatic change of the electrode surface by pitting was noticeable; by contrast, however, surface roughness did not change appreciably when sonication was carried out in an organic solvent such as dioxane [47]. The explanation for this lies in the difference in cavitation collapse energy between the two solvents. Marken *et al.* analysed the surface state of glassy carbon electrodes immersed in 0.1 M NaOH [48], applying an ultrasonic intensity of 30 W cm^{-2} at 20 kHz and for 1 cm electrode surface–horn distance, no detectable damage was observed after 30 minutes of sonication. Later work [33] carried out with an ultrasound bath at 30 kHz applying 100 W with the glassy carbon electrodes 6 cm from the transducer showed that the surface of the electrode did not exhibit any signs of superficial pitting or damage after 20 minutes of sonication.

Electrodes made from boron-doped diamond have been widely used in sonoelectrochemistry [49], especially in sonoelectroanalytical applications [50], due to its attractive properties including a wide potential window, low background currents and high resistance to electrode fouling. These electrodes exhibit remarkable resistance to etching and are mechanically stable in sonoelectrochemical applications.

In summary, it has been shown that ultrasound can damage the electrode surface and that the extent of damage depends on the electrode material, intensity of sonication, the distance between the electrode and the source of ultrasound and the duration of the experiment. In electrochemical work it is important that the electrode surface remains undamaged throughout the experiment and this is an even more critical criterion in sonoelectrochemistry where cavitation damage is a possibility and must be avoided. For this reason, it is important to ascertain whether there is any surface damage to the electrodes during sonication in order to make sure that the results are reproducible and quantitative.

1.7.3 Some Applications of Sonoelectrochemistry

Table 1.2 summarises a range of examples of applications of sonoelectrochemistry and some of these will be treated in more depth in the following chapters of this book.

1.7.3.1 Electro- and Electroless Deposition

The benefits of the application of an ultrasound field to electrodeposition processes are well established and it has been widely used. Several researchers have demonstrated that employing ultrasound in electroplating baths enhances the deposition rate and the quality of the deposit, increasing thickness, hardness and brightness of the deposited films. Several metallic deposits such as silver oxide [68], silver [66] and tungsten [69], together with bioactive calcium phosphate coating [70] and CdSe film semiconductors [71], have been obtained under ultrasound.

Studies have also been made on the influence of ultrasonic irradiation on the electroless plating of nickel [57], copper [72] and cobalt [73].

1.7.3.2 Electroanalysis

Several studies have reported the use of ultrasound for the electroanalytical determination of metals [74] and organic compounds [75] using stripping techniques on different types of electrode and in various complex matrices. For example Banks *et al.* [76] reported the deposition and detection of cadmium using boron-doped diamond electrodes with the aid of anodic stripping with square wave voltammetry in water. The detection limit was 2.5×10^{-8} M in silent conditions and in the presence of ultrasound the detection limit was improved to 3.9×10^{-9} M. The improvement in the detection limit reflected ultrasonically enhanced mass transport of the electroactive species to the electrode surface during the deposition process.

Table 1.2 Examples of applications of sonoelectrochemistry.

	Advantages over silent conditions	Examples
Sonoelectroorganic synthesis	Changes the product yield and the selectivity Increases current efficiency Lower voltage applied Avoids electrode fouling	Electroreduction of benzophenone [51] and acrylonitrile [52] Electrooxidation of diacetone-*L*-sorbose [53]
Sonoelectroanalysis	Limiting currents increase due to increased mass transport and electrode surface passivation is prevented	Vitamin C analysis [54] Determination of manganese [55]
Electrodeposition and electroless	Uniform deposits with superior quality are obtained Increases deposit hardness and brightness Decreases of the internal stress and the grain size of the deposit	Electrodeposition of copper [56] and lead [40] Deposition of nickel by electroless [57] Aniline polymerisation [58]
Production of nanomaterials	Avoids the use of reducing agents and surfactants Allows the use of water as a solvent Size of the nanoparticles can be adjusted by varying the operating parameters (ultrasonic power, current density, deposition potential and the ultrasonic vs electrochemical pulse times)	Synthesis of CdSe nanotubes [59] Synthesis of copper [60], platinum [61] and gold nanoparticles [62]
Environmental protection	Enhances mass transport of the pollutant to the electrode surface Avoids electrode fouling Possibility of new paths of degradation	Water disinfection [63] Dye [64] and organochlorine compound degradation [65] Recovery of silver [66] Cyanide degradation [67]

1.7.3.3 Environmental Protection

The application of sonoelectrochemistry to the protection of the environment is an emerging field of research that has been applied successfully for the degradation of a range of important contaminants in water [34, 64, 65]. This technology can be used in combination with other methods including biological treatment, ozonation, photolysis and Fenton processes [77] in order to achieve synergistic effects and reduce the energy cost of the treatment [14].

Sonoelectrochemistry has several advantages over more conventional technologies for the removal of contamination in that it can be carried out under milder conditions, at room temperature and without the need for additional chemical treatment.

1.7.3.4 Electrosynthesis

The benefits of the application of ultrasound to electroorganic synthesis includes activation of the electrode surface, increased yield and the possibility of alteration in the mechanistic pathway together with changed product distribution [53]. A well-known example of product selectivity change induced by sonication is the Kolbe reaction for oxidation of carboxylate anions to hydrocarbons [10].

1.7.3.5 Nanomaterials

Currently the synthesis of nanomaterials is one of the most active fields of research in sonoelectrochemistry [13]. Pulsed sonoelectrochemical techniques [78] have been employed to produce numerous pure metals [79, 80], alloys [81], semiconductors [82] and polymer nanoparticles [83]. The particular attributes of metal powders obtained using this technology are that they are produced in a very finely divided state with high surface area, high chemical purity and with an average particle size of 100 nm [84].

References

(1) Wood, R.W. and Loomis, A.L. (1927) The physical and biological effects of high frequency sound waves of great intensity. *Philos. Mag.*, **4**, 417–436.
(2) Richards, W.T. and Loomis, A.L. (1927) The chemical effects of high frequency sound waves I. A preliminary survey. *J. Am. Chem. Soc.*, **49**, 3086–3100.
(3) Richards, W.T. (1929) The chemical effects of high frequency sound waves II. A study of emulsifying action. *J. Am. Chem. Soc.*, **51**, 1724–1729.
(4) Weissler, A. (1948) Ultrasonics in chemistry. *J. Chem. Educ.*, **25**, 28–30.
(5) Neppiras, E.A. (1972) Macrosonics in industry 1. Introduction. *Ultrasonics*, **10**, 9–13.
(6) Lorimer, J.P. and Mason, T.J. (1987) Sonochemistry Part 1 – The physical aspects. *Chem. Soc. Rev.*, **16**, 239–274.
(7) Lindley, J. and Mason, T.J. (1987) Sonochemistry Part 2 – Synthetic applications. *Chem. Soc. Rev.*, **16**, 275–311.
(8) Moriguchi, N. (1934) The effect of supersonic waves on chemical phenomena (III). The effect on the concentration polarization. *J. Chem. Soc. Jpn.*, **55**, 749–750.
(9) Walker, R. (1990) Ultrasound and electroplating. *Chemistry in Britain*, **26**, 251–254.
(10) Mason, T.J., Lorimer, J.P. and Walton, D.J. (1990) Sonoelectrochemistry. *Ultrasonics*, **28**, 333–337.
(11) Compton, R.G., Eklund, J.C. and Marken, F. (1997) Sonoelectrochemical Processes: A review. *Electroanalysis*, **9**, 509–522.
(12) Compton, R.G., Eklund, J.C., Marken, F., Rebbitt, T.O., Akkermans, R.P. and Waller, D.N. (1997) Dual activation: coupling ultrasound to electrochemistry – an overview. *Electrochim. Acta*, **42**, 2919–2927.
(13) Sáez, V. and Mason, T.J. (2009) Sonoelectrochemical synthesis of nanoparticles. *Molecules*, **14**, 4284–4299.
(14) González-García, J., Esclapez, M.D., Bonete, P., Vargas Hernández, Y., GaeteGarretón, L. and Sáez, V. (2010) Current topics on sonoelectrochemistry. *Ultrasonics*, **50**, 318–322.
(15) Birkin, P.R., Offin, D.G., Joseph, P.F. and Leighton, T.G. (2005) Cavitation, shock waves and the invasive nature of sonoelectrochemistry. *J. Phys. Chem. B*, **109**, 16997–17005.
(16) Klima, J. and Bernard, C. (1999) Sonoassisted electrooxidative polymerisation of salicylic acid: Role of acoustic streaming and microjetting. *J. Electroanal. Chem.*, **462**, 181–186.

(17) Leighton, T.G. (1994) *The Acoustic Bubble*, Academic Press, London.
(18) Frenkel, V., Gurka, R., Liberzon, A., Shavit, U. and Kimmel, E. (2001) Preliminary investigations of ultrasound induced acoustic streaming using a particle image velocimetry. *Ultrasonics*, **39**, 153–156.
(19) Lighthill, S.J. (1978) Acoustic streaming. *J. Sound Vibration*, **61**, 391–418.
(20) Klima, J. (2011) Application of ultrasound in electrochemistry. An overview of mechanisms and design of experimental arrangement. *Ultrasonics*, **51**, 202–209.
(21) Kumar, A., Kumaresan, T., Pandit, A.B. and Joshi, J.B. (2006) Characterization of flow phenomena induced by ultrasonic horn. *Chem. Eng. Sci.*, **61**, 7410–7420.
(22) Mason, T.J. (1999) *Sonochemistry*, Oxford University Press, New York.
(23) Lauterborn, W. (1982) Cavitation bubble dynamics – New tools for an intricate problem. *Appl. Sci. Res.*, **38**, 165–178.
(24) Plesset, M.S. and Chapman, R.B. (1971) Collapse of an initially spherical vapour cavity in the neighbourhood of a solid boundary. *J. Fluid Mech.*, **47**, 283–290.
(25) Lauterborn, W. and Bolle, H. (1975) Experimental investigations of cavitation-bubble collapse in the neighbourhood of a solid boundary. *J. Fluid Mech.*, **72**, 391–399.
(26) Elder, S.A. (1959) Cavitation microstreaming. *J. Acoust. Soc. Am.*, **31**, 54–64.
(27) Mason, T.J. (1989) *Sonochemistry: The Uses of Ultrasound in Chemistry*, Royal Society of Chemistry, Cambridge, UK.
(28) Rayleigh, L. (1917) On the pressure developed in a liquid during the collapse of a spherical cavity. *Philos. Mag.*, **34**, 94–98.
(29) Riesz, P. and Kondo, T. (1992) Free radical formation induced by ultrasound and its biological implications. *Free Rad. Biol. Med.*, **13**, 247–270.
(30) Suslick, K.S. and Doktycz, S.J. (1990) *Advances in Sonochemistry*, Vol. **1** (ed. T.J. Mason), JAI Press, London.
(31) Sáez, V., Frías-Ferrer, A., Iniesta, J., González-García, J., Aldaz, A. and Riera, E. (2005) Characterization of a 20 kHz sonoreactor. Part II: Analysis of chemical effects by classical and electrochemical methods. *Ultrason. Sonochem.*, **12**, 67–72.
(32) Marchante, E., Lana-Villarreal, T., Sáez, V., González-García, J. and Gomez, R. (2009) Sonopotential: A new concept in electrochemistry. *Chem. Commun.*, **27**, 4127–4129.
(33) González-García, J., Iniesta, J., Aldaz, A. and Montiel, V. (1998) Effects of ultrasound on the electrodeposition of lead dioxide on glassy carbon electrodes. *New J. Chem.*, **22**, 343–347.
(34) Trabelsi, F., Ait-Lyazidi, H., Ratsimba, B., Wilhelm, A.M., Delmas, H., Fabre, P.-L. and Berlan, J. (1996) Oxidation of phenol in wastewater by sonoelectrochemistry. *Chem. Eng. Sci.*, **51**, 1857–1865.
(35) Walton, D.J., Phull, S.S., Chyla, A., Lorimer, J.P., Mason, T.J., Burke, L.D., Murphy, M., Compton, R.G., Eklund, J.C. and Page, S.D. (1995) Sonovoltammetry at platinum electrodes: surface phenomena and mass transport processes. *J. Appl. Electrochem.*, **25**, 1083–1090.
(36) Marken, F., Akkermans, R.P. and Compton, R.G. (1996) Voltammetry in the presence of ultrasound: the limit of acoustic streaming induced diffusion layer thinning and the effect of solvent viscosity. *J. Electroanal. Chem.*, **415**, 55–63.
(37) Compton, R.G., Eklund, J.C., Page, S.D., Mason, T.J. and Walton, D.J. (1996) Voltammetry in the presence of ultrasound: mass transport effects. *J. Appl. Electrochem.*, **26**, 775–784.
(38) Kumbhat, S. (2000) Potentialities of power ultrasound in electrochemistry: An overview. *Bull. Electrochem.*, **16**, 29–32.
(39) Walton, D.J., Burke, L.D. and Murphy, M.M. (1997) Sonoelectrochemistry: Chlorine, hydrogen and oxygen evolution at platinised platinum. *Electrochim. Acta*, **41**, 2747–2751.
(40) Agullo, E., González-García, J., Exposito, E., Montiel, V. and Aldaz, A. (1999) Influence of an ultrasonic field on lead electrodeposition on copper using a fuoroboric bath. *New J. Chem.*, **23**, 95–101.

(41) Brett, C. (2008) *Sonoelectrochemistry. Piezoelectric Transducers and Applications* (ed. A. Arnau), Springer-Verlag, Berlin Heidelberg, Germany.
(42) Reisse, J., Vandercammen, J., Yang, D.H., Broeckaert, L., Caulier, T., Fabre, O., Maerschalk, C., Lepoint, Th. and Mullie, F. (1992) *Current Trends in Sonochemistry* (ed. G.J. Price), Royal Society of Chemistry, Cambridge.
(43) González-García, J. (2010) *Sonoelectrochemical Synthesis of Materials. Theoretical and Experimental Sonochemistry involving Inorganic Systems* (eds M. Ashokkumar and A. Kumar), M/S Springer, UK.
(44) Reisse, J., Francois, H., Vandercammen, J., Fabre, O., Kirsch-de Mesmaeker, A., Maerschalk, C. and Delplancke, J.L. (1994) Sonoelectrochemistry in aqueous electrolyte: A new type of sonoelectroreactor. *Electrochim. Acta*, **39**, 37–39.
(45) Esclapez, M.D., Sáez, V., Milán-Yáñez, D., Tudela, I., Louisnard, O. and González-García, J. (2010) Sonoelectrochemical treatment of water polluted with trichloroacetic acid: From sonovoltammetry to pre-pilot plant scale. *Ultrason. Sonochem.*, **17**, 1010–1020.
(46) Compton, R.G., Eklund, J.C., Page, S.D., Sanders, G.H.W. and Booth, J. (1994) Voltammetry in the presence of ultrasound. sonovoltammetry and surface effects. *J. Phys. Chem.*, **98**, 12410–12414.
(47) Madigan, N.A., Hagan, C.R.S., Zhang, H. and Coury, L.A. Jr. (1996) Effects of sonication on electrode surfaces and metal particles. *Ultrason. Sonochem.*, **3**, S239–S247.
(48) Marken, F., Kumbhat, S., Sanders, G.H.W. and Compton, R.G. (1996) Voltammetry in the presence of ultrasound: surface and solution processes in the sonovoltammetric reduction of nitrobenzene at glassy carbon and gold electrodes. *J. Electroanal. Chem.*, **414**, 95–105.
(49) Sáez, V., González-García, J., Kulandainathan, M.A. and Marken, F. (2007) Electro-deposition and stripping of catalytically active ion metal nanoparticles at boron-doped diamond electrodes. *Electrochem. Commun.*, **9**, 1127–1133.
(50) Compton, R.G., Hardcastle, J.L. and del Campo, J. (2007) *Sonoelectroanalysis: Applications. Encyclopedia of Electrochemistry*, Wiley-VCH Verlag GmbH & Co.
(51) Atobe, M., Tonoi, T. and Nonaka, T. (2002) Ultrasonic effects on electroorganic processes. 21. A limiting current study on electroreduction of benzophenone in the presence of a proton donor. *Ultrason. Sonochem.*, **9**, 71–77.
(52) Atobe, M., Sasahira, M. and Nonaka, T. (2000) Ultrasonic effects on electroorganic processes: Part 17. Product selectivity control in cathodic reduction of acrylonitrile. *Ultrason. Sonochem.*, **7**, 103–107.
(53) Cognet, P., Wilhelm, A.M., Delmas, H., Ait Lyazidi, H. and Fabre, P.L. (2000) Ultrasound in organic electrosynthesis. *Ultrason. Sonochem*, **7**, 163–167.
(54) Akkermans, R.P., Wu, M. and Compton, R.G. (1998) A comparison between pulsed sonovoltammetry and low power laser activated voltammetry for the electroanalysis of ascorbic acid in a commercial fruit drink. *Electroanalysis*, **10**, 814–820.
(55) Saterlay, A.J., Foord, J.S. and Compton, R.G. (1999) Sono-cathodic stripping voltammetry of manganese at a polished boron-doped diamond electrode: Application to the determination of manganese in instant tea. *Analyst*, **124**, 1791–1796.
(56) Hong, B., Jiang, C.H. and Wang, X.J. (2008) Effects of ultrasound on morphology of copper electrodeposited on titanium in aqueous and organic solutions. *Mater. Trans.*, **49**, 275–277.
(57) Abyaneh, M.Y., Sterritt, A. and Mason, T.J. (2007) Effects of ultrasonic irradiation on the kinetics of formation, structure, and hardness of electroless nickel deposits. *J. Electrochem. Soc.*, **154**, D467–D472.
(58) Atobe, M., Kaburagi, T. and Nonaka, T. (1999) Ultrasonic effects on electroorganic processes XIII. A role of ultrasonic cavitation in electro-oxidative polymerization of aniline. *Electrochemistry*, **67**, 1114–1117.

(59) Shen, Q., Jiang, L., Miao, J., Hou, W. and Zhu, J.J. (2008) Sonoelectrochemical synthesis of CdSe nanotubes. *Chem. Commun.*, **14**, 1683–1685.
(60) Haas, I., Shanmugam, S. and Gedanken, A. (2006) Pulsed sonoelectrochemical synthesis of size-controlled copper nanoparticles stabilized by Poly(N-vinylpyrrolidone). *J. Phys. Chem. B*, **110**, 16947–6952.
(61) Zin, V., Pollet, B.G. and Dabala, M. (2009) Sonoelectrochemical (20 kHz) production of platinum nanoparticles from aqueous solutions. *Electrochim. Acta*, **54**, 7201–7206.
(62) Aqil, A., Serwas, H., Delplancke, J.L., Jerome, R., Jerome, C. and Canet, L. (2008) Preparation of stable suspensions of gold nanoparticles in water by sonoelectrochemistry. *Ultrason. Sonochem.*, **15**, 1055–1061.
(63) Joyce, E., Mason, T.J., Phull, S.S. and Lorimer, J.P. (2003) The development and evaluation of electrolysis in conjunction with power ultrasound for the disinfection of bacterial suspensions. *Ultrason. Sonochem.*, **10**, 231–234.
(64) Lorimer, J.P., Mason, T.J., Plattes, M. and Phull, S.S. (2000) Dye effluent decolourisation using ultrasonically assisted electro-oxidation. *Ultrason. Sonochem.*, **7**, 237–242.
(65) Sáez, V., Esclapez, M.D., Tudela, I., Bonete, P., Louisnard, O. and González-García, J. (2010) 20 kHz sonoelectrochemical degradation of perchloroethylene in sodium sulphate aqueous media: Influence of the operational variables in batch mode. *J. Hazard. Mater.*, **183**, 648–654.
(66) Pollet, B., Lorimer, J.P., Phull, S.S. and Hihn, J.Y. (2000) Sonoelectrochemical recovery of silver from photographic processing solutions. *Ultrason. Sonochem.*, **7**, 69–76.
(67) Iordache, I., Nechita, M.T., Rosaca, I. and Aelenei, N. (2004) Ultrasound assisted electrochemical degradation of cyanides: Influence of electrode type. *Turkish J. Eng. Env. Sci.*, **28**, 377–380.
(68) Saterlay, A.J., Wilkins, S.J., Goeting, Ch.H., Foord, J.S., Compton, R.G. and Marken, F. (2000) Sonoelectrochemistry at highly boron-doped diamond electrodes: silver oxide deposition and electrocatalysis in the presence of ultrasound. *J. Solid State Electrochem.*, **4**, 383–389.
(69) Goeting, Ch.H., Foord, J.S., Marken, F. and Compton, R.G. (1999) Sonoelectrochemistry at tungsten-supported boron-doped CVD diamond electrodes. *Diam. Relat. Mater.*, **8**, 824–829.
(70) Han, H.M., Phillips, G.J., Mikhalovsky, S.V., FitzGerald, S. and Lloyd, A.W. (2008) Sonoelectrochemical deposition of calcium phosphates on carbon materials: effect of current density. *J. Mater. Sci.: Mater. Med.*, **19**, 1787–1791.
(71) Murali, K.R. and Sasindran, P. (2004) Structural and optical properties of sonoelectrochemically deposited CdSe films. *J. Mater. Sci.*, **39**, 6347–6348.
(72) Touyeras, F., Hihn, J.Y., Bourgoin, X., Jacques, B., Hallez, L. and Branger, V. (2005) Effects of ultrasonic irradiation on the properties of coatings obtaiend by electroless plating and electro plating. *Ultrason. Sonochem.*, **12**, 13–19.
(73) Matsuda, H. and Takano, O. (1993) Influence of ultrasonic radiation on electroless cobalt phosphorus deposition. *Trans. Inst. Met. Finish.*, **71**, 138–141.
(74) Madigan, N.A., Murphy, T.J., Fortune, J.M., Hagan, C.R.S. and Couty, L.A. Jr (1995) Sonochemical stripping voltammetry. *Anal. Chem.*, **67**, 2781–2786.
(75) Hardcastle, J.L., Paterson, C.J. and Compton, R.G. (2001) Biphasic sonoelectroanalysis: Simultaneous extraction from, and determination of vanillin in food flavoring. *Electroanalysis*, **13**, 899–905.
(76) Banks, C.E., Hyde, M.E., Tomcik, P., Jacobs, R. and Compton, R.G. (2004) Cadmium detection via boron-doped diamond electrodes: surfactant inhibited stripping voltammetry. *Talanta*, **62**, 279–286.
(77) Ai, Z., Li, J., Zhang, L. and Lee, S. (2010) Rapid decolorization of azo dyes in aqueous solution by an ultrasound-assisted electrocatalytic oxidation process. *Ultrason. Sonochem.*, **17**, 370–375.

(78) Durant, A., Delplancke, J.-L., Winand, R. and Reisse, J. (1995) A new procedure for the production of highly metal powders by pulsed sonoelectrochemical reduction. *Tetrahedron Lett.*, **36**, 4257–4260.
(79) Aqil, A., Serwas, H., Delplancke, J.L., Jerome, R., Jerome, C. and Canet, L. (2008) Preparation of stable suspensions of gold nanoparticles in water by sonoelectrochemistry. *Ultrason. Sonochem.*, **15**, 1055–1061.
(80) Shen, Q., Jiang, L., Zhang, H., Min, Q., Hou, W. and Zhu, J.J. (2008) Three-dimensional dendritic Pt nanostructures: Sonoelectrochemical synthesis and electrochemical applications. *J. Phys. Chem. C*, **112**, 16385–16392.
(81) Dabala, M., Pollet, B.G., Zin, V., Campadello, E. and Mason, T.J. (2008) Sonoelectrochemical (20 kHz) production of $Co_{65}Fe_{35}$ alloy nanoparticles from Aotani solutions. *J. Appl. Electrochem.*, **38**, 395–402.
(82) Qiu, X., Lou, Y., Samia, A.C.S., Devadoss, A., Burgess, J.D., Dayal, S. and Burda, C. (2005) PbTe Nanorods by sonoelectrochemistry. *Angew. Chem., Int. Ed. Engl.*, **44**, 5855–5857.
(83) Atobe, M., Ishikawa, K., Asami, R. and Fuchigami, T. (2009) Size-controlled synthesis of conducting polymer microspheres by pulsed sonoelectrochemical polymerization. *Angew. Chem., Int. Ed. Engl.*, **48**, 6069–6072.
(84) Reisse, J., Caulier, T., Deckerkheer, C., Fabre, O., Vandercammen, J., Delplancke, J.L. and Winand, R. (1996) Quantitative sonochemistry. *Ultrason. Sonochem.*, **3**, S147–S151.

2

The Use of Electrochemistry as a Tool to Investigate Cavitation Bubble Dynamics

Peter R. Birkin

2.1 Introduction

Although electrochemistry is dominated by the processes that occur at the solid/liquid interface of an electrode material, it offers many powerful tools for the analysis of cavitation and bubble dynamics within complex environments. This chapter is aimed at illustrating the role that electrochemistry can play in providing evidence for different mechanisms triggered by pressure changes within a liquid environment. However, although electrochemical techniques are powerful and elucidative in their ability to provide key information on many bubble processes, it is important to recognise that due to the complex nature of the systems involved, the electrochemical data provides complementary evidence to the analysis of the system. Full elucidation of the mechanism and changes that are occurring in the media requires the combination of electrochemical methods with other experimental techniques. Once this has been recognised, and as it will be demonstrated here, useful insights into the processes involved within the media are possible. In addition to this experimental approach, an understanding of the basic acoustics of the media and the vessels employed for its deployment is recommended to the reader.

In the next section an outline of the general classes of bubble dynamics is given, which will be used throughout the chapter as a reference point for the electrochemical experimental data and interpretation given.

Power Ultrasound in Electrochemistry: From Versatile Laboratory Tool to Engineering Solution, First Edition. Edited by Bruno G. Pollet.

2.2 An Overview of Bubble Behaviour

In order to understand the behaviour of cavitation bubbles within a particular environment, and the subsequent chemical and physical effects which are observed from such systems, an appreciation of the interaction of gas bubbles within the liquid and an applied pressure field (e.g. a ultrasonic field within a reactor of some design) is required. This is a complex subject and the reader is directed to comprehensive texts which eloquently describe in detail the dynamics and processes involved [1, 2]. In this section we will give an overview of some of the important factors which are relevant to the understanding of the electrochemical data obtained from a variety of experiments reported here. In the first instance gas bubbles are extremely active in a sound field. However, this activity is dominated by the physical parameters of the system. This will include the local pressure, the solution properties, surface tension, viscosity, the dynamics of the sound field imposed on the system and the bubble size and so on. It is illustrative to consider the behaviour of a single gas bubble (note this consideration will have to be modified when bubble clusters [3–5] or other multibubble cavitation environments are generated) as a sinusoidal pressure field is imposed on the system in addition to the base hydrostatic pressure value. Here we impose amplitudes which induce bubble motion of various magnitudes. Initially, at low pressure amplitudes (e.g. 10s Pa zero-to-peak amplitude), the bubble oscillates in its so-called breathing mode in a linear fashion related to the frequency of the sound applied. In addition, the amplitude of this oscillation is maximised if the sound field matches the bubble's resonant frequency. The resonant frequency of a bubble in an infinite medium is conveniently described by the Minnaert equation [1]. Either side of this frequency, the magnitude of motion falls (and the phase relationship between the bubble motion and the imposed sound field alters). However, as the pressure is increased further, surface oscillations (e.g. Faraday waves) will also be generated. These are larger in displacement amplitude and, as will be subsequently demonstrated, have a significant impact on local mass transfer to an electrode surface placed in close proximity to the driven gas bubble, but have a weak acoustic signature (whereas the breathing mode has a strong acoustic signature). Increasing the pressure amplitude further will cause non-linear behaviour of the bubbles to be generated. Two pressure amplitude thresholds become relevant to the explanation of the final and most dramatic stage of bubble behaviour; inertial cavitation. These thresholds are the Blake pressure and the inertial/non-inertial boundary. The former refers to the condition where rapid growth of a bubble is encountered. The later refers to the conditions where the pressure conditions are sufficient for the collapse phase of the bubbles cycle to be dominated by the inertial forces of the liquid rather than the pressure within the bubble interior.

Once these inertial forces [1, 6, 7] dominate the collapse, the dramatic effects of cavitation (e.g. shock wave emission, surface damage, sonochemistry, sonoluminescence and so on [8, 9]) are expected. Hence, this last scenario, termed inertial cavitation, [10–12] can be exploited in, for example, dentistry [13, 14] and lithotripsy [15]. In an effort to decide on which type of cavitation, whether inertial (sometimes referred to as 'transient') or non-inertial (sometimes referred to as 'stable'), are likely to be produced by a particular experimental arrangement, Apfel and Holland produced a relatively simple model based on the response of a single bubble in relation to a sinusoidal pressure wave [10–12, 16, 17]. This study also linked experimental data used to infer temperatures [18–23] within the core of the collapsing bubble (~5000 K) to the minimum pressure amplitude necessary to

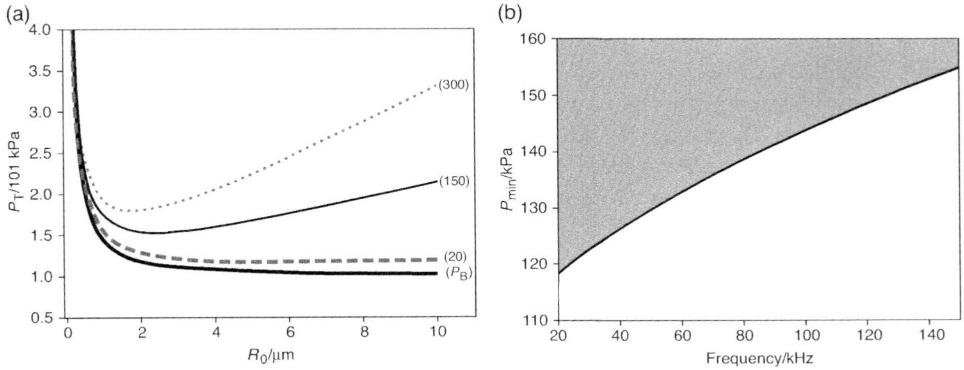

Figure 2.1 *(a) Plot showing the Blake threshold for explosive growth (P_B) and the inertial pressure threshold (P_T) for three different ultrasonic frequencies (shown in parenthesis in kPa) as a function of initial bubble size. (b) Plot showing the minimum zero-to-peak pressure required to generate an inertial event under ideal (e.g. initial bubble size) conditions as a function of the ultrasonic frequency employed. The shaded region indicates where inertial events may occur. Both plots where calculated for water (density = 1000 kg m^{-3}, surface tension = 0.072 N m^{-1} and viscosity 8.9 × 10^{-4} Pa s) and relied on the expansion criteria of $R_{max} = 2.3R_o$ where R_o indicates the initial bubble size.*

generate such conditions. This pressure is then used to mark the threshold between the two forms of cavitation event. However, it should be noted that other temperature criteria, for example, the 1550 K of Šponer [24–26], can also be deployed in such a model [27, 28]. In addition other assessment methods can also be employed. One such method is the expansion ratio (defined as the ratio of the maximum expansion radius to the initial radius of a given bubble), such that, for example, a value of this ratio in the range 2–2.3 appears to predict the inertial/non-inertial border [27]. Using this particular border, bubbles with greater expansion ratios than the set value are assumed to undergo inertial cavitation, and those with smaller expansion ratios are assumed to undergo non-inertial cavitation. Figure 2.1(a) shows how this expansion ratio can be used to predict the minimum zero-to-peak pressure amplitude required to generate an inertial cavitation event. Note that Figure 2.1(a) shows that the pressure threshold increases for the same initial bubble size as the frequency of ultrasound is increased. Figure 2.1(b) shows how the minima in the pressure amplitude required to generate an inertial event under ideal conditions[1] varies as function of frequency over an extended frequency range.

It turns out that the choice of expansion or temperature criteria is not that vital. This is a result of the dramatic change in both these parameters as bubbles cross from non-inertial to inertial in character. Hence broader tolerances in such criteria can be tolerated than if temperatures and expansion ratios scaled linearly with driving pressure, for example. This makes the model of Apfel and Holland an extremely useful guide in predicting the type of

[1]Note that this implies that the optimum initial bubble size is present within the medium. This initial bubble size, here ~1–5 µm in radii, will vary according the conditions employed.

cavitation event expected. However, it should be noted that the model refers to the behaviour of a single bubble in response to the local pressure field. In addition, only the first cycle of the bubble dynamic is considered. Nevertheless the distinction between inertial and non-inertial cavitation environments is useful for the interpretation of the electrochemical data that can be gathered from such systems. In the subsequent sections of this chapter, a discussion of the electrochemical data for many of the dynamics discussed above will be presented. In each case the appropriate conditions employed and experimental arrangements used will be discussed as relevant to the isolation of the desired mechanistic detail.

2.3 Mass Transfer Effects of Cavitation

The effects of cavitation generated by ultrasound on mass transport to electrode surfaces has been reported in the literature in a variety of different experimental arrangements [29–39]. In many of these investigations macroelectrodes (with critical dimensions typically greater than 50 μm) were deployed. In addition varying geometries and ultrasonic sources have been used. A relatively early example by Perusich *et al.* employed the effect of a focused ultrasonic field on mass transport to an electrode [40]. Alternatively Dewald *et al.* employed a pulsed 20 kHz ultrasonic source to generate a hydrodynamically modulated voltammetric experiment [41] with a detection limit for $[Fe(CN)_6]^{4-}$ of the order of 10^{-7} mol dm^{-3} reported. Coury *et al.* [30] showed that a high pseudo steady state mass transfer rate could be achieved by positioning an electrode within the plume of an operating ultrasonic horn [30]. They compared their results to the rotating disc electrode and calculated a theoretical rotation rate of circa 2.7 kHz would be required to generate equivalent mass transfer conditions. However, many of these reports have relied on a time-averaged measurement rather than focusing on the individual transient nature of the cavitation events present. Many of these literature examples employ ultrasonic fields that generate a number of different phenomena. Hence these systems will produce a broad range of different physical and chemical effects. Consequently, an appreciation of the different forced convection mechanisms that can be generated is vital to the interpretation of the experimental data (see Section 2.2) gathered to monitor transport of material. It is also illustrative if a study isolates one particular mass transfer enhancement mechanism. A discussion of such an approach is undertaken in the next section.

2.4 Isolating Single Mechanisms for Mass Transfer Enhancement

The general effects of ultrasound on mass transfer to the solid/liquid interface of an electrode surface can be associated with four major mechanisms. These are acoustic streaming, microstreaming, bubble motion and microjetting. Briefly, acoustic streaming arises from attenuation of the sound field within the liquid resulting in subsequent liquid motion. This mechanism is highly frequency, geometry and intensity dependent. In addition it can be expected that a steady flow of liquid (and hence time independent enhancement to mass transfer) will be generated under appropriate conditions. The remaining three mechanisms are non-steady state in that the motion of the liquid is transient in nature.

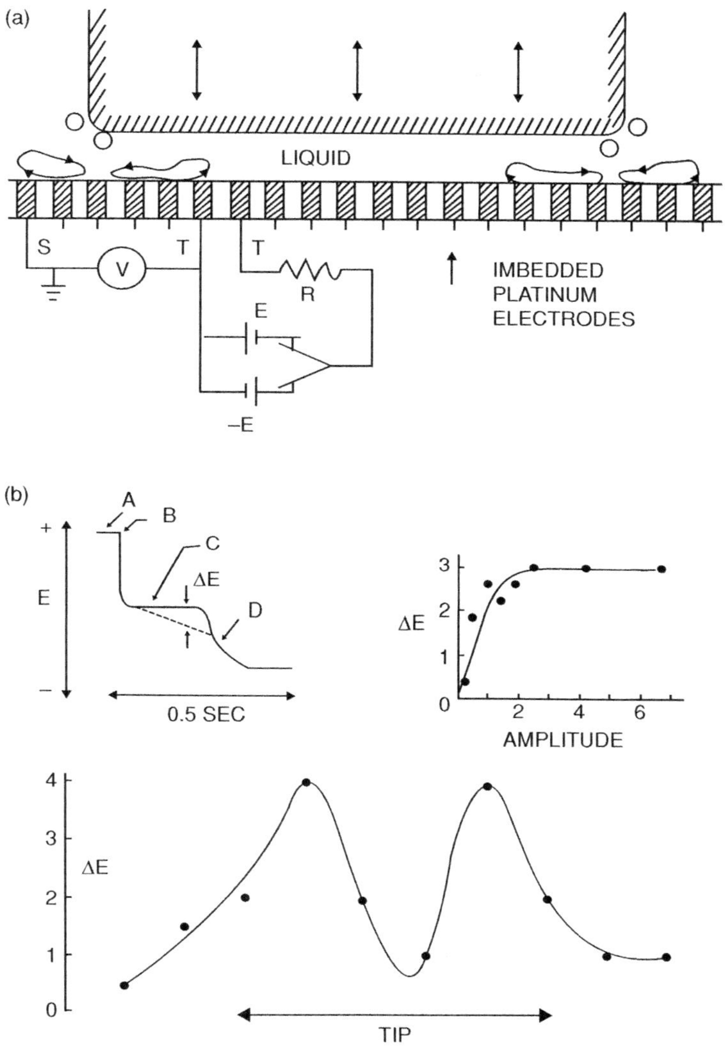

Figure 2.2 *(a) Schematic of electrode array used to determine the spatial extent of microstreaming from an oscillating piston. (b) Cathodic depolarisation waveform and results presented by Nyborg* et al. *(see reference [42] for further details) (Reprinted with permission from [42] Copyright (1960) Elsevier Ltd).*

As an example, the isolation of one particular forced convection mechanism, microstreaming, was reported in a novel and ground breaking study using an electrochemical approach by Nyborg *et al.* [42,44,45]. Figure 2.2 shows the experimental setup employed and some of the results obtained from this work. Here an array of band electrodes (down to 200 μm in width) were used. The mass transfer of material (through cathodic depolarisation) to a surface below an oscillating body was then monitored as a function of

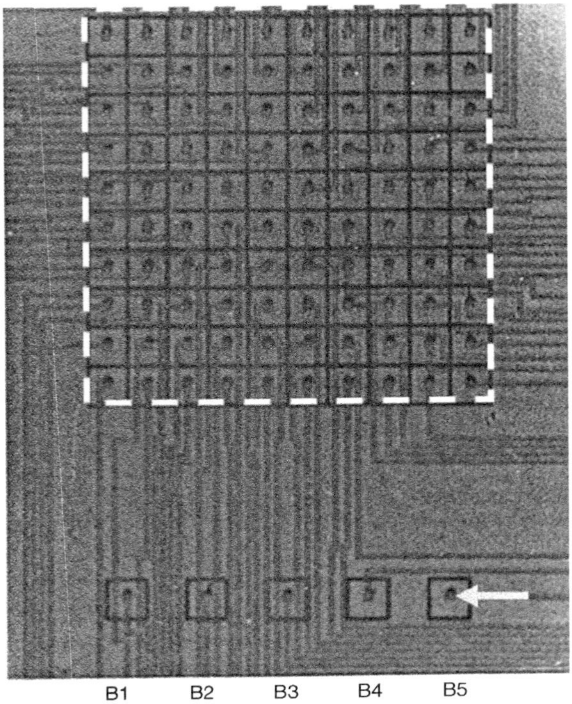

Figure 2.3 *Scanned image of a micromosaic electrode used by Tobias* et al., *to investigate bubble contributions to mass transfer (taken from reference [43]). The arrow indicates the generator 'satellite' electrodes while the highlighted box shows the array of 100 microelectrode detectors (Reprinted from [43] Copyright (1994) Electrochemical Society).*

electrode position. In addition to the study of microstreaming around an oscillating body, the application of electrochemical technology to study the effects of microstreaming from an oscillating gas bubble is also of relevance. This is detailed in a subsequent section of this chapter. The study of mass transfer enhancement through bubble motion (driven by buoyancy forces, for example) is a particularly important area when considered with the importance of a variety of large-scale industrial processes. As such, many comprehensive studies on the motion of gas bubbles and their effects on mass transfer to solid electrode surfaces can be found in the literature [43,46–48]. A particularly eloquent set of studies has been reported by Tobias *et al.* These authors used a variety of techniques to understand the electrochemical discharge and nature of electrolytic bubbles generated in solution. As part of these investigations Tobias *et al.* fabricated and employed a microelectrode (micromosaic) array for the tracking and assessment of the mass transfer coefficient associated with bubbles produced electrochemically. Figure 2.3 shows the microelectrode array used to detect bubbles generated by 'satellite' electrodes at the base of the structure. Other important investigations of bubble generation and behaviour are noteworthy. An eloquent example reported by Kelsall *et al.* investigated the influence of bubble skin effects (in comparison to truly clean bubbles) on the characteristics of gas bubbles within liquids [49, 50].

The last class of commonly encountered mass transfer mechanisms is termed microjetting. Here the collapse phase of the bubble is influenced by the presence of a solid/liquid boundary close to the collapsing bubble. The ensuing jet (or microjet) [51, 52] that forms is known to travel at high velocity through the gas phase (of the order of 100 m s^{-1}) and be at least partly responsible for damage to the solid/liquid interface [53, 54].

Clearly all the physical mechanisms described above will contribute to the mass transfer of material to a solid/liquid interface. However, although many studies have been focused on these effects, most have concentrated on the employment of large macroelectrodes. These electrodes have critical dimensions >50 μm and as such the diffusion pattern generated cannot be described as 'steady state'. An alternative approach and, as will be shown here, highly useful strategy, is the deployment of microelectrodes [55] for the characterisation of many of these mass transfer mechanisms [56–59]. The advantages of employing microelectrodes within these environments are many but can be summerised in two major characteristics: their small size enabling them to interrogate individual bubble events and their diffusion characteristics, which allow the electrode to recover quickly from perturbations of their diffusion field (for example as a result of transient convection). Note, we restrict ourselves to 'events' as it turns out to be extremely difficult to assign a mechanism directly to an electrochemical mass transfer measurement generated using ultrasonic irradiation of a fluid. This is because in many instances all mechanisms will be generated together and the dissection of the current time data often recorded in these experiments is difficult to achieve without simplifying measures (e.g. raised overpressure to eliminate bubble effects, the generation of single inertial cavitation events through laser discharge or the study of single bubbles driven into oscillation by an appropriate sound field). In the next sections we focus on the use of microelectrodes to study mass transfer enhancement generated by single bubble phenomena. In particular we start by discussing an investigation of mass transfer enhancements produced by a tethered gas bubble (held by buoyancy forces on a solid surface or in a wire loop) driven into oscillation by a low pressure amplitude (10s Pa) at or close to the resonant frequency of the bubble employed.

2.5 Electrochemistry Next to a Tethered Permanent Gas Bubble

In order to investigate the dynamics of bubble motion induced by sound irradiation, a particularly useful approach is the investigation of the current recorded at a microelectrode placed in close proximity to a stationary bubble. Here it is an advantage to hold a gas bubble on a solid support and approach the gas/liquid interface of the bubble with a microelectrode in a controlled manner. This approach employs the concepts developed in scanning electrochemical microscopy (SECM) [60, 61]. There are two possible ways in which this experiment can be undertaken. In the first method a solution-phase redox-active molecule is added to the liquid and the current recorded under mass transfer controlled conditions[2]. Note that the current is monitored both as a function of time and position with respect to the gas/liquid interface of the bubble employed. Far from the gas/liquid interface of the bubble

---

[2]Mass transfer control refers to the situation where the electrode potential is chosen so as a redox active species deployed in the solution is completely reduced or oxidised at the surface of an electrode so that its concentration at the electrode/electrolyte interface is zero at all times. The current recorded at the electrode is then a function of the mass transfer regime employed in the system. Changes in this regime, caused by perturbation of the available mass transfer stimuli, are subsequently recorded as changes in the current recorded at the electrode. In addition, the mass transfer coefficient (k_m) can be calculated using $i_{mt} = nFAk_mC$ where i_{mt} represents the current recorded under mass transfer limiting conditions, n the number of electrons exchanged, A the electrode area, F Faraday's constant and C the concentration of the redox system in question in the bulk of the liquid.

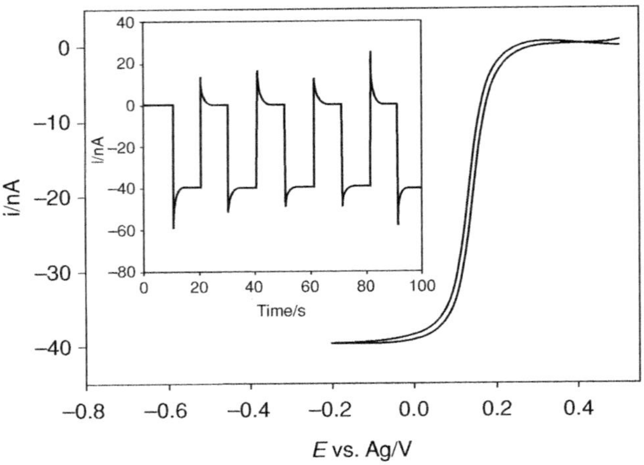

Figure 2.4 *Plot showing the cyclic voltammogram of the microelectrode tip (25 μm diameter Pt) recorded at 10 mV s*$^{-1}$*. The solution contained 10 mmol dm*$^{-3}$ *Fe(CN)*$_6^{3-}$ *in 0.2 mol dm*$^{-3}$ *Sr(NO*$_3$*)*$_2$*. The solution was sparged with argon to ensure anaerobic conditions. The temperature of the solution was 18–22 °C. The insert in the plot shows a series of potential steps from −200 mV versus Ag to +500 mV versus Ag (Reprinted with permission from [65] Copyright (2002) American Chemical Society).*

(distance, d, > 10a where a is the microdisk electrode radius), the electrode operates as if it were in bulk solution (normal conditions). Here the voltammetric and amperometric response of the electrode is largely as expected. Figure 2.4 shows a typical cyclic voltammetric response and amperometric response of a 25 μm diameter Pt microdisk electrode in solution under these conditions. Note the sigmoidal shape of the voltammetry and the steady state nature of the chronoamperometry are characteristic a microelectrode under these conditions.

The insert in Figure 2.4 shows the current/time response of the microelectrode as the potential was stepped between +0.5 V versus Ag to −0.2 V versus Ag in a repetitive manner. A clear steady state current (of ~ −40 nA) was observed when the electrode potential was set at −0.2 V versus Ag (this corresponds to a mass transfer limiting potential under these conditions). In order to investigate the effect of the gas/liquid interface of a bubble on the electrochemistry at the microelectrode (and in turn the effect of the motion of the bubble interface on the mass transfer to a microelectrode in the presence of sound irradiation), it is useful to employ an electrochemical cell with the ability to control the distance between the electrode and the gas/liquid interface. In addition the cell must be able to of apply a suitable sound field and allow visual inspection of the corresponding bubble motion generated [62–67]. Figure 2.5 shows an example of such a cell. Probing the electrode response held at a mass transfer limiting potential as a function of distance reveals that the gas/liquid interface can be viewed as an inert surface under the conditions employed. Figure 2.6 shows the response of the microelectrode as a function of distance normalised to the response in the bulk and fitted to the appropriate theory describing negative feedback for an SECM [68].

The observed reduction in the current magnitude as a function of distance between the electrode tip and the bubble gas/liquid interface is a result of the inert nature of the

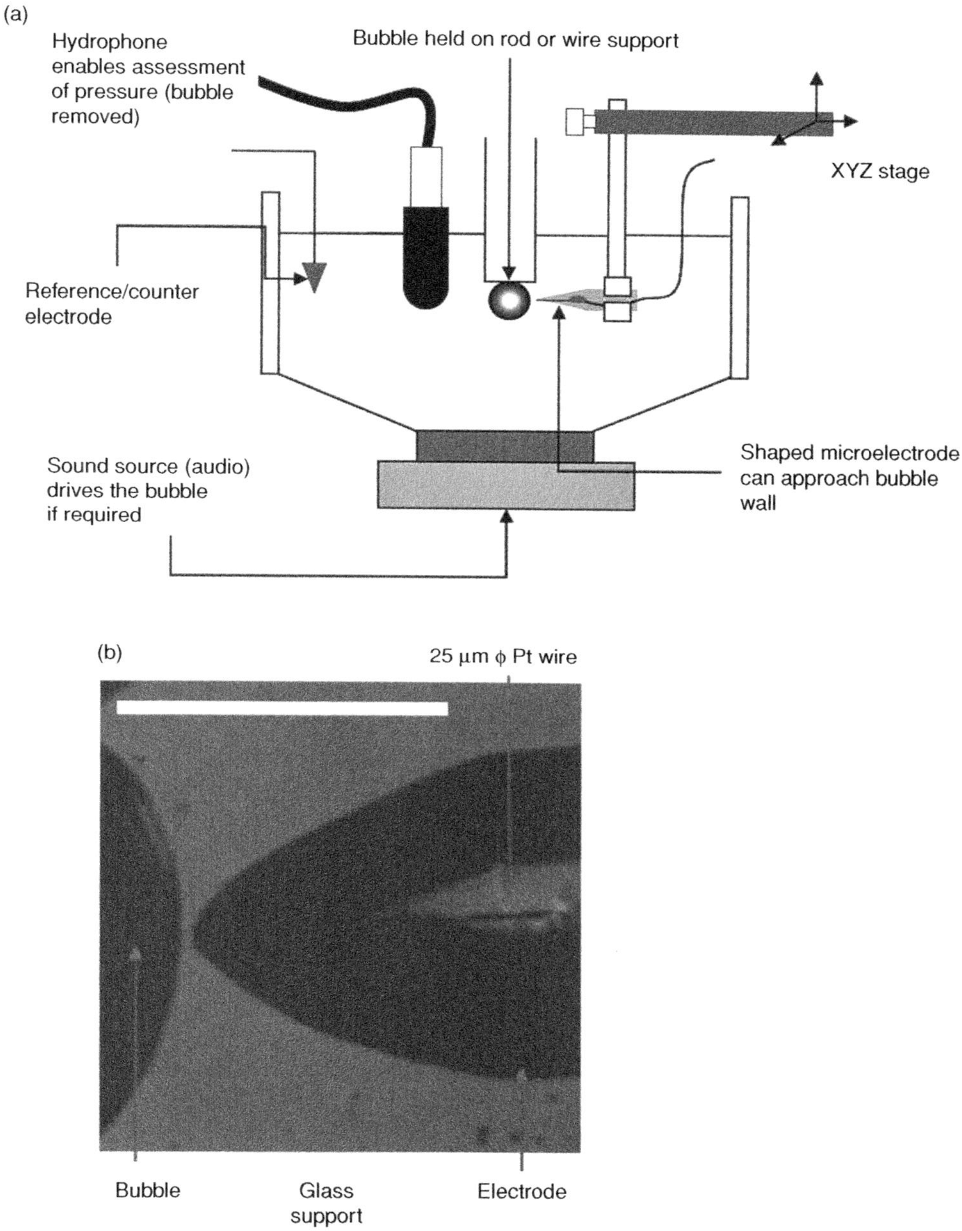

Figure 2.5 *(a) Schematic representation of the acoustoelectrochemical cell employed to study gas exchange reactions at air/liquid interfaces and surface waves on the bubble wall. (b) Image taken of the shaped microelectrode as it approaches the gas/liquid interface of an air bubble held on a solid support. The scale bar represents 0.75 mm.*

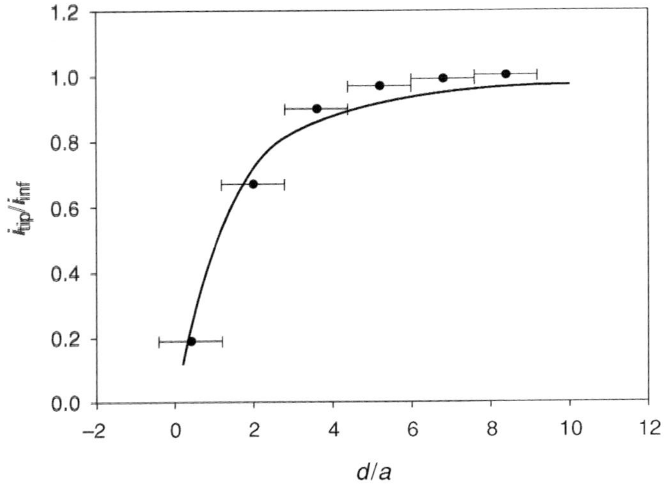

Figure 2.6 *Graph of i_{tip}/i_{inf} (normalised current) versus d/a (distance, d, from the bubble wall divided by electrode radius, a) for a 25 μm diameter Pt electrode in 10 mmol dm^{-3} $Fe(CN)_6^{3-}$/0.2 mol dm^{-3} $Sr(NO_3)_2$. The solid line represents the theoretical response of the microelectrode under the conditions assumed for negative feedback. The points represent the positions employed in the experiment each 20 μm apart. The error bars represent the estimated accuracy of the positioning system employed (Reprinted with permission from [65] Copyright (2002) American Chemical Society).*

interface under these conditions. The bubble wall effectively inhibits the mass transfer of the $Fe(CN)_6^{3-}$ species to the microelectrode surface (a process that can be described as hindered diffusion) and reduces the magnitude of the current recorded. Under these conditions a 75–50% reduction in the current recorded at the microelectrode can be detected. This corresponds to a distance of 5–10 μm between the bubble wall and the microelectrode surface. It is also interesting to observe the response of the microelectrode when the redox species in solution is replaced with molecular oxygen and a stationary air bubble is approached in the same fashion. This represents a different scenario as the redox active material is both in the solution and in the interior of the air bubble. Figure 2.7 shows the current/time response of a Pt microelectrode (held at the mass transfer limiting potential for the reduction of molecular oxygen under the conditions employed) as the electrode was progressively moved towards the gas/liquid interface of an air bubble. Clearly under these conditions the gas/liquid interface can no longer be viewed as inert (the current increases as the microelectrode approaches the bubble wall). This can be understood if one considers that as the diffusion field of the microelectrode encroaches on the gas/liquid interface of the bubble wall, oxygen transfer from the gas phase into the liquid phase is induced. This effect results in an enhanced current detected at the microelectrode (here depicted as an increase in the normalised current). This experimental observation is supported by the previous work of Unwin *et al.* [69] who suggested that reduction of oxygen at a microelectrode creates a depleted region next to a microelectrode that, if adjacent to the interface between a gas and a liquid, provides a thermodynamic driving force for the transfer of oxygen from the air to liquid phase of the system. In Unwin's experiment a planar air/liquid interface was created

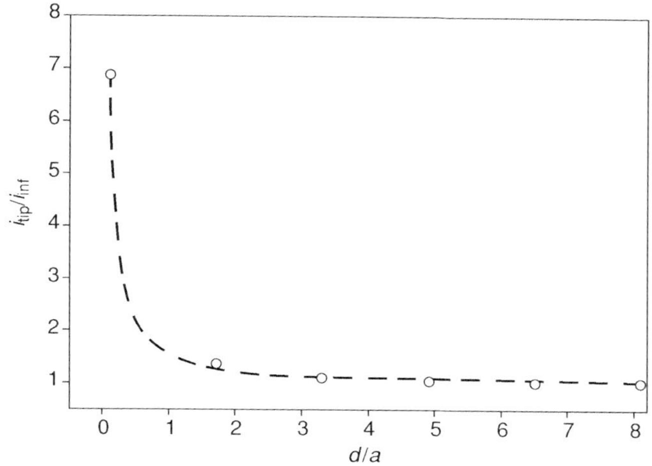

Figure 2.7 *Plot of the normalised current (○) for a 25 µm Pt electrode in 0.1 M KCl approaching the surface of an air bubble. The electrode potential was held at −0.7 V versus Ag. The symbols show where the electrode was moved in increments of 20 µm using an SECM apparatus. The dotted line (— —) represents the fit of the data to the appropriate positive feedback model [68] (Reprinted with permission from [62] Copyright (2004) Elsevier Ltd).*

using a Langmuir trough. In the results shown in Figure 2.7, the interface is the gas/liquid barrier of an air bubble. The addition of surfactants to the media also influences the rate of the transfer between the gas and liquid phase of the system considered.

2.6 Mass Transfer from Forced Permanent Gas Bubble Oscillation

While the proceeding section discussed the study of a gas/liquid interface of a bubble in the absence of sound, the next section discusses the observations obtained when a gas bubble is irradiated with sound close to the bubble's resonant frequency. Here the electrochemical response of a microelectrode placed in a controlled manner with respect to the gas/liquid interface of the bubble is employed in conjunction with other experimental observations (e.g. high-speed imaging and acoustic measurements) in order to support the conclusions drawn. Figure 2.8(a) shows the current time history for a 25 µm diameter Pt microelectrode placed close to the gas/liquid interface of an air bubble held in solution by a support [67]. In this experiment the solution contains a suitable redox species (here $Fe(CN)_6^{3-}$) and the microelectrode was positioned close to (~5–10 µm) the gas/liquid interface of a bubble using the SECM method described in Section 2.5. In the absence of sound irradiation, a steady state signal was recorded at the microelectrode (here ~ −7.5 nA).

However, if sound irradiation of the bubble is introduced (in this example at a frequency close to bubble resonance, 1.167 kHz) a number of experimental observations can be made. First, the cathodic (negative) current increases as a result of forced convection to the microelectrode induced by microstreaming around the bubble bringing the redox species to the surface of the electrode at a higher rate. Second, as the zero-to-peak pressure

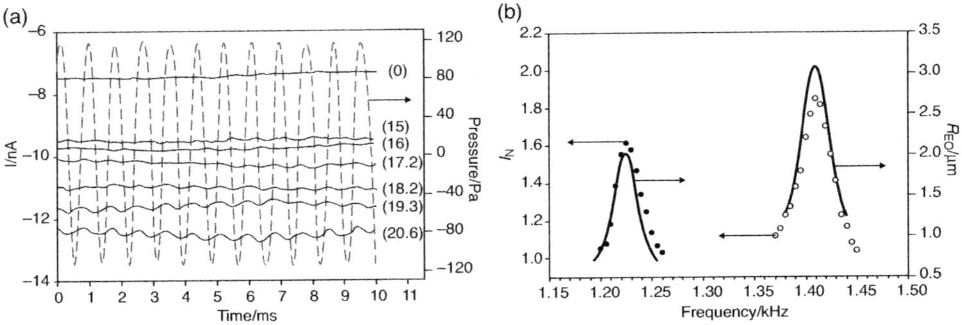

Figure 2.8 *(a) Plot showing the current (——) recorded as a function of time as the acoustic pressure was increased. The actual pressure amplitude at which each current was recorded is given in the parenthesis in the figure next to the current time trace. (– – –) represents an example of the acoustic pressure measured simultaneously in the presence of the bubble for reference. The acoustic driving signal was maintained at 1.167 kHz. (b) Plot showing time-averaged normalised current recorded for two air bubbles driven into oscillation below the threshold for Faraday wave excitation. In each case the excitation pressure was 5.8 and 10.4 Pa for (•) and (○) respectively. The solid lines indicate the estimated bubble wall displacement amplitude calculated under the conditions stated. The currents were normalised to the current recorded in the absence of sound irradiation (6.82 and 8.85 nA for • and ○ respectively) (Reprinted with permission from [67] Copyright (2003) Elsevier Ltd).*

amplitude was increased, the cathodic current increased in magnitude and possessed an AC component. In this case the AC component of the current (see Figure 2.8(a) 20.6 Pa current time trace) matches the frequency of the applied sound field (see Figure 2.8(a) – – –). In this case we attributed the mass transfer enhancement to forced convection caused by the breathing mode of the bubble itself. Figure 2.8(b) shows how the enhancement in current to the microelectrode varies as a function of frequency for two different bubbles driven into oscillation. In this case the pressure field is <10 Pa zero-to-peak. Figure 2.8(b) shows that the current enhancement varies as a function of frequency around a value which is assumed to be the resonant frequency of each bubble employed. This type of behavior is only observed at relatively low zero-to-peak pressure amplitudes. If the pressure amplitude was increased further above those shown in Figure 2.8(a), a large step change increase in the cathodic current was observed. In addition, the AC nature of the current changed to a subharmonic of the driving field. This is attributed to the onset of surface waves (in this case a Faraday wave) at the gas/liquid interface of the bubble. This is further supported by visual observation of a 'shimmer' on the bubble wall. Figure 2.9(a) shows the time-averaged current recorded at the microelectrode as a function of the drive zero-to-peak pressure amplitude. Note in this case the sound frequency was held at a constant value while the pressure amplitude was gradually increased.

Clearly a steady increase in the current was observed up until the zero-to-peak pressure amplitude exceeded a critical threshold value (in this experiment ~21 Pa). At zero-to-peak pressure amplitudes greater than this value a surface wave on the bubble wall could be observed by eye (this is seen as a 'shimmer' of the interface). In addition the cathodic current above this threshold possess a subharmonic component and is assigned to the excitation of a Faraday wave at the gas/liquid interface (labelled FW on Figure 2.9(a) with

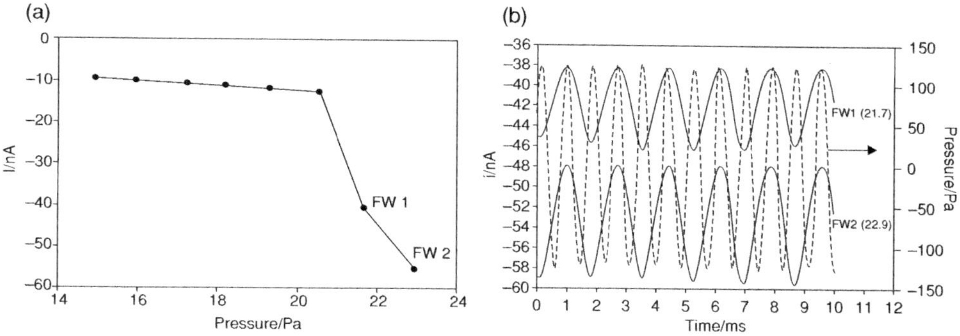

Figure 2.9 (a) Plot showing the average current recorded for a single air bubbles driven into oscillation as a function of the acoustic driving pressure. The points labelled with 'FW' represent those experiments where Faraday waves were observed. (b) Plot showing the current (———) recorded as a function of time as the acoustic pressure was increased. The actual pressure amplitude at which each current was recorded is given in parenthesis in the figure next to the current time trace. (— —) represents the acoustic pressure measured simultaneously in the presence of the bubble for reference. The acoustic driving signal was maintained at 1.167 kHz (Reprinted with permission from [67] Copyright (2003) Elsevier Ltd).

the respective zero-to-peak pressure amplitude in parenthesis). The subharmonic nature of the current recorded at the microelectrode in relation to the driving pressure field is shown in Figure 2.9(b) for two pressure amplitudes (note these are measured in the presence of the bubble under the same conditions).

Figure 2.10(a) shows an image of a bubble (the image was taken from below) driven with an appropriate acoustic field to generate a visual 'shimmer' on the gas/liquid interface. This image clearly shows formation of a surface wave (with 15 peaks seen and labelled on the image) on the gas/liquid interface of the bubble wall. It is possible to explore the effect of the driving frequency on the exact mode of bubble oscillation (i.e. breathing versus surface wave oscillation) using this electrochemical method. Figure 2.10(b) shows the results of such an experiment. Here the time-averaged current (•) recorded at the microelectrode varies as a function of the acoustic drive frequency employed. Note that in this case the bubble was irradiated with a zero-to-peak pressure amplitude of ~41 Pa over the entire frequency range studied. At the extremes of the frequency range employed (1.23 to 1.46 kHz) the current enhancement is at a minimum, indicating relatively little enhancement in the convective flow to the microelectrode surface at those frequencies. However, as the frequency approaches ~1.38 kHz (from either frequency limit) a steady increase in the current magnitude was observed. This is attributed to the progressive increase in the motion of the bubble wall in the breathing mode oscillation (see Figure 2.8). This trend in increasing current continues until a step change in the current was observed (at either 1.35 kHz or 1.41 kHz). This coincides with the visual observation of shimmer caused by a surface wave on the gas/liquid interface of the bubble wall. This is accompanied with a subharmonic component in the current time history again characteristic of a Faraday wave on the bubble wall. The threshold for the onset of Faraday wave generation (determined from the current/time history of the microelectrode) is also shown in Figure 2.10(b) (see ○).

In order to determine the pressure threshold for the Faraday wave on the bubble wall, the zero-to-peak pressure amplitude was increased at each frequency employed until a

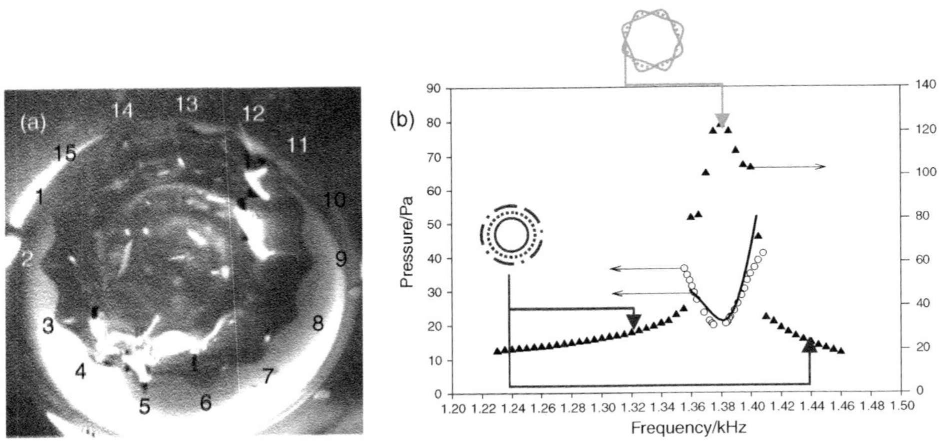

Figure 2.10 *(a) Image of a Faraday wave generated on a bubble. The image was recorded using a high-speed camera. The bubble was driven at 1.297 kHz. The scale bar represents 2 mm. The image is annotated to indicate the mode excited (n = 15). (b) Plot showing time averaged current recorded at a microelectrode positioned close (ca. 5–10 μm) from the gas/liquid interface of an air bubble plotted as a function of frequency (•). The aerobic aqueous solution contained 5 mmol dm^{-3} $K_3[Fe(CN)_6]$ in 0.2 mol dm^{-3} $Sr(NO_3)_2$. The 25 μm diameter Pt electrode was held at −0.1 V vs. Ag. The solution temperature was 20–23 °C and the acoustic pressure was 41 Pa. The pressure threshold for Faraday wave observation is also shown for interest (▲). (—) represents the predicted threshold (n = 14, σ = 0.0685 N m^{-1}). The mode of excitation including the breathing mode (—) and Faraday wave (—) are shown for illustration purposes and are not drawn to scale (Reprinted with permission from [67] Copyright (2003) Elsevier Ltd).*

subharmonic component in the current/time history was observed. This demarcation point is assigned to the particular Faraday wave threshold under the conditions reported. This process was repeated for the all the frequencies shown. This experimental approach shows a parabolic well-like threshold for the minimum zero-to-peak pressure amplitude required to generate the Faraday wave. This threshold pressure amplitude can be compared to a simple theory for this phenomenon. Equation (2.1) shows how the pressure threshold varies with frequency:

$$P_A = P_T = R_0{}^2 \rho C_t \left[\left(\omega_0{}^2 - \omega^2 \right)^2 + \left(\omega^2 d_{tot} \right)^2 \right]^{1/2} \tag{2.1}$$

where P_A represents the zero-to-peak pressure amplitude required to generate the surface wave, ω the driving frequency, R_0 is the equilibrium radius, ρ the density of the liquid, ω_0 the resonant frequency of the bubble, C_t is a dimension less term defined elsewhere (see [67]), and d_{tot} the total dimensionless damping coefficient. Figure 2.10(b) shows the predicted threshold pressure (—) as a function of frequency under the conditions employed. The agreement between the theoretical pressure amplitude and the experimental measurements is reasonable considering some of the assumptions made and the experimental arrangement employed. Further more-complex theory associated with the generation of a Faraday wave on a bubble wall can be found elsewhere [70–73].

If the pressure amplitude is increased further above this critical pressure amplitude required to generate the first surface wave observed with the lowest threshold, multiple

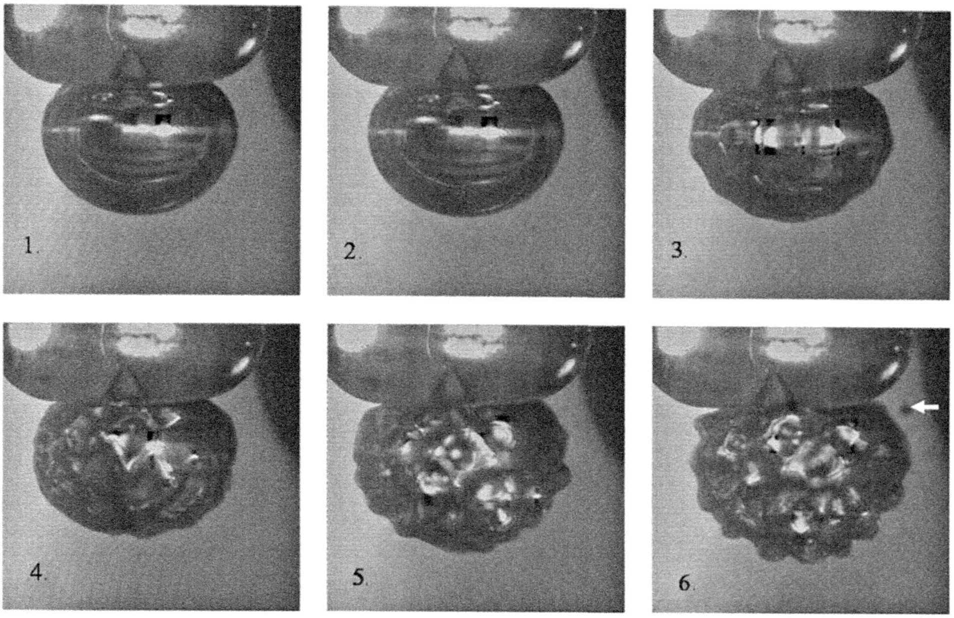

Figure 2.11 *A set of images of the same bubble resonant at 1.5006 kHz at zero-to-peak driving pressure amplitudes of (1) 10.9 Pa, (2) 24 Pa, (3) 44 Pa, (4) 79 Pa, (5) 94 Pa and (6) 139 Pa. At 139 Pa the bubble fragments denoted by a white arrow on frame 6. Note the glass rod used as a support was 6 mm in diameter.*

modes can be generated on the bubble wall. Under these circumstances the dynamics of the gas/liquid interface can become complex. In order to illustrate this Figure 2.11 shows images recorded with a high-speed camera of a single air bubble driven into oscillation with progressively increasing zero-to-peak pressure amplitudes. The onset of a surface wave is clearly seen in Figure 2.11 frame 3. As the zero-to-peak pressure amplitude increased, multiple modes of excitation are clearly evident on the bubble wall (see images 4, 5 and 6). Finally fragmentation can be observed, with bubble ejection evident in frame 6 (highlighted with a white arrow). To the naked eye this behaviour is observed as a 'frosted' bubble ejecting smaller non-resonant bubbles from its surface.

While the generation of surface waves and their interrogation with a microelectrode is of fundamental interest, the practical applications of this phenomenon are also noteworthy. One such application is the local enhancement in mass transfer generated by the motion of the bubble wall. Figure 2.12 shows the enhancement in mass transfer to a microelectrode detected as a function of distance between the gas/liquid interface and the electrode surface. In this case, the bubble was driven into oscillation at a pressure sufficient to generate surface waves on the gas/liquid interface. This was again observed in two ways. First, the presence of surface waves on the bubble wall was observed optically as a 'shimmer'. Second, when the microelectrode was positioned close to the bubble wall (circa 5–10 μm) the periodicity of the motion of the bubble wall can be resolved from the current time trace recorded. This is shown as an insert in Figure 2.12.

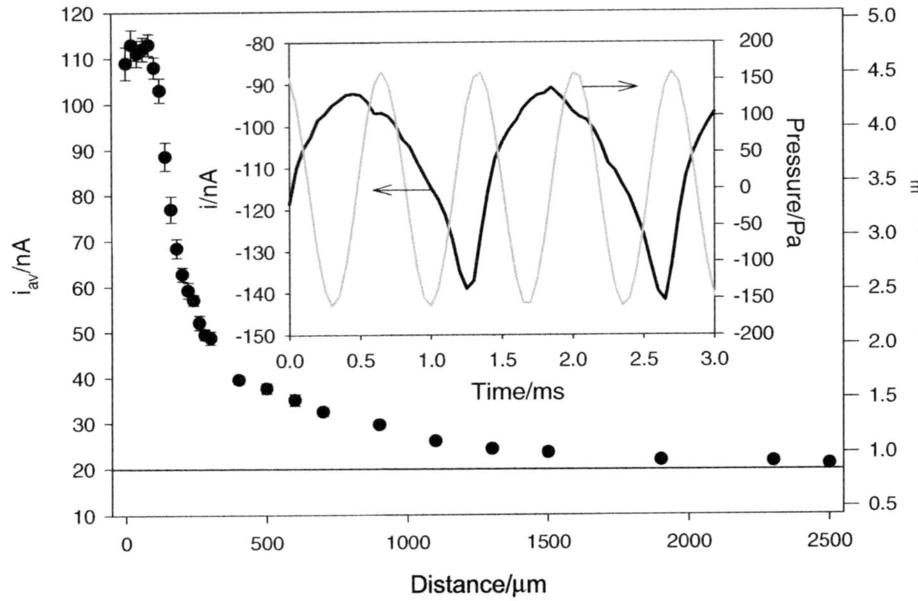

Figure 2.12 Plot showing the average current (•) and associated mass transfer coefficient for a single air bubble driven into oscillation by a sound field operating at 1.46 kHz. The solution contained 5 mmol dm^{-3} $[Fe(CN)_6]^{3-}$ in 0.2 mol dm^{-3} $Sr(NO_3)_2$. The error bars show the 95% confidence interval. The insert shows the oscillation in the current (——) and pressure[2] (——) as a function of time employing high temporal resolution equipment. The experiment was performed at ca. 20–23 °C under aerobic conditions. The solid horizontal line represents the steady state current or mass transfer coefficient for the microelectrode in a stagnant solution (Reprinted with permission from [64] Copyright (2001) Royal Society of Chemistry).

The insert also shows the pressure *vs.* time[3] traces for the bubble driven to oscillate with surface motion. This clearly shows that the current vs. time trace has a sub-harmonic component (here 0.5f where f represents the drive frequency, in this case 1.46 kHz). Figure 2.12 shows that within the first ~100 μm, the mass transfer enhancement remains approximately constant although a small reduction is apparent at the distance of closest approach (estimated to be ~5–10 μm microelectrode to gas/liquid interface separation using the SECM method, see Section 2.5).

If the distance between the bubble wall and the electrode is increased above 100 μm the current falls away as the distance was increased. However, the perturbation in the mass transfer coefficient can be detected at extended distances away form the gas/liquid interface. Clearly a significant enhancement was observed up to distances of circa 2500 μm. This is particularly significant considering that the mass transfer coefficient of the 25 μm diameter platinum microelectrode is already high in stagnant solution (in this case 0.008 cm s^{-1}). The maximum enhancement in mass transfer as a result of the bubble motion exceeds this

[3]Note that the absolute pressure will not be the same as that shown in the insert of Figure 2.12 as the presence of the oscillating bubble will alter the measured pressure. The absolute pressure can only be measured when the bubble has been removed from the glass support. This is the case for all zero-to-peak pressures quoted as thresholds etc. within this chapter.

value (up to 0.0477 cm s^{-1} in this case) but is brought about by a relatively small pressure amplitude (<100 Pa). This is a significant point as it clearly demonstrates that enhancing the mass transfer of material to an electrode in this manner would be a more efficient way when compared to the employment of ultrasound to induce inertial cavitation (see Section 2.2) and other high-energy phenomena. For example, although it has been shown that ultrasound can enhance mass transfer significantly [13–15], the magnitude of the pressure field required to achieve these inertial (transient) cavitation events is considerable. In order to generate cavitation in water under standard conditions a zero-to-peak pressure amplitude in excess of ~120 kPa at 23 kHz is estimated (see Section 2.2 for a discussion on this threshold) [4]. This is over a factor of 1000 higher than the pressures employed in the experimental results shown in Figure 2.12 and in turn a factor of 10^6 in intensity. Clearly the generation of mass transfer enhancements using acoustically oscillated bubbles, rather than inertial cavitation, should be significantly more efficient. This increase in efficiency is due to the differing mechanisms responsible for the forced convection enhancements observed. The process of inertial cavitation requires that small (e.g. of the order of micron radius) bubbles expand against atmospheric pressure and surface tension forces to a critical radius, before collapsing. However, exciting a large bubble (e.g. mm radius in this case) at resonance takes advantage of their exceptionally good acoustic coupling [4]. In turn, the pressure amplitude required to generate surface waves is relatively small, but the enhancement in forced convection is still significant. In the results shown in Figure 2.12, mass transfer coefficients are ~1 order of magnitude lower than those reported for inertial cavitation (~0.1 to 1 cm s^{-1}) [58]. However, the mass transfer coefficients shown in Figure 2.12 required pressure amplitudes which are 3 orders of magnitude less than those required to generate inertial cavitation events. This is because the excitation of a surface wave does not require a large overall volume change of the bubble, compared to the generation of inertial cavitation. An additional advantage of these surface wave systems is that the generation of inertial cavitation also introduces the possibility of erosive damage [74–76]. These erosiove mechanisms, under certain circumstances where valuable or fragile substrates are present within the media, can be a significant disadvantage.

An example of where bubble oscillations may be of significant benefit is in the deposition of metals onto solid substrates. Metal deposition (as part of the electrowinning process [77]) is a vital stage in the refining of a number of metals, for example copper [78], nickel [79] and cobalt [80]. The use of targeted bubble oscillation has the potential to increase the efficiency of such processes, leading to significant reduction in costs and environmental impact. Figure 2.13 shows the results of a study of bubble oscillation on the electrodeposition of copper metal.

In this experiment an argon bubble (~1.5 mm radius) was placed on a Cu surface (4 cm^2). The effect of different (controlled) bubble oscillation regimes was then investigated. Figures 2.13(a) and (b) show the enhanced deposition of copper around the gas bubble. In this case the bubble was driven to excite multiple mode surface waves (up to a zero-to-peak pressure amplitude of ~70 Pa). The SEM images clearly show the enhanced deposition rate in the region close to the bubble wall while the region directly below the bubble (shown as a white dotted circle) was obscured by the bubble itself and is relatively smooth. The enhanced deposition rate extends from the bubble wall to ~1500 μm from the gas/liquid interface in agreement with the results shown in Figure 2.12 (note the bubble deployed in Figure 2.13 is smaller than that used in Figure 2.12). However, although effective this still represents a relatively small surface area (13.43 mm^2). Figure 2.13(c) shows that

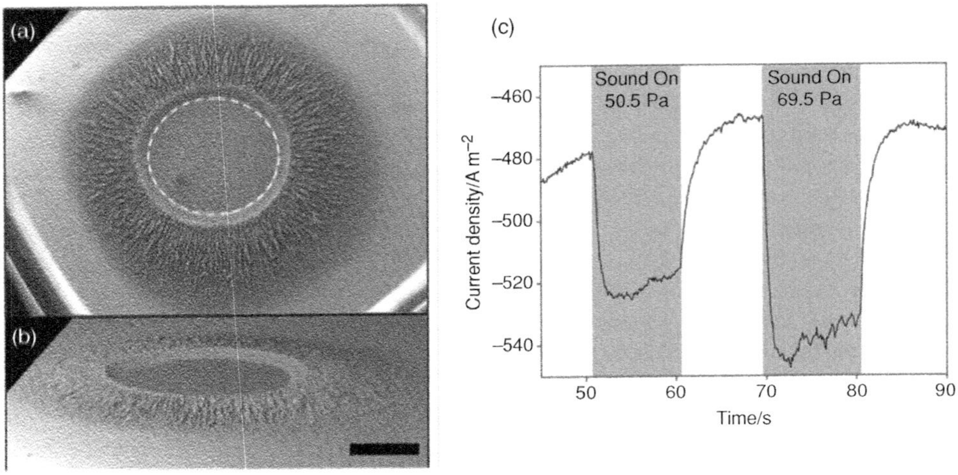

Figure 2.13 (a) and (b) SEM images showing copper deposited on an electrode in the presence of an argon bubble driven into oscillation at 69.5 Pa and a frequency of 2.0677 kHz. The electrode was held at −0.5 V versus SCE in a solution of 0.5 M $CuSO_4$ in 2 M H_2SO_4. A total charge of 84.9 C was passed at an average current density of 607 A m^{-2}. The scale bar represents 1 mm. (c) Plot showing the current density at a 4 cm^2 copper electrode held at −0.5 versus SCE as a function of time under various experimental conditions. Shaded regions indicate times at which sound (2.0677 kHz, see figure for zero-to-peak pressure) was on (Reprinted with permission from Electrochemistry Communications, Electrodeposition of copper in the presence of an acoustically excited gas bubble by D.G. Offin et al., **9** (5), 1062–1068 Copyright (2007) Elsevier Ltd).

the electrochemical data also indicates that sound irradiation induces a modest (spatially averaged) increase in mass transfer as a result of bubble oscillation. Clearly in order for this technology to be scaled up arrays of bubbles are necessary. Nevertheless this technological approach is of merit and deserves further investigation.

In the preceding sections the emphasis has been placed on the investigation of mass transfer effects of permanent gas bubbles driven into oscillation using an appropriate acoustic field. This has clear advantages over inertial cavitation studies in terms of the sound intensities required and the absence of damaging erosive mechanisms. However, it is also necessary to understand the effects of inertial cavitation as it is of significance in industrial flow systems, the medical arena and in research focused on the understanding of these fascinating but complex phenomena. In the next section we look at approaches to the investigation of cavitation bubbles using electrochemical techniques designed to investigate their effects at the solid/liquid interface.

2.7 Mass Transfer Effects from Single Inertial Cavitation Bubbles

The study of cavitation dynamics has been significantly aided by the interrogation of single cavitation bubble events. One method whereby a large (and hence easily imaged using high-speed camera techniques) cavitation event can be controllably generated is by the deployment of focused laser energy (with high local intensities $>10^{10}$ W cm^{-2}) [81] into a liquid medium [53, 54, 82, 83]. These large (of the order of mm in radii) single cavitation events can, as well as being generated by laser pulses, be produced through spark discharge into a liquid [84]. One such study of laser-induced cavitation events was reported by Philip and Lauterborn [54]. They detailed an extensive investigation of the damage to a surface produced by single cavitation bubbles (with a maximum bubble radius, R_{max}, of the order of 1.45 mm) produced by laser action [54]. They showed, using high frame rate photography techniques (frame rates up to 10^6 frames per second) and *ex situ* microscopy of various solid surfaces, that the damage caused by one or more (up to 5000) single bubble collapses depended strongly on the dimensionless distance between the bubble and the solid/liquid interface (this is defined as s/R_{max} where s is the distance between the bubble centre and the solid/liquid interface). These authors showed that most surface damage associated with single bubble collapse originates from shock wave emission. This emission originates from the collapse phase of the bubble and is often associated with the generation of a toroidal bubble as the result of microjet formation. Surface damage, as the result of microjet impingement on the surface, was not found to be extensive. Philip and Lauterborn also showed that, even though the scale of the damage they observed was of the order of μm in depth, the surface damage patterns were very complex in shape. However, this and many previous studies have relied on the *ex situ* investigation of surface damage as the result of either microjet or shock wave impact. Furthermore these *ex situ* investigations were time integrated such that they examine only the cumulative result of the entirety of each erosion event. While a great deal of important information has been gathered in this manner, *in situ* investigations would be clearly useful in understanding the actual processes occurring close to these large single cavitation bubbles. Hence the following results obtained from an investigation of cavitation produced by a laser pulse are of direct relevance. In these experiments a large (maximum bubble radius, R_{max}, ~0.5–1.0 mm) single bubble was formed by a single Nd : YAG laser pulse (5 ns). This was focused inside bespoke electrochemical cell close to a microelectrode surface [85–87] under electrochemical control. Figure 2.14(a) shows the challenging experimental setup required to generate a single cavitation bubble, control the separation between the bubble centre and the electrode surface (this was achieved using stages) and image the resulting bubble motion.

Figure 2.14(b) shows the shape and size of an individual bubble event created in this experimental arrangement. In this set of frames it can be seen that the bubble is generated below the electrode substrate [highlighted in Figure 2.14(b) frame (0)]. Note in this case a Au microelectrode imbedded in glass was employed. Figure 2.14(a) shows that the bubble can grows over a period of ~60 μs and then collapse (after ~110 μs). Under these conditions a bubble rebound was also observed (see frame 121). Eventually the bubble fragments leaving a cloud of small bubbles close to the electrode surface. In addition to these growth and collapse phases of the bubble, the bubble moves towards the microelectrode. Jetting (highlighted with a white arrow in frame 121) onto the surface can also be observed during this set of images.

Figure 2.15 shows electrochemical, photodiode (used to trigger the recording apparatus) and acoustic data recorded for a single laser-induced cavitation event (note that this was generated under different experimental conditions to Figure 2.14) in this experimental arrangement.

Figure 2.15(a) shows that a pressure shock was emitted as the liquid was broken down by the focused laser action (denoted LS). In this case the magnitude of the LS is impressive

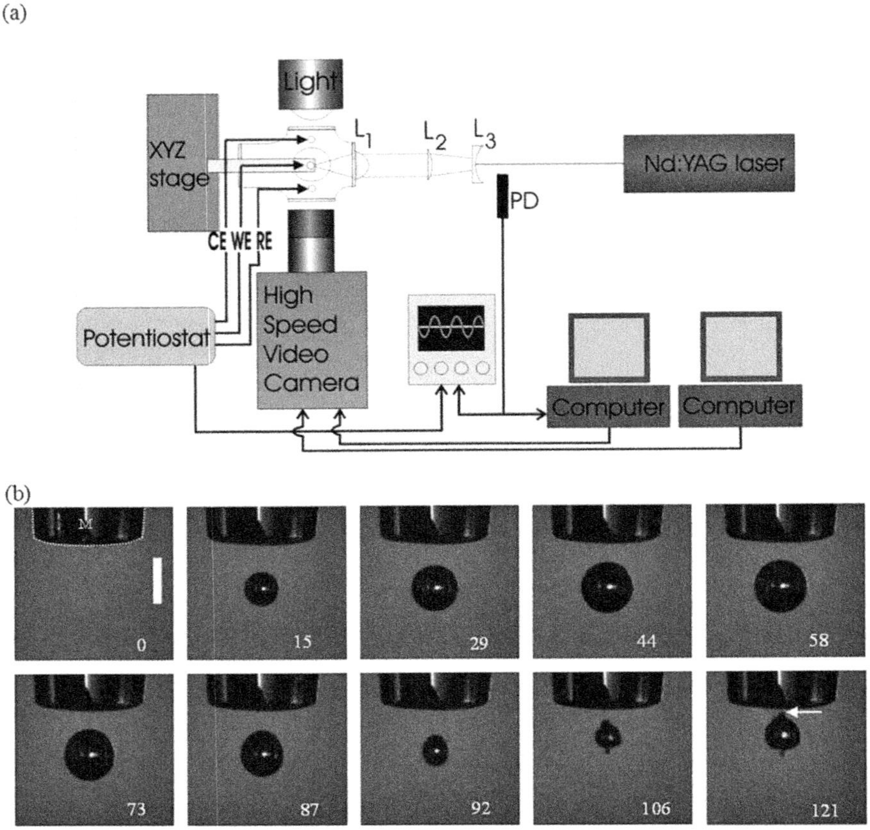

Figure 2.14 *(a) A schematic representation of the lens arrangement described in the experimental section. Note PD is the photodiode, WE the working electrode, CE the counter electrode and RE the reference electrode. (b) High-speed video image sequence of a single laser induced bubble generated below a microelectrode. Frame rate = 68 000 fps. The laser power was ~9 mJ. The solution consisted of 1 mM KI in 0.2 M Na_2SO_4. The experiments were performed at room temperature (~20 °C). The scale bar (see frame (a)) represents 1 mm. The bubble maximum radius was measured at 535 μm with an s/R_{max} value of 1.76. Note the microelectrode (denoted by 'M') boundary is highlighted by the dotted line in frame (0). The numbers on the frames indicate the time after the laser pulse (in μs) (Reprinted with permission from [85] Copyright (2006) Elsevier Ltd).*

and >2.5 MPa. In addition two further strong acoustic signatures are seen (labelled BS_1 and BS_2). These are attributed to the pressure signals generated by the bubble dynamics and are also considerable (note BS_1 ~ 0.9 MPa). A second collapse after rebound (BS_2 ~ 0.55 MPa) can also be detected by the acoustic measurement. Figure 2.15(b) shows the electrochemical data for a microelectrode held under mass transfer limiting conditions and placed close (~1.5 mm) to the inertial laser-generated bubble centre. Several key features are of interest. First, Figure 2.15(b) shows two large mass transfer enhancement

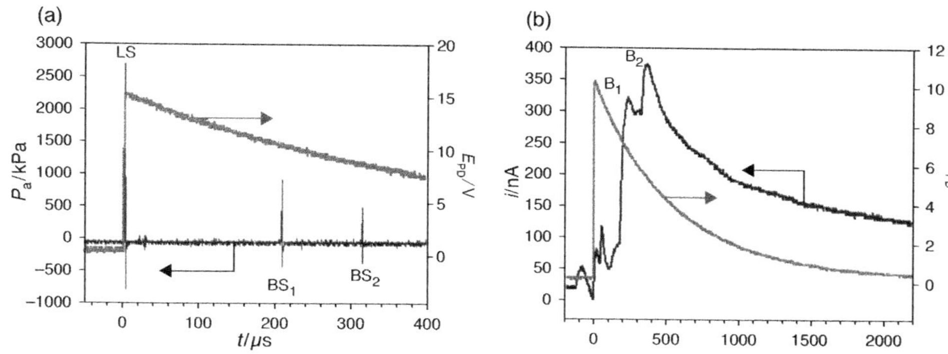

Figure 2.15 *(a) Plots showing the acoustic trace detected with a G.R.A.S. 10DB needle hydrophone (——) in comparison to the laser pulse (shown as the photodiode output, E_{PD} ——). The laser power was ~7 mJ. The electrolyte consisted of a filtered and degassed[4] 5 mM $K_4Fe(CN)_6$/0.1 M $Sr(NO_3)_2$ solution. (b) Plots showing the simultaneous measurement of the response of the photodiode and the electrochemical signal (——) recorded by a microelectrode (25 μm diameter Au) following a laser pulse into the solution. The electrode was held at +0.4 V vs. SCE. The microelectrode was positioned ~1.5 mm from the focal point of the laser. The experiments were performed at room temperature (~20 °C) (Reprinted with permission from [85] Copyright (2006) Elsevier Ltd).*

events (denoted as B_1 and B_2) are detected after the events associated with the laser pulse. Second, the separation in time between these two mass transfer events was found to be of the order of 120 ± 9 μs. This is in close agreement with the separation between primary collapse and rebound of the bubble (~106 μs) determined from the acoustic pressure time history (see Figure 2.15(a)). This evidence suggests that the electrochemical mass transfer enhancement (events labelled B_1 and B_2) are associated with primary bubble collapse and rebound events respectively.

While these experiments are illustrative of the processes (e.g. forced convection enhancement) that occur close to a single bubble, it is desirable to gather information on multibubble systems (e.g. those found in ultrasonic reactors) as these offer the largest extent of potential applications for sonochemistry and processing and so on. This is discussed in the following section.

2.8 Investigating Non-inertial Cavitation Under an Ultrasonic Horn

The same concept of employing a single microelectrode to monitor the physical and chemical effects generated within a multibubble environment, where cavitation bubbles are excited by acoustic driving forces, is possible. This was first reported in 1995 and has been used by several groups to investigate cavitation effects within an ultrasonic environment [37, 38, 56–58]. However, care should be applied to the interpretation of the data from this

[4] Although the electrolyte was degassed prior to use, some gas reintroduction may occur over the time period of the experiment.

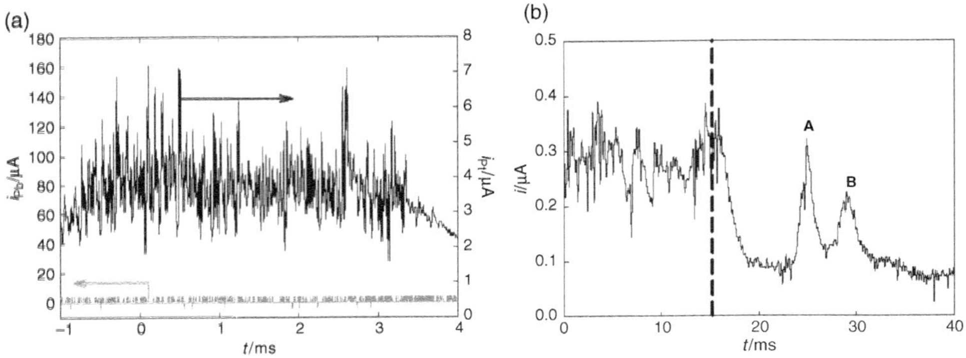

Figure 2.16 (a) Plot showing the current traces recorded simultaneously from lead (125 µm diameter, ——) and platinum (50 µm diameter, ——) microdiscs of a dual electrode under exposure to ultrasound (22.83 kHz, 56 ± 5 W cm^{-2}). The electrode-to-horn separation was 4.5 mm. The solution contained 20 mmol dm^{-3} $K_4Fe(CN)_6$ and 0.75 mol dm^{-3} Na_2SO_4 and the experiment was performed under aerobic conditions. The potentials of both discs was held at +0.8 V versus Ag. The initial solution temperature was 20–25 °C (b) Plot showing the current recorded as a function of time at a 25 µm diameter platinum electrode in a solution of 2 mM KI/ 0.1 M KCl during a ring-down experiment. The potential of the electrode was held at +0.6 V versus SCE corresponding to the mass transfer limited oxidation of I^- within the system.

approach as, although it is useful, cavitation exists in a complex and interactive environment and many mechanisms can produce enhanced mass transfer. As a result, current time history on its own has to be used with some caution. Combining the electrochemical mass transfer data with other techniques (e.g. high-speed imaging and surface erosion measurements) can provide further insight into the processes that occur. For example, Figure 2.16 shows the response of a dual electrode[5] used within a cavitation environment generated by an ultrasonic horn (or piston-like emitter).

Figure 2.16(a) shows the current time history of both electrodes recorded simultaneously. Under the conditions used the Pt electrode detects mass transfer enhancement of a redox species and shows a repetitive event. However, the erosion sensor (a passivated Pb electrode in this case), which is located close to the mass transfer sensor (~50 µm away) shows no surface erosion has occurred. This absence of surface erosion indicated that inertial cavitation is unlikely to be present and non-inertial bubble motion is a likely candidate for the forced convection generated under these conditions. This is supported by a ring-down experiments performed with simultaneous high-speed imaging of the liquid above the electrode. In this experiment the sound source was terminated and the current time history of the Pt microelectrode recorded for a period of 100 ms. Figure 2.16(b) shows the results of such an experiment. Initially the current is high, indicating mass transfer enhancement due to forced convection.

[5]This comprises of an electrode designed to measure both surface erosion effects through the use of a passivated Pb electrode in this case and a Pt microelectrode for mass transfer monitoring.

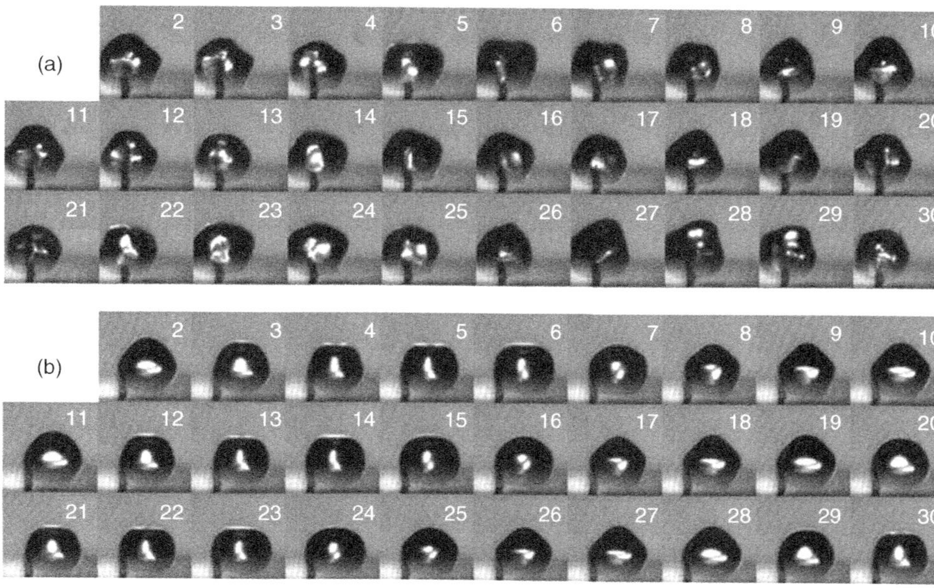

Figure 2.17 *High-speed images showing the resident surface bubble behaviour during a ring-down experiment. The images were taken from a 'side-on' perspective with respect to the electrode/solution interface. The inter-frame time is 0.5 ms (every 50th frame from data recorded at 100 000 fps). Note: sequence (a) was taken at an earlier time in the ring-down and multiple modes are seen on the gas/liquid interface of the bubble while (b), recorded later, appears to only show a single mode.*

Interestingly, when the excitation of the acoustic source is terminated (denoted by a vertical dashed line on Figure 2.16(b) and leading to a decay or 'ring-down' in the acoustic field) although the current at the microelectrode falls, several current transients are detected (labelled A, B, Figure 2.16(b)). This data indicates that the 'event' that causes these current enhancements remains present near the microelectrode surface and is still in motion. Further evidence for the presence of a residual event was obtained from high-speed imaging of the region above the electrode surface. Figure 2.17 shows high-speed images of the event which can be seen to be a permanent gas bubble trapped on the surface of the electrode. Initially this bubble can be observed to move around the surface and oscillate dramatically within the ultrasonic field. However, the ring-down experiment shows that the motion of the bubble wall becomes less dramatic as the pressure field dies away with a single surface wave present at later times within the ring-down experiment (see Figure 2.17(b)). Eventually a stationary bubble is seen left behind on the surface of the electrode. These types of events may be responsible for the repetitive multiple current/time transients reported in the literature [37, 38].

The type of mass transfer event, reported in Figure 2.16, only occurs under certain conditions. In particular under the conditions employed, they only occurred when the electrode to sound source separation was ~5 mm. This is a key point as the sound field developed by an ultrasonic horn under the conditions employed is predicted to have the highest zero-to-peak pressure amplitude at the surface of the piston-like emitter. Hence, if inertial events (note the event shown in Figure 2.17 is non-inertial) are to be investigated, the electrode must be placed within the region of the liquid where the zero-to-pressure amplitude is greater than ~120 kPa (the threshold for inertial events under the conditions employed – see Section 2.2). This is most likely to be encountered close to the surface of the piston-like emitter. As the electrode is moved away from the source, one would expect that a transition from inertial events to non-inertial events would be encountered. Detecting this boundary using mass transfer measurements on their own is extremely difficult due to the number of mechanisms that lead to transient mass transfer enhancements. Hence, a technique which is able to isolate effects of inertial cavitation is advantageous. This is discussed in the next section.

2.9 Measuring Individual Erosion Events from Inertial Cavitation

One characteristic of inertial cavitation is its association with surface erosion [88, 89]. In order to follow surface erosion using an electrochemical technique it is necessary to employ an electrochemical system where the electrode material itself is passive unless erosion takes place [74]. Consider the case where a passive electrode is then exposed to inertial cavitation (and hence erosive mechanisms). Figure 2.18 shows the sequence of events which can occur under these conditions. Prior to erosion of the surface, the electrode potential and the solution conditions are chosen to maintain a stable thin passive layer of material on the surface of the electrode. Under these conditions no current is observed to pass at the electrode.

However, erosion of the surface by an inertial event (or cluster of events) damages the passive protecting layer and causes the underlying electrode to be exposed to the solution. Under the potential/solution conditions employed immediate oxidation of the substrate would be expected and an anodic current will pass. This anodic current can be detected and measured and corresponds to the 'healing' of the damage caused by the erosive mechanism. Once the exposed area has been re-passivated the current returns to zero. The size and charge passed in each transient relates to the amount of damage caused to the passive surface by the erosive mechanism. This approach to the study of erosion by single cavitation events was first reported in 1997–8 for $Pb/PbSO_4$, aluminium and stainless steel [59, 74].

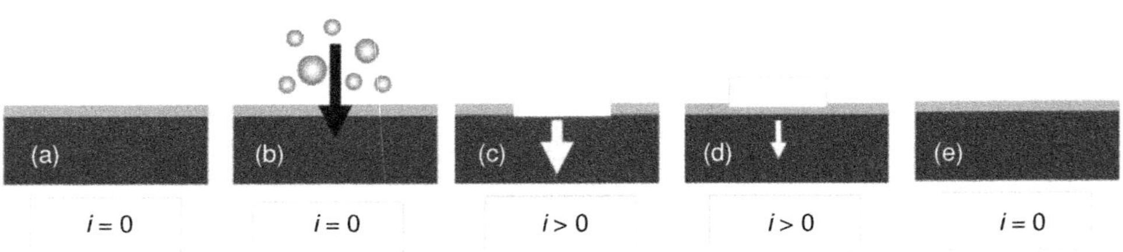

Figure 2.18 *Sequence of events associated with the erosion/corrosion measurements described here. The passive layer on the electrode surface and the underlying electrode substrate are shown. The current (i) recorded at the electrode is also shown. Note the erosion stage is indicated by a black arrow while the magnitude and direction of electron flow is represented by a white arrow.*

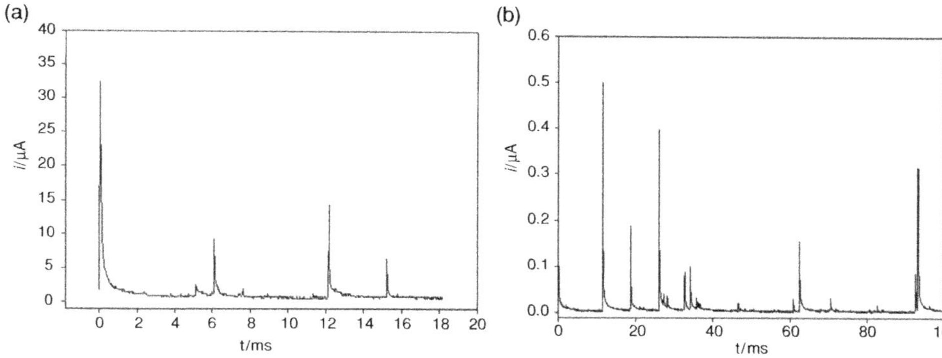

Figure 2.19 *(a) Plot showing the current time trace recorded for a passivated 125 μm diameter Pb electrode exposed to continuous ultrasound (~23 kHz). The cell contained 1 mol dm*$^{-3}$ H_2SO_4*. The electrode to source separation was 1 mm. The electrode was held at +500 mV versus SCE (Reprinted with permission from [74] Copyright (1998) Royal Society of Chemistry) (b) Plot showing the current time data recorded as a function of time for a stainless steel microelectrode (25 μm diameter) exposed to inertial cavitation. The tip/electrode distance was 0.3 mm and the potential was held at +250 mV versus SCE. In this case the solution contained 0.5 mol dm*$^{-3}$ K_2CO_3 *(Reprinted with permission from [90] Copyright (2005) Elsevier Ltd).*

Figure 2.19 shows typical individual erosion events for two different electrode materials exposed to inertial cavitation by continuous ultrasound [74, 90]. Note here the electrode was positioned accurately within the ultrasonic field close to the surface of the piston-like emitter (<2 mm axial electrode-to-horn separation).Monitoring the location and number of inertial erosion events detected using this electrochemical approach is useful in a number of ways [91–95]. First, the location of surface erosion is an indicator of acoustic pressure amplitudes greater than the minimum required for inertial events (this can be related to the inertial boundary discussed in section 2.2). Second, these measurements can be used with other techniques (e.g. high-speed imaging and luminescence) to investigate the timing and processes present within the complex media. In this second class, cluster collapse effects [3–5,96] have been interrogated and correlated to a variety of other characteristics of the ultrasonic field employed. It is also possible to determine boundaries between the presence and absence of inertial cavitation. Figure 2.20 shows an example of such a measurement [97].

Here a passivated lead electrode is positioned in the centre of the cavitation plume generated by an ultrasonic horn. In order to do this, the mass transfer enhancement recorded for a Pt electrode sealed in the same electrode body was mapped as a function of position in the XY plane at a Z displacement greater than that expected for inertial cavitation in the experimental arrangement used. The electrode was subsequently positioned in the centre of the plume and then moved progressively towards the sound source. This allowed for the determination of the boundary between the regions of the liquid where inertial events where present (denoted as the observation of current time transients at the Pb electrode as a result of surface erosion) and where it was not. This is an important boundary as it is thought to be an indication of where the acoustic pressure amplitude crosses the minimum required for inertial cavitation. Figure 2.20 shows five repeat measurements performed for a

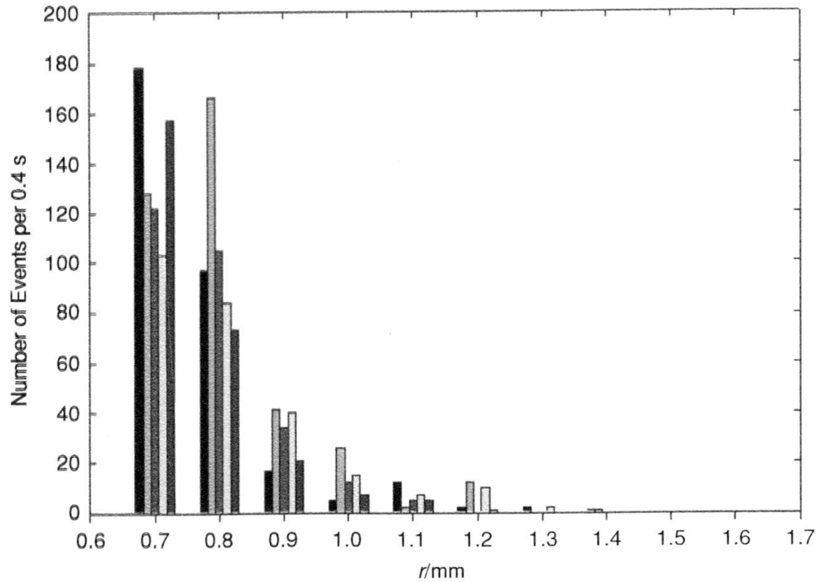

Figure 2.20 *Plot showing the number of erosion events per 0.4 s determined as a function of displacement (denoted as r) in the centre of the cavitation plume. The figure shows the results from five repeat experiments. This was recorded using a dual electrode to first map the mass transfer produced by the horn then the Pb electrode (housed in resin) was moved on progressively towards the sound source [97] (Reprinted with permission from [97] Copyright (2005) Royal Society of Chemistry).*

piston-like emitter source. This indicates that this boundary occurs around a displacement of 1.3 mm under these conditions. These electrochemical measurements can be combined with luminescence, acoustic modelling and acoustic measurements. Modelling the sound field developed by such a piston-like emitter indicates that the direct field (assuming at this stage that contributions from the reverberant and scattered field are small) falls dramatically with distance.

This demarcation between inertial and non-inertial cavitation behaviour has been investigated using a number of techniques the results of which are reported in Figure 2.21. Figure 2.21 shows a collection of results combining electrochemistry (specifically surface erosion), multibubble sonoluminescence imaging (MBSL [98]) and acoustic measurements all designed to investigate the 'inertial cavitation threshold' [10, 12, 97, 99]. Figure 2.21 shows the predicted[6] acoustic pressure amplitude plotted as a function of distance under the conditions employed. Over the distance considered (0.2 mm to 3 mm), the pressure amplitude falls considerably from ~130 kPa at 0.2 mm to 30 kPa at 3 mm. Under the conditions employed (22.85 kHz in water) the theoretical pressure threshold for inertial

[6] Note this prediction was made using acoustic measurements and modelling.

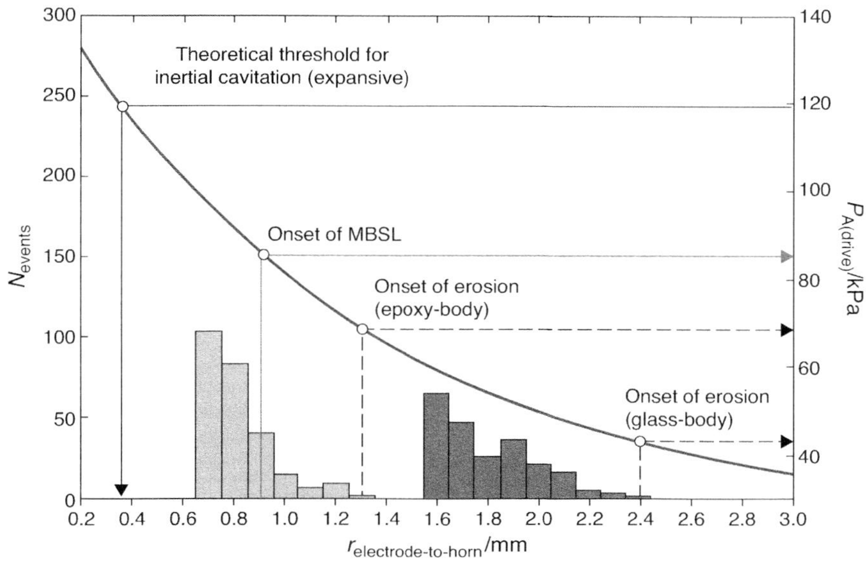

Figure 2.21 *Collection of results from a study to investigate the inertial cavitation threshold [96]. The figure shows the average number of erosion corrosion events detected using a electrochemical sensor [90, 96, 97] as a function of electrode-to-horn distance ($r_{electrode-to-horn}$). Two different electrode supports were employed; epoxy (▢) and glass (■) bodied electrodes. The predicted pressure amplitude (——) from the piston-like emitter at 22.85 kHz. The theoretical threshold pressure for inertial cavitation (——) is shown along with the experimentally determined pressures considering MBSL (——) and erosion/corrosion (– – –) from different electrode supports.*

cavitation is ~120 kPa (see Section 2.2). Hence, if correct, this predicts that inertial events should only be observed up to ~0.35 mm from the face plate of the piston-like emitter.

However, the experimental investigations of this system revealed that the effects of inertial cavitation (in this case MBSL luminescence and surface erosion) were detected at considerably greater distances than this prediction. In addition the materials used in the construction of the electrochemical sensor also altered the experimental threshold prediction [96]. Consider the MBSL output of the system first. Lowlight imaging was used to measure the extent of MBSL in this system. Figure 2.22 frames (a) and (b) show the sound source under normal and blackout conditions respectively. A clear zone of MBSL luminescence can be seen in Figure 2.22(b). In the absence of an electrode the boundary between luminescent bubbles (considered inertial) and non-inertial events (considered non-luminescent) was measured at ~0.9 mm. This corresponds to a pressure amplitude of ~85 kPa. Further measurements of surface erosion suggested that the inertial boundary was even further from the sound source and dependent on the materials employed in body of the sensor itself [96]. For example when an electrode supported in epoxy resin (see Figure 2.21 ▢) was employed, the boundary between inertial and non-inertial events (determined as the demarcation in distance between the presence of erosion/corrosion and the absence

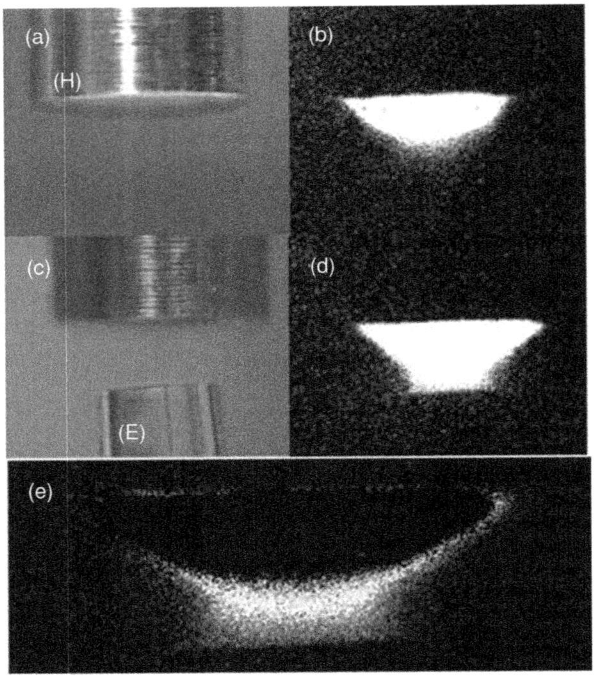

Figure 2.22 *Images (a–d) show images of ultrasonic horn (labelled H – note 3 mm diameter for scale purposes) and electrode (labelled E) taken with a cooled CCD camera. Frames (a) and (c) are taken under normal daylight conditions while (b) and (d) are taken in a dark room. Frame (a) shows the ultrasonic horn on its own and (b) the corresponding light emission (exposure time 10 min) in the presence of continuous ultrasonic irradiation (22.85 kHz, 56 ± 5 W cm^{-2}). Frame (c) shows the horn and an electrode and (d) the corresponding light emission (exposure time 10 min) in the presence of continuous ultrasonic irradiation. Frame (e) shows the result of image subtraction of frame (b) from frame (d) (Reprinted with permission from [96] Copyright (2005) American Chemical Society).*

of this effect) was determined to be 1.3 mm (corresponding to a pressure threshold of ~70 kPa) while when a glass-bodied electrode (see Figure 2.21 ■) was employed, the same boundary was found to be extended to 2.4 mm (corresponding to a pressure threshold of only 45 kPa). It is clear that these different measurements suggest that the system is considerably more complex than the simple pressure threshold predicts. It should be noted that the theoretical prediction (based on the work of Apfel *et al.* [10, 12]) assumes a sinusoidal excitation of a pre-existing bubble nuclei. However, the sound field produced by this sound source is considerably more complex (see reference [97]) containing not only the direct 23 kHz component from the piston-like emitter but also a shock wave-like component (in this system found to occur at a frequency of 7.6 kHz). This shock wave signal was associated with a cluster collapse model [3–5]. The presence of these shock waves has a number of interesting consequences. First, it is proposed that these shocks can cause bubbles to be inertial which, in their absence, would be classed as non-inertial [100]. This leads to the observation of MBSL output further from the surface of the sound source than expected (in agreement with the images shown in Figure 2.22). Second, these shocks can interact with objects (e.g. electrodes and other immersed probes) placed within this region of the sound field. This is a consequence of the high-frequency nature of a pressure shock which results in reflection of this pressure pulse from small (~2 mm) diameter disk-like objects [101, 102] (e.g. the electrode housing). As a consequence of this reflection (and the physics involved at the surface of interest), the pressure at the surface of the electrode is increased. In order to quantify these effects the scattering of sound by an object may be modelled as a function of the dimensionless variable ka where k represents the wave number and a the radius of the object in question [102]. For low ka values the object does not scatter strongly while for higher ka values strong scattering should be observed. For an electrode of a fixed radius this dependence manifests itself as a relationship in frequency whereby low frequency sound waves (the driving pressure 23 kHz field in this case) are not perturbed by the electrode. However, the high-frequency pressure shocks are efficiently reflected and the electrode becomes invasive. Experimental evidence for the invasive nature of the electrode is shown in Figure 2.22 frames (c) and (d). These images show the effect of the presence of an electrode on the MBSL emission of the system. Comparison of Figure 2.22 frame (b) with frame (d) clearly shows that the electrode has the effect of increasing the spatial extent of MBSL. This is highlighted by the image subtraction shown in Figure 2.22 frame (e). The precise increase in pressure caused by the electrode is also a function of the materials employed [96]. This is supported by the material dependence of the erosion limit (see Figure 2.21) and the MBSL emission. Glass has different acoustic properties when compared to epoxy resin and is found to be more reflecting (owing to their respective densities and speed of sound [96, 97]), which leads to a greater pressure increase in the case of the glass-bodied electrode. As a consequence there is an increased range over which erosion is detectable using the glass-bodied electrode when compared with the epoxy-bodied electrode (see Figure 2.21). These observations, along with the theoretical treatment of the scattering from the disk [96,101], suggested that pressure shocks generated by a cluster collapse mechanism were largely responsible for the effects observed. Further evidence for the presence of a cluster collapse model can be obtained from high-speed imaging of the system. These results suggest that cluster collapse effects and the associated pressure shocks generated have a significant influence on the physical and electrochemical measurements within the inertial region of an operating ultrasonic horn.

2.10 Conclusions

This chapter has detailed the investigation of bubble dynamics within a liquid containing different bubble phenomena. This ranges from the large single events that can be generated by a laser pulse to permanent bubbles physically held within the media. In all cases electrochemical data is valuable in assessing the mass transfer or surface erosion effects associated with the systems employed. However, an understanding of the acoustics and effects of the electrodes on these systems is vital to the full understanding of this fascinating but complex subject. Clearly in order to gain this understanding, combining the electrochemical measurements with complementary techniques (e.g. high-speed imaging, acoustic measurements, acoustic modelling, luminescence etc.) is of utmost importance.

Acknowledgements

The work presented in this chapter originating in Southampton has had a number of significant contributions. In particular I would like to thank my friend and colleague Professor Tim Leighton who has been invaluable in understanding and developing this research area. In addition I would like to thank Douglas Offin, Christopher Vian, Hanne Hirsimaki and Yvonne Watson for their resilience in the face of 'how hard can it be' ideas. Finally I would like to acknowledge financial support from the EPSRC, NPL, The Nuffield Foundation and the University of Southampton.

References

(1) Leighton, T.G. (1994) *The Acoustic Bubble*, Academic Press, London.
(2) Young, F.R. (1999) *Cavitation*, Imperial College Press, London.
(3) Vyas, B. and Preece, C.M. (1976) Stress produced in a solid by cavitation. *Journal of Applied Physics*, **47**, 5133.
(4) Hansson, I., Kedrinskii, V. and Morch, K.A. (1982) On the dynamics of cavity clusters. *Journal of Physics D: Applied Physics*, **15**, 1725.
(5) Hansson, I. and Morch, K.A. (1980) The dynamics of cavity clusters in ultrasonic (vibratory) cavitation erosion. *Journal of Applied Physics*, **51**, 4651.
(6) Flynn, H.G. (1975) Cavitation Dynamics II. Free pulsations and models for cavitation bubbles. *Journal of the Acoustics Society of America*, **58**, 1160.
(7) Flynn, H.G. (1975) Cavitation Dynamics I. A mathematical formulation. *Journal of the Acoustics Society of America*, **57**, 1379.
(8) Weissler, A. (1959) Formation of hydrogen peroxide by ultrasonic waves: free radicals. *Journal of the American Chemical Society*, **81**, 1077.
(9) Weissler, A., Cooper, H.W. and Snyder, S. (1950) Chemical effect of ultrasonic waves: oxidation of potassium iodide solution by carbon tetrachloride. *Journal of the American Chemical Society*, **72**, 1769.
(10) Apfel, R.E. (1980) Some new results on cavitation threshold prediction and bubble dynamics. Cavitation and Inhomogeneities in underwater acoustics, Gottingen.
(11) Apfel, R.E. and Holland, C.K. (1991) Gauging the likelihood of cavitation from short-pulse, low-duty cycle diagnostic ultrasound. *Ultrasound in Medicine and Biology*, **17**, 179.
(12) Holland, C.K. and Apfel, R.E. (1989) An improved theory for the prediction of microcavitation thresholds. *IEEE Transactions Ultrasonics Ferroelectrics and Frequency Control*, **36**, 204.
(13) Lea, S.C., Price, G.J. and Walmsley, A.D. (2005) A study to determine whether cavitation occurs around dental ultrasonic scaling instruments. *Ultrasonics Sonochemistry*, **12**, 233.
(14) Oleary, R., Sved, A.M., Davies, E.H., Leighton, T.G., Wilson, M. and Kieser, J.B. (1997) The bactericidal effects of dental ultrasound on Actinobacillus actinomycetemcomitans and Porphyromonas gingivalis – An in vitro investigation. *Journal of Clinical Periodontology*, **24**, 432.
(15) Leighton, T.G. and Cleveland, R.O. (2010) Lithotripsy. *Journal of Engineering in Medicine*, **224**, 317.
(16) Apfel, R.E. (1981) *Methods in Experimental Physics*, Vol. **19** (ed. P.D. Edmonds), Academic Press, New York.
(17) Apfel, R.E. and Holland, C.K. (1991) *Ultrasound in Medicine and Biology*, **17**, 179.
(18) Didenko, Y.T. and Suslick, K.S. (2002) The energy efficiency of formation of photons, radicals and ions during single bubble cavitation. *Nature*, **418**, 394.

(19) Eddingsaas, N.C. and Suslick, K.S. (2007) Evidence for a plasma core during multibubble sonoluminescence in sulfuric acid. *Journal of the American Chemical Society*, **129**, 3838.
(20) Flint, E.B. and Suslick, K.S. (1991) The temperature of cavitation. *Science*, **253**, 1397.
(21) Suslick, K.S. (1990) Sonochemistry. *Science*, **247**, 1439.
(22) Suslick, K.S., Doktycz, S.J. and Flint, E.B. (1990) On the origin of sonoluminescence and sonochemistry. *Ultrasonics*, **28**, 280.
(23) Suslick, K.S., Hammerton, D.A. and Cline, R.E. (1986) The sonochemical hotspot. *Journal of the American Chemical Society*, **108**, 5641.
(24) Šponer, J. (1990) Dependance of ultrasonic cavitation threshold on raised ambient pressure. *Studia Biophysica*, **137**, 91.
(25) Šponer, J., Davadorzh, C. and Mornstein, V. (1990) Influence of viscosity on the ultrasonic cavitation threshold for sonoluminescence at low megahertz region. *Studia Biophysica*, **137**, 81.
(26) Šponer, J. (1991) Theoretical estimation of the cavitation threshold for very short pulses of ultrasound. *Ultrasonics*, **29**, 376.
(27) Leighton, T.G. (1994) *The Acoustic Bubble, section 4.3.1*, Academic Press, London.
(28) Leighton, T.G. (1995) Bubble population phenomena in acoustic cavitation. *Ultrasonics Sonochemistry*, **2**, S123.
(29) Cooper, E.L. and Coury, L.A. (1998) Mass transport in sonovoltammetry with evidence of hydrodynamic modulation from ultrasound. *Journal of the Electrochemical Society*, **145**, 1994.
(30) Hagan, C.R.S. and Coury, L.A. (1994) Comparison of hydrodynamic voltammetry implemented by sonication to a rotating-disk electrode. *Analytical Chemistry*, **66**, 399.
(31) Madigan, N.A. and Coury, L.A. (1997) Measurement of heterogeneous electron transfer rates. *Analytical Chemistry*, **69**, 5.
(32) Zhang, H.H. and Coury, L.A. (1993) Effects of high-intensity ultrasound on glassy-carbon electrodes. *Analytical Chemistry*, **65**, 1552.
(33) Banks, C.E. and Compton, R.G. (2003) Sonoelectroanalysis: A review. *Chemical Analysis*, **48**, 159.
(34) Banks, C.E. and Compton, R.G. (2003) Ultrasonically enhanced voltammetric analysis and applications: An overview. *Electroanalysis*, **15**, 329.
(35) Banks, C.E. and Compton, R.G. (2004) Ultrasound: promoting electroanalysis in difficult real world media. *Analyst*, **129**, 678.
(36) Banks, C.E., Kruusma, J., Hyde, M.E., Salimi, A. and Compton, R.G. (2004) Sonoelectroanalysis: investigation of bismuth-film-modified glassy carbon electrodes. *Analytical and Bioanalytical Chemistry*, **379**, 277.
(37) Maisonhaute, E., Brookes, B.A. and Compton, R.G. (2002) Surface acoustic cavitation understood via nanosecond electrochemistry. 2. The motion of acoustic bubbles. *Journal of Physical Chemistry B*, **106**, 3166.
(38) Maisonhaute, E., White, P.C. and Compton, R.G. (2001) Surface acoustic cavitation understood by nanosecond electrochemistry. *Journal of Physical Chemistry B*, **105**, 12087.
(39) Marken, F., Eklund, J.C. and Compton, R.G. (1995) Voltammetry in the presence of ultrasound: can ultrasound modify heterogeneous electron transfer kinetics? *Journal of Electroanalytical Chemistry*, **395**, 335.
(40) Perusich, S.A. and Alkire, R.C. (1991) Ultrasonically induced cavitation studies of electrochemical passivity and transport mechanisms, part I. *Journal of the Electrochemical Society*, **138**, 700.
(41) Dewald, H.D. and Peterson, B.A. (1990) Ultrasonic hydrodynamic modulation voltammetry. *Analytical Chemistry*, **62**, 779.

(42) Rowe, W.E. and Nyborg, W.L. (1966) Changes in the electrode process brought about by small scale acoustic streaming. *Journal of the Acoustics Society of America*, **39**, 965.
(43) Whitney, G.M. and Tobias, C.W. (1988) Mass transfer effects of bubble streams rising near vertical electrodes. *AIChE Journal*, **34**, 1981.
(44) Kolb, J. and Nyborg, W. (1956) Small-scale acoustic streaming in liquids. *Journal of the Acoustics Society of America*, **28**, 1237.
(45) Nyborg, W.L. and Seegall, M.I.L. (1960) *Effects of Acoustic Microstreaming at Electrodes.* Proceedings of the 3rd International Congress on Acoustics, 1960.
(46) Deslouis, C. and Tribollet, B. (1990) Flow modulation techniques in electrochemistry, in *Advances in Electrochemical Science and Engineering*, Vol. **2** (eds G. Gerischer and C.W. Tobias), VCH, New York, p. 205.
(47) Sutija, D.P. and Tobias, C.W. (1994) Mass-transport enhancement by rising bubble curtains. *Journal of the Electrochemical Society*, **141**, 2599.
(48) Tobias, C.W. (1959) Effect of gas evolution on current distribution and ohmic resistance in electrolysers. *Journal of the Electrochemical Society*, **106**, 833.
(49) Brandon, N.P. and Kelsall, G.H. (1985) Growth kinetics of bubbles electrogenerated at micro-electrodes. *Journal of Applied Electrochemistry*, **15**, 475.
(50) Kelsall, G.H., Tang, S.Y., Smith, A.L. and Yurdakul, S. (1996) Measurement of rise and electrophoretic velocities of gas bubbles. *Journal of the Chemical Society Faraday Transactions*, **92**, 3879.
(51) Klima, J., Bernard, C. and Degrand, C. (1995) Sonoelectrochemistry: transient cavitation in acetonitrile in the neighbourhood of a polarized electrode. *Journal of Electroanalytical Chemistry*, **399**, 147.
(52) Klima, J., Bernard, C. and Degrand, C. (1994) Sonoelectrochemistry: effects of ultrasound on voltammetric measurements at a solid electrode. *Journal of Electroanalytical Chemistry*, **367**, 297.
(53) Tomita, Y. and Shima, A. (1990) High speed photographic observations of laser induced cavitation bubbles. *Acustica*, **71**, 161.
(54) Philip, A. and Lauterborn, W. (1998) Cavitation erosion by single laser-produced bubbles. *Journal of Fluid Mechanics*, **361**, 75.
(55) Denuault, G. (1996) Microelectrodes. *Chemistry in Industry*, **18**, 678.
(56) Birkin, P.R. and Martinez, S.S. (1997) A study on the effect of ultrasound on electrochemical phenomena. *Ultrasonic Sonochemistry*, **4**, 121.
(57) Birkin, P.R. and Silva-Martinez, S. (1995) The effect of ultrasound on mass transfer to a microelectrode. *Journal of the Chemical Society, Chemical Communications*, 1807.
(58) Birkin, P.R. and Silva-Martinez, S. (1996) A study of the effect of ultrasound on mass transfer to a microelectrode. *Journal of Electroanalytical Chemistry*, **416**, 127.
(59) Silva-Martinez, S. (1997) Applications of Ultrasound in Electrochemistry. PhD, University of Southampton.
(60) Engstrom, R.C. and Pharr, C.M. (1989) Scanning electrochemical microscopy. *Analytical Chemistry*, **61**, A1099.
(61) Kwak, J. and Bard, A.J. (1989) Scanning electrochemical microscopy – theory of the feedback mode. *Analytical Chemistry*, **61**, 1221.
(62) Birkin, P.R., Leighton, T.G. and Watson, Y.E. (2003) The use of acoustoelectrochemistry to investigate rectified diffusion. *Ultrasonics Sonochemistry*, **10**, 65.
(63) Birkin, P.R., Leighton, T.G. and Watson, Y.E. (2003) The use of Acoustoelectrochemistry to investigate bubble phenomena – Rectified diffusion. Applications of Power Ultrasound in Physical and Chemical Processing 4, Besançon.
(64) Birkin, P.R., Watson, Y.E., Leighton, T.G. (2001) Efficient mass transfer from an acoustically oscillated gas bubble. *Journal of the Chemical Society, Chemical Communications*, 2650.

(65) Birkin, P.R., Watson, Y.E., Leighton, T.G. and Smith, K.L. (2002) Electrochemical detection of Faraday waves on the surface of a gas bubble. *Langmuir*, **18**, 2135.
(66) Watson, Y.E. (2003) Electrochemical investigation of bubble wall motion during non-inertial cavitation. PhD, University of Southampton.
(67) Watson, Y.E., Birkin, P.R. and Leighton, T.G. (2003) Electrochemical detection of bubble oscillation. *Ultrasonics Sonochemistry*, **10**, 65.
(68) Amphlett, J.L. and Denuault, G. (1998) Scanning electrochemical microscopy (SECM): An investigation of the effects of tip geometry on amperometric tip response. *Journal of Physical Chemistry B*, **102**, 9946.
(69) Slevin, C.J., Ryley, S., Walton, D.J. and Unwin, P.R. (1998) A new approach for measuring the effect of a monolayer on molecular transfer across an air/water interface using scanning electrochemical microscopy. *Langmuir*, **14**, 5331.
(70) Maksimov, A.O. and Leighton, T.G. (2001) Transient processes near the threshold of acoustically driven bubble shape oscillations. *Acta Acustica*, **87**, 322.
(71) Maksimov, A.O., Leighton, T.G. and Birkin, P.R. (2006) Dynamics of a tethered bubble. *Innovations in Nonlinear Acoustics*, **838**, 512.
(72) Maksimov, A.O., Leighton, T.G. and Birkin, P.R. (2008) Self focusing of acoustically excited Faraday ripples on a bubble wall. *Physics Letters A*, **372**, 3210.
(73) Maksimov, A., Winkels, K., Birkin, P.R. and Leighton, T.G. (2008) Hopf bifurcation in acoustically excited Faraday ripples on a bubble wall, in *Nonlinear Acoustics: Fundamentals and Applications. 18th International Symposium on Nonlinear Acoustics – ISNA*, Vol. **18** (eds B.O. Endlo, C.M. Hedberg and L. Kari). Springer, Melville, New York, pp. 229–232.
(74) Birkin, P.R., O'Connor, R., Rapple, C. and Silva-Martinez, S. (1998) Electrochemical measurement of erosion from individual cavitation generated from continuous ultrasound. *Journal of the Chemical Society, Faraday Transactions*, **94**, 3365.
(75) Garcia, R., Hammitt, F.G. and Nystrom, R.E. (1966) *Correlation of Cavitation Damage with Other Material and Fluid Properties*, ASTM, Philadelphia.
(76) Vian, C.J.B., Birkin, P.R. and Leighton, T.C. Cluster collapse in a cylindrical cell: Correlating multibubble sonoluminescence, acoustic pressure, and erosion. *Journal of Physical Chemistry C*, **114**, 16416.
(77) Rigby, G.D., Grazier, P.E., Stuart, A.D. and Smithson, E.P. (2001) Gas bubble induced mixing in electrowinning baths. *Chemical Engineering Science*, **56**, 6329.
(78) Cooper, W.C. (1985) Reviews of applied electrochemistry 11. Advances and future prospects in copper electrowinning. *Journal of Applied Electrochemistry*, **15**, 789.
(79) Moskalyk, R.R. and Alfantazi, A.M. (2002) Nickel laterite processing and electrowinning practice. *Minerals Engineering*, **15**, 593.
(80) Moskalyk, R.R. and Alfantazi, A.M. (2000) Review of present cobalt recovery practice. *Minerals and Metallurgical Processing*, **17**, 205.
(81) Vogel, A., Nahen, K., Theisen, D. and Noack, J. (1996) Plasma formation in water by picosecond and nanosecond Nd:YAG laser pulses – part I: Optical breakdown at threshold and superthreshold irradiances. *IEEE Journal of Selected Topics in Quantum Electronics*, **2**, 847.
(82) Lauterborn, W. and Ohl, C.D. (1997) Cavitation bubble dynamics. *Ultrasonics Sonochemistry*, **4**, 67.
(83) Lindau, O. and Lauterborn, W. (2003) Cinematographic observation of the collapse and rebound of a laser-produced cavitation bubble near a wall. *Journal of Fluid Mechanics*, **479**, 327.
(84) Mellen, R.H. (1956) An experimental study of the collapse of a spherical cavity in water. *Journal of the Acoustical Society of America*, **28**, 447.
(85) Birkin, P.R., Hirsimaki, H.M., Frey, J.G. and Leighton, T.G. (2006) Mass transfer enhancement produced by laser induced cavitation. *Electrochemistry Communications*, **8**, 1603.

(86) Hirsimäki, H. (2007) *An Electrochemical Investigation of Laser Cavitation*, University of Southampton.

(87) Hirsimäki, H., Birkin, P.R., Frey, J.G. and Leighton, T.G. (2006) *The Laser-Induced Formation of Plasma Bubbles in Water – Electrochemical Measurements*, IOA Spring Conference, 2006, Southampton.

(88) Whillock, G.O.H. and Harvey, B.F. (1997) Ultrasonically enhanced corrosion of 304L stainless steel II: The effect of frequency, acoustic power and horn to specimen distance. *Ultrasonics Sonochemistry*, **4**, 33.

(89) Whillock, G.O.H. and Harvey, B.F. (1997) Ultrasonically enhanced corrosion of 304L stainless steel I: The effect of temperature and hydrostatic pressure. *Ultrasonics Sonochemistry*, **4**, 23.

(90) Birkin, P.R., Offin, D.G. and Leighton, T.G. (2005) The study of surface processes under electrochemical control in the presence of inertial cavitation. *Wear*, **258**, 623.

(91) Birkin, P.R., Leighton, T., Offin, D.G. and Vian, C. (2010) Towards the understanding of a complex cavitation environment – electrochemistry, acoustics and imaging observations (Invited Paper). Internoise – Noise and Sustainability, Lisbon.

(92) Birkin, P.R., Leighton, T., Offin, D.G. and Vian, C. (2010) *Observations of Bubble Clusters Within Different Cavitation Environments – Electrochemistry, Acoustics and Imaging Studies*, ICA, Sydney.

(93) Vian, C., Birkin, P. and Leighton, T. (2010) Cluster collapse in a cylindrical cell: Correlating multibubble sonoluminescence, acoustic pressure, and erosion. *Journal of Physical Chemistry C.*, **114**, 16416.

(94) Vian, C.J.B. (2007) *A Comparison of Measurement Techniques for Acoustic Cavitation*, University of Southampton.

(95) Vian, C.J.B., Birkin, P.R. and Leighton, T.G. (2009) Opto-isolation of electrochemical systems in cavitation environments. *Analytical Chemistry*, **81**, 5064.

(96) Birkin, P.R., Offin, D.G., Joseph, P.F. and Leighton, T.G. (2005) Cavitation, shock waves and the invasive nature of sonoelectrochemistry. *Journal of Physical Chemistry B*, **109**, 16997.

(97) Birkin, P.R., Offin, D.G. and Leighton, T.G. (2005) Experimental and theoretical characterisation of sonochemical cells. Part 2: cell disruptors (Ultrasonic horns) and cavity cluster collapse. *PCCP*, **7**, 530.

(98) Walton, A.J. and Reynolds, G.T. (1984) Sonoluminescence. *Advances in Physics*, **33**, 595.

(99) Roy, R.A., Atchley, A.A., Crum, L.A., Fowlkes, J.B. and Reidy, J.J. (1985) A precise technique for the measurement of acoustic cavitation thresholds and some preliminary results. *Journal of the Acoustical Society of America*, **78**, 1799.

(100) Wolfrum, B., Kurz, T., Mettin, R. and Lauterborn, W. (2003) Shock wave induced interaction of microbubbles and boundaries. *Physics of Fluids*, **15**, 2916.

(101) Rayleigh, L. (1896) *Theory of Sound*, 2nd edn, Macmillan, London.

(102) Morse, P.M. and Ingard, K.U. (1986) *Theoretical Acoustics*, Princeton University Press, New York.

3

Sonoelectroanalysis: An Overview

Jonathan P. Metters, Jaanus Kruusma and Craig E. Banks

3.1 Introduction

The application of power ultrasound is shown to facilitate electroanalysis at ultra-trace levels through acoustic streaming when applied during an electrochemical accumulation step. Yet the 'real' power of the technique lies upon applying insonation into electrochemical processes which are inhibited in complex media but yet are transformed into large quantifiable signatures, which is impossible using conventional methods via *in situ* cavitational cleaning processes. In this chapter we provide an up-to-date overview of this exciting area known as **sonoelectroanalysis**.

The use of stripping voltammetry is a highly sensitive methodology for the analytical detection of a wide range of metals which can achieve nano- and pico-molar concentration in certain cases. The methodology is divided into three categories; anodic, cathodic or adsorptive stripping voltammetry [1]. In each case there is first a pre-conditioning step – for anodic stripping the process may be described by:

$$M^{n+}(aq) + ne^- \rightarrow M(electrode) \tag{3.1}$$

while for cathodic stripping either

$$2M^{n+}(aq) + mH_2O \rightarrow M_2O_m(electrode) + 2mH^+ + 2(m - n)e^- \tag{3.2}$$

or

$$M + L^{m-} \rightarrow ML^{(n-m)-}(ads) + ne^- \tag{3.3}$$

Power Ultrasound in Electrochemistry: From Versatile Laboratory Tool to Engineering Solution, First Edition. Edited by Bruno G. Pollet.

and for adsorptive stripping:

$$A(aq) \rightarrow A(ads) \tag{3.4}$$

Following a suitable accumulation of the target species at a suitable electrode substrate, a stripping step is applied in which the electrode potential is swept to induce a Faradaic current arising from the oxidation (or reduction) of the accumulated material, which can be described by:

$$M(electrode) \rightarrow M^{n+}(aq) + ne^- \tag{3.5}$$

for anodic, while cathodic either

$$M_2O_m(electrode) + 2mH^+ + 2(m-n)e^- \rightarrow 2M^{n+}(aq) + mH_2O \tag{3.6}$$

or

$$ML^{(n-m)-}(ads) + ne^- \rightarrow M + L^m \tag{3.7}$$

and for adsorptive stripping:

$$A(ads) \pm ne^- \rightarrow B(aq) \tag{3.8}$$

In the stripping step, a characteristic voltammetric signature is produced allowing quantification of the target ion. Figure 3.1 depicts a typical voltammetric signal for a target metal, M^{n+} which has undergone the electrochemical transitions as described by Equations (3.1) and (3.5).

Stripping voltammetric methods are sensitive and reproducible with ease of computerisation and automation of analytical procedures, utilising instrumentation which is simple and easy to use and is economical in comparison to alternative analytical methodologies [1].

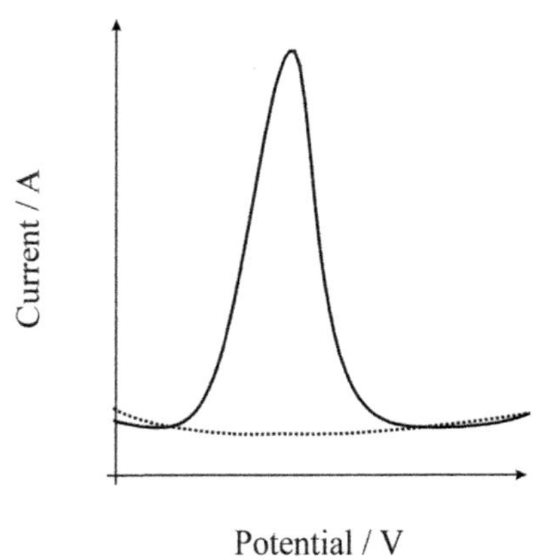

Figure 3.1 *Schematic representation of a typical voltammetric signal (solid line) obtained from the electrochemical stripping of a target metal from a suitable electrode substrate. Shown for clarity is the response of the electrode substrate when the target metal has not been suitably accumulated via the electrodeposition step.*

Mercury was traditionally used in anodic stripping voltammetry (ASV) where a thin mercury film is formed *in situ* on a suitable electrode substrate via the addition of a mercuric salt to the sample solution where the electrodeposition of mercury and target trace metals occur simultaneously. The target metals are then anodically stripped from the mercury film which typically has a thickness of 0.001–0.1 μm. After the analysis, the mercury film can be completely removed from the electrode by electrochemical stripping or simply wiping with a tissue or cloth [1]. This technique is highly sensitive, giving excellent resolution of voltammetric signals towards target metal species for instance, and, because of the method's simplicity and reliability, is particularly suitable for routine analysis, with the methodology allowing the electroanalytical sensing of target metals ions in a diverse range of samples. However, mercury has a drawback of being restricted in light of recent legislation and its inherent toxicology [1, 2]. Note, however, that there are some target analytes which still require the use of mercury.

In the application of stripping voltammetry, the analytical range and detection limit can be tailored through careful choice of the applied deposition potentials and duration. However, the former is limited by accessible voltammetric windows where at high applied potentials, solution decomposition results in the formation of gaseous species resulting in the analytical methodology becoming irreproducible. In the latter case for trace detection, deposition times can approach many tens of minutes, resulting in the overall time of analysis becoming excessive.

A common approach is to apply convection during the accumulation step, increasing the transport of material towards the electrode surface and thus allowing higher analytical sensitivities and greatly reduced accumulation times [3–5]; a typical cell assembly is shown in Figure 3.2.

A common approach is to apply power ultrasound during the accumulation step. When applying power ultrasound to a solution, a process known as **acoustic streaming** arises

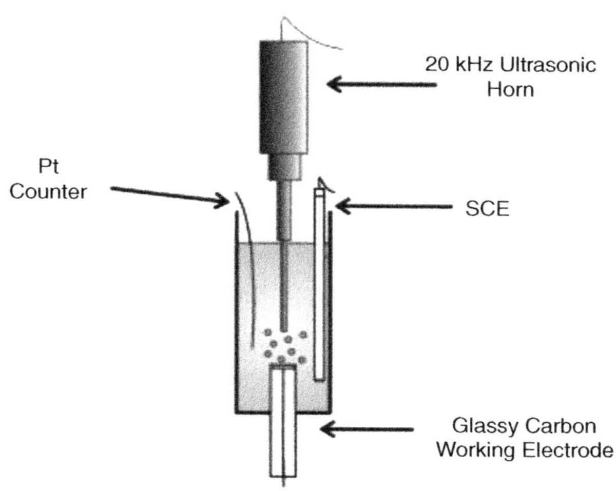

Figure 3.2 *Typical cell assembly used in sonoelectroanalysis (Reprinted with permission from [78] Copyright (2003) Wiley).*

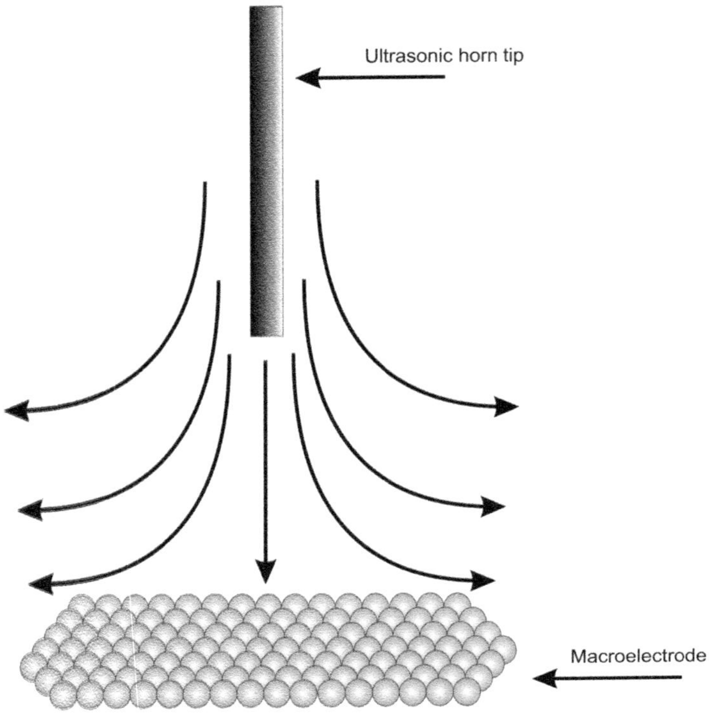

Figure 3.3 *Representation of acoustic streaming indicating a typical flow pattern induced by applying power ultrasound into a solution in 'face-on' mode opposite to a macroelectrode (Reprinted with permission from [6] Copyright (2003) Wiley).*

where forced convection induces a typical flow pattern as shown in Figure 3.3 [6]. This is a non-linear effect [7], which has been suggested to lead to local flow rates of more than circa 10 m s^{-1} [4, 8].

This forced convection can be parameterised by assuming that ultrasonic application results in a truncated or thinned Nernst diffusion layer. This diffusion layer model allows a naive description of the mass transport at the electrode interface by assuming a laminar sub-layer close to the surface and an approximately linear concentration gradient across a thin layer adjacent to the electrode [8, 9], Moriguchi [10] is thought to have first drawn attention to the fact that ultrasound can decrease the thickness of the diffusion layer, although Marken *et al.* [11] and Cooper *et al.* [12] proposed independently that acoustic streaming is the major contribution of the possible ultrasonic mechanisms on the rate of mass transport to an electrode. Marken *et al.* also elegantly demonstrated that under extreme conditions the minimum diffusion layer achievable in aqueous solution to be as thin as 0.7 μm (± 0.1 μm) [8].

When ultrasound is applied into electroanalytical methodologies, the term **sonoelectroanalysis** is utilised. The unique effect of applying acoustic streaming on the anodic stripping voltammetry is the transformation of a small analytical signal obtained under silent conditions (no power ultrasound) into a large, easily quantifiable signature from a result of

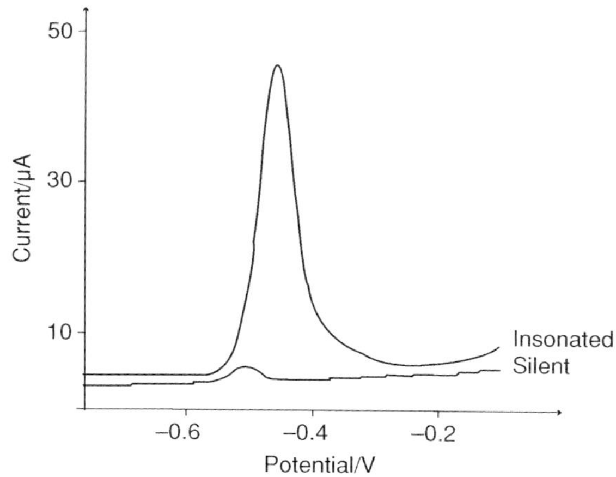

Figure 3.4 *Comparison of the effect of using ultrasound (insonated) during the accumulation step with that observed under quiescent conditions. Conditions: 5 × 10^{-7} mol L^{-1} lead(II) utilising a 60 s deposition time at −1.5 V at a* in situ *bismuth film coated boron-doped diamond electrode (Reprinted with permission from [13] Copyright (2004) Springer Science + Business Media).*

insonation [1, 4, 8]. This is shown in Figure 3.4 for the anodic stripping voltammetry of lead, as described by Equations (3.1) and (3.5) above. Here the anodic stripping voltammetry of lead is performed under silent conditions (no ultrasound) at an *in situ*-formed bismuth film modified boron-doped diamond electrode and compared with that of applying ultrasound during the accumulation step. Clearly a large increase in the magnitude of the voltammetric peak is evident, which has resulted in an increase of ∼ 14-fold. Table 3.1 represents a non-exhaustive list of various analytes that can be determined using sonoelectroanalysis – here in this chapter we highlight the prominent examples.

The majority of methodologies make use of the 'face-on' approach, as depicted in Figure 3.3 which is one of the most widely used approaches; a second variant is the 'side-on' approach but it is not widely utilised due to the reduction in acoustic streaming [9, 11]. Later, we will discuss a more preferred approach, the **sonotrode** in Section 3.8. Note that a critical parameter that needs to be monitored and recorded in order for the electroanalytical response to be produced, is the ultrasonic intensity (ultrasonic power/ultrasonic probe surface area) delivered into the solution medium, and also the horn-tip to electrode distance [8, 9].

The benefits of acoustic streaming can be applied into other areas of electroanalysis. Ward-Jones [14] has shown that the high rate of mass transport from acoustic streaming applied to metallic micron particles suspended in octane within an electrochemical cell in 'face-on' mode results in immobilisation of the particles onto an electrode surface. It is believed that the acoustic streaming forces the metallic particles into the electrode surface causing defects such as scratches. The immobilised particles are electrically conducting and can be used as a random ensemble of microelectrodes [14]. Microelectrodes are attractive tools for a plethora of electrochemical explorations such as in the study of ultra-fast kinetic

Table 3.1 *A non-exhaustive list of various analytes that can be determined using sonoelectroanalysis.*

Element Analysed	Detection Limit/ M	Ultrasonic Intensity/ W cm^{-2}	Comments	Reference
Copper	Ω	200	Mercury thin-film and glassy carbon electrodes; proof of concept shown through analysis of beer	[34]
Cadmium	3×10^{-11}	30	Nafion-coated mercury thin-film glassy carbon working electrode	[93]
Lead	3×10^{-11}	30	Nafion-coated mercury thin-film glassy carbon working electrode	[93]
	9.65×10^{-9}	3.5	ASV sonotrode mercury modified electrode	[94]
	Ψ	52	Mercury modified platinum; proof of concept shown	[95]
	Ω	26	In wine; determined ~20 μL levels; mercury platinum electrode	[96]
	2.41×10^{-9}	40	Human saliva; *in situ* plated mercury thin film GCE	[64]
	1×10^{-10}	34	Bismuth film modified BDD electrode; applied to diluted blood	[13]
Arsenic	1×10^{-8}	57	Gold electrode; proof of concept shown	[97]
Zinc	4×10^{-6}	140	Proof of concept shown within difficult surfactant media	[79]
Vanadium	Ω	17.7	Adsorptive stripping; mercury-plated copper electrode	[98]
Ascorbic acid	Ψ	25	Pulsed sonoelectroanalysis; proof of concept for detection in Ribena drinks compared favourably with standard HPLC	[99]
Nitrite	9×10^{-6}	50	Utilised in phloroglucinol as a chemical marker	[48]

(continued)

Table 3.1 (Continued)

Element Analysed	Detection Limit/M	Ultrasonic Intensity/ W cm^{-2}	Comments	Reference
Manganese	1×10^{-11}	14	Boron-doped-diamond electrode; instant tea; cathodic stripping voltammetry	[50]
Copper	6.29×10^{-8}	200	Applied to rainbow trout gill mucous	[100]
Nitrate	Ω	Ω	Applied to treated sewage; copper modified electrode	[101]
Cadmium	$5 \times 1\ 0^{-8}$	50	In presence of surfactants (titrant X-100)	[102]
5-aminosalicylic acid	3×10^{-7}	Ω	N/A	[103]
Cadmium	9×10^{-9}		Human saliva; *in situ* plated mercury thin film GCE	[64]
Lead	1×10^{-10}	34	bismuth film modified BDD electrode; applied to diluted blood	[13]

Ψ = not examined.
Ω = not provided.

processes or for use in electroanalysis [15–17]. Microelectrode arrays can be fabricated by photolithography or laser ablation, but these are costly approaches. Other methodologies include sonochemical ablation of thin-polymer films to reveal an underlying metal layer [18, 19] deposition of gold into the pores of a nano-porous material [20] and forming screen-printed arrays [21]. Another method is to seal thousands of carbon wires randomly into epoxy resin producing random assemblies of microdiscs, which are known as RAM electrodes [22]. Note that the word 'assembly' is used to indicate that the microdiscs are not regularly spaced. A possible method to create a random microelectrode array would be to 'mechanically attach' metal particles to a bare carbon electrode surface, which may be achieved by abrasively rubbing the electrode surface into the metal particles [23–25]. This can be applied to an analytical sample where the metals are present and are electrochemically stripped from the surface for use in their quantification.

Returning to the acoustically fabricated assemblies, proof-of-concept for use within electroanalysis was demonstrated for the sensing of arsenic(III) and cadmium(II) at silver and bismuth random microelectrode assemblies, respectively. This approach suggests a simple generic methodology for the construction of microelectrode assemblies via abrasive attachment induced by insonation with power ultrasound [14].

Last we note that by acoustically accelerating metallic particles towards an electrode surface, single impacts of the solid particles can be monitored by ultra-fast chronoamperometry [26]. Such an approach produces current spikes of 1–5 μs duration only observable using a potentiostat with an upper scan rate limit of ~2.2 MV^{-1} [26]. The electrode at which the acoustically agitated solid particles are aimed is held over a range of potentials in an aqueous solution, where the observed current spikes are monitored, with the polarity of current spikes found to invert at potentials corresponding to the potential of zero charge (PZC). Such an approach allows a simple methodology for the determination of the number and size of the particles suspended in solution [26].

While acoustic streaming is a very effective methodology for applying forced convection into an electroanalytical process, there are of course many other approaches that can be utilised such as a high speed rotating electrode [27–29] and microfluidics [30]. So the pertinent question is: *what is so attractive about utilising power ultrasound?* The answer to this arises from the other phenomenon, acoustic cavitation, which is produced through the application of power ultrasound into a liquid media. Cavitation arises from the alternating longitudinal sound waves creating presumed variations in the liquid media leading to the breaking of the cohesive forces which hold the liquid molecules together [4, 6, 9, 31, 32]. Since this itself alone is insufficient to break the intermolecular forces, gas-filled bubbles grow near in-homogeneities such as solid impurities and/or any tiny gas bubbles to provide a cavitation nucleus in which acoustic cavitation is initiated. These cavitational bubbles grow over the timescales of the order of an acoustic cycle and then undergo an energetic collapse occurring in less than one microsecond. On collapse, localised high temperatures and pressures are produced, which can lead to highly reactive species within the gas bubble [1, 4, 6, 9, 31, 32].

It was originally thought that cavitation bubbles at the electrode's surface were undergoing asymmetric collapse, which leads to the formation of a microjet. It has been demonstrated that the microjet development is not responsible for the intense and periodic current spikes observed when fast chronoamperometry at a microelectrode was carried out [31–33]. It has recently been shown that the shear flow involved during bubble oscillation is likely the case; Figure 3.5 presents a schematical representation [31–33]. The cavitational activity at the electrode surface creates an *in situ* cleaning mechanism permitting the electroanalysis in otherwise passivating media [1]. The efficiency of the ultrasonic cleaning strongly depends upon the size of the adsorbed particle to be removed and on the electrode to horn distance [33]. This clearly puts limitations on the application of power ultrasound into electrochemical processes in real media samples, but it is one of the best approaches in overcoming the detrimental affects that would otherwise be observed.

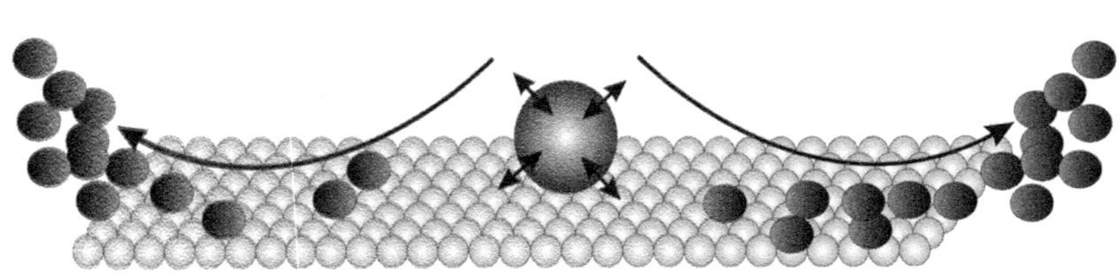

Figure 3.5 *Schematic representation of* in situ *cavitational cleaning via oscillating (20 kHz) bubbles generating a strong shear flow across a macroelectrode surface (Reprinted with permission from [64] Copyright (2002) Royal Society of Chemistry).*

The field of sonoelectroanalysis has been authoritatively overviewed in recent years [1, 4–6, 8, 9, 34] and in this chapter we wish to build on these, providing an update on recent developments and highlighting prominent examples.

3.2 Analysis of Pesticides

As food production intensifies, the greater use of agrochemicals prevails, and unfortunately, as a result, environmental pollution caused by pesticides and their degradation products occur, which is both an ecological issue and a health issue. Chromatographic approaches to detection are laborious and require lengthy preparative clean-up steps [35–37]. Electroanalytical protocols for quantifying pesticides in a food have been reported using mercury and other electrode substrates, but only after lengthy pre-treatment of the samples [38–40]. Problems are also faced when using an electroanalytical approach, as adsorption from both the target species and interferents are involved which results in irreproducible voltammetric signals. Garbellini *et al.* have reported that the sonoelectroanalytical quantification of 4-nitrophenol on a boron-doped diamond electrode, where improvements in the protocol were facilitated by the application of power ultrasound through cavitational cleaning and enhancements in mass transport, allowed detection limits of ~4 μg L^{-1} and ~3μg L^{-1} for the oxidation and reduction processes respectively [6, 41]. This has been extended further by Garbellini [42] for sensing methylparathion in potato and corn extracts, and also 4-nitrophenol in fruit juices using sonoelectroanalysis at boron-doped diamond electrodes. Analysis of these real samples required only simple dilution with aqueous solutions and filtration. Limits of detection of ~5 and ~10 μg L^{-1} were obtained for methylparathion in water and corn extract respectively. Interestingly, these detection limits for methylparathion in water and corn extract are 55 and 72% lower than those obtained using silent voltammetric analysis, indicating the enhancement by acoustic streaming [42].

3.3 Quantifying Nitrite

Nitrite is found in a plethora of samples and there is a requirement for its continued monitoring due to its toxicity to human health [43]. Nitrite is metabolised to carcinogenic *N*-nitroso compounds [43]. For a detailed overview of detection strategies see Moorcroft *et al.* [43]. In the realms of electroanalysis it was widely reported within literature that bare, unmodified electrodes were unsuitable due to surface passivation which would decrease the electroanalytical sensitivity and reproducibility. However, recent work by Kozub *et al.* [44] has reported that the traditional use of chemically modified electrodes and metallic modified electrodes [45, 46] do not need to be applied, and that in fact bare unmodified carbon electrodes can be utilised instead [44].

This has been extended to the use of screen-printed microelectrode arrays allowing low detection limits [47]. While these approaches will work in environmental samples, the quantification of nitrite in complex media, such as in food, for example eggs, would be extremely taxing, since the high concentrations of protein and fat foul the surface of the electrode preventing anything useful to be derived. Davis and Compton [48] demonstrated

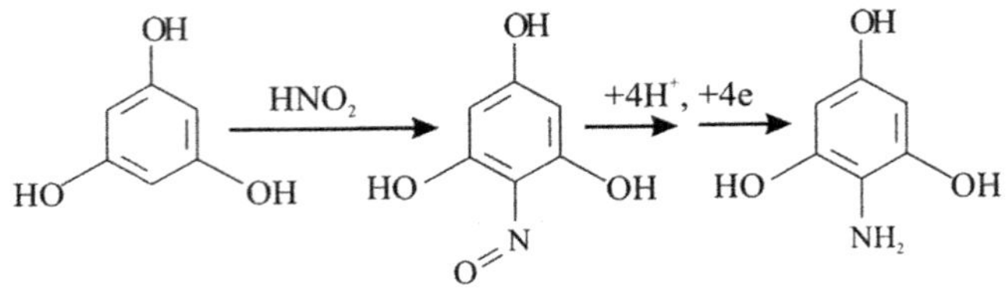

Scheme 3.1 Reaction of phloroglucinol with nitrite and its subsequent reduction to the corresponding aminophloroglucinol [47].

that nitrite sensing could be achieved in homogenised egg via sonoelectroanalysis. The egg sample was homogenised (white and yolk) and phloroglucinol (in aqueous solution) was added. This addition serves as to dilute the sample and provide electrolyte, while the phloroglucinol acts as an assay; see Scheme 3.1. The solution is then acidified and is ready for use via sonoelectroanalysis.

The use of power ultrasound ensures efficient mixing of the egg sample and *in situ* cavitational processes provides electrode cleaning, meaning that a signal is obtained in highly passivating media, which would otherwise not yield any voltammetric response. While the application might be questionable, the example of the use of power ultrasound in highly passivating media is not.

3.4 Biogeochemistry

The biogeochemistry of metals is an important parameter in the ecological status of the natural waters [49]. The total concentration of trace materials is an indicator of pollution and their distribution in the water is equally important [49]. A recent report of the sonoelectroanalysis of manganese(II) at a boron-doped diamond electrode by Saterlay *et al.* [50] has been used in the analysis of natural waters [49, 51]. This is based on cathodic stripping voltammetry where the electrodeposition/stripping is due to the following process:

$$MnO_2 + 2e^- + 4H^+ \rightarrow Mn^{2+} + 2H_2O \tag{3.9}$$

This electroanalytical approach has been reported previously by Rodrigues *et al.* [52], but to enhance the detection limits, the power ultrasound has been applied allowing ultra-trace levels, $\sim 10^{-11}$ M, to be readily achieved using only a deposition time of 120 seconds [50]! The enhanced mass transport from applying power ultrasound is clearly evident. This approach is favourable where other metal species (ions), likely to be found within real samples, are present. Zinc(II), copper(II), lead(II) and iron(III) ions are found to have no effect on the voltammetric signature of the manganese detection. The protocol is also tolerant to Hg^{II} at levels of at least 50-fold those of Mn^{II} and fivefold excess of Al^{3+}. The only downside to this approach is the presence of iron(II) which at similar levels affects the Mn^{II} signal. This may be overcome by chemical reagents or oxidising the Fe^{2+} to Fe^{3+} via application of the potential prior to analysis [50]. Saterlay *et al.* demonstrated that the sonoelectroanalytical approach could be used for sensing in tea samples [50]. Godwin *et al.* demonstrated that this could be applied to sensing Mn^{II} in seawater near a former ferromanganese factory in Šibenik, Croatia [51]. The sonoelectroanalytical approach was verified with the analysis of a certified seawater reference material and was then diligently applied to water samples from the bay of Šibenik in Croatia. Interestingly, the levels of Mn^{II} near the former ferromanganese factory were found to be similar to that in natural seawater, which was suggested to be due to the speciation of the Mn^{II} [51]. Lastly, Omanović *et al.* [49] have extended this to an in-depth study of Mn^{II} as well as Cd, Pb and Cu ions in Šibenik bay (Krka river estuary) [49], where the low detection limits facilitated by ultrasound allow for the monitoring of trace metal species in natural water samples.

3.5 Quantifying Metal in 'Life or Death' Situations

The World Health Organization (WHO) has reported that the contamination of water supplies in Bangladesh is the largest 'poisoning' of a population in history [53–56]. This is a geological problem which not only affects Bangladesh but also the USA, Taiwan, New Zealand, China and many other nations [57].

Ingesting arsenic has many adverse health influences, especially long-term exposure that results in mutagenic and carcinogenic effects [57]. As a consequence, the analysis of arsenic in drinking water wells is imperative, with many portable kits such as the Merck glow test based on the Marsh/Gutzeit test [58]. These tests do, however, have the drawbacks of not meeting the WHO limit of 10 $\mu g\ L^{-1}$ and also interpreting the colour response is needed. This approach was used to paint water wells 'green' or 'red', with the former indicating that the water is safe to drink. Note that in the case of Bangladesh, the guidelines for arsenic within drinking water is higher than that of the WHOs. The introduction of the Arsenator [59] has largely met this, but it still does not meet the desired WHO levels.

Interestingly, electroanalysis of the water wells revealed that an improved detection limit was viable through the application of anodic stripping voltammetry. The electroanalysis of samples resulted in a system which was not only sensitive, rapid and below the WHO guidelines, but also cheap and portable. The only drawback to the system is the possible interference of copper, which was determined to overlap the signal response. Consequently there is the requirement of sensing arsenic using economical and sensitive measurement techniques, yet still allowing for portability. Other approaches have reported that the possible limits of detection at levels are 0.5 $\mu g\ L^{-1}$ on platinum electrodes and 0.02 $\mu g\ L^{-1}$ on gold electrodes [60]. However, the detection times are excessive, approaching 10 minutes for each addition! To overcome this Simm *et al.* have reported the sensing of ultra-trace arsenic using a gold macroelectrode under insonation. A detection limit of $\sim 10^{-8}$ M using a deposition time and potential of 60 s at -1.5 V versus SCE respectively is possible [53]. The methodology was shown to be possible in river samples containing copper contamination and passivating organic materials. It was shown that the presence of surfactants (and humic acid), well known to passivate electrode surfaces/processes, is overcome through *in situ* cavitational activity, and additionally the acoustic streaming permits low detection levels, where gas bubbles formed at the electrode surface are removed, which would otherwise provide irreproducible voltammetric signals [53].

While such approaches are advantageous, the implementation in the field is limited. To this end, Simm *et al.* [53] report the use of a low frequency (250 Hz) sound source integrated with an electrochemical sensor. This truly hand-held sensor, where the whole device fits

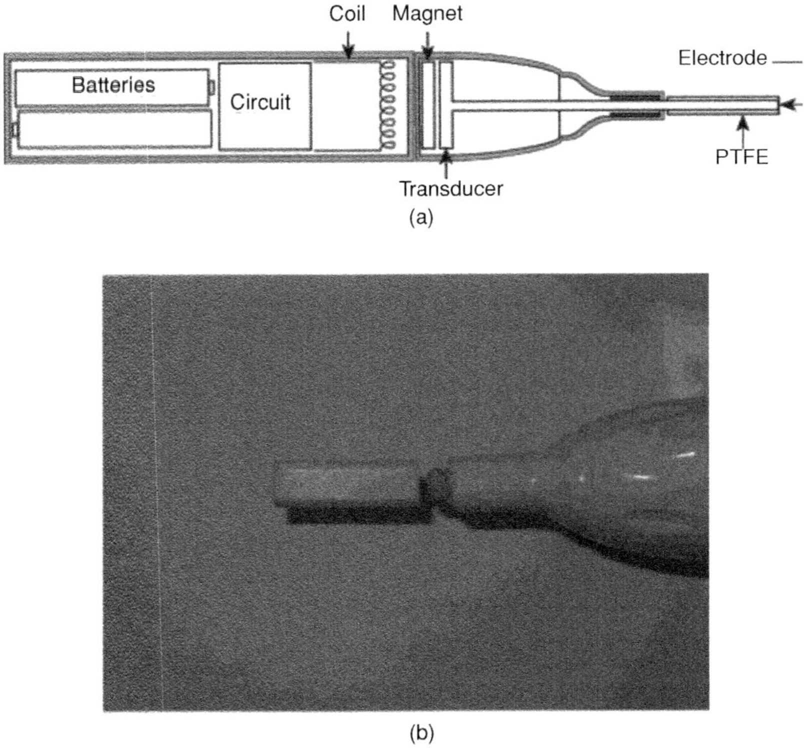

Figure 3.6 (a) Schematic diagram showing the dimensions and layout of the sonotrode. (b) A photograph of the output arm of the sonotrode with a gold microdisc electrode attached in the vertical orientation (Reprinted with permission from [61] Copyright (2004) American Chemical Society).

into one's hand, is akin to a sonotrode (see later) allowing detection limits of ~4 μM with a sensitivity of ~27 A/M (120 second deposition) which represents a ~50 times increase in sensitivity and reduction in the limit of detection by two orders of magnitude compared to a silent approach [57]. Figure 3.6 demonstrates the layout of the sonotrode, which enabled the portability of sensor through its highly compact design; it is battery powered, further increasing its 'in the field' applicability. This methodology was also later shown to be applicable to river water [61].

3.6 Analysis of Trace Metals in Clinical Samples

The monitoring of metal ions concentration within the body is essential where deficiency is associated with a range of diseases and excess indicates direct exposure [62, 63]. One area is the exposure to lead, where the toxicity is well documented [64]. Levels of lead within the body are usually around the 10^{-7} M level and below, with exposure to lead leading to concentrations in the range of $\sim 10^{-6}$ M [65, 66] and higher. Leading work by Jagner described the first approach using a mercury modified glassy carbon electrode [65], which was extended with a Nafion modified mercury film [66] with only simple dilution of the blood sample [66]. Mercury was usually utilised since the constituents within blood are highly passivating and greatly reduce the sensitivity of the electroanalytical measurement. However, given the toxicity of mercury and recent bans in many countries, such an approach for analysis of clinical samples is no longer viable.

One approach to overcome this is to employ the use of bismuth-modified electrodes, which have been described as environmentally friendly alternatives to mercury, with their greatly reduced levels of toxicity [67, 68]. Wang pioneered the use of bismuth-modified electrodes [69, 70]. Where mercury utilises amalgam formation for its stripping performance [69], bismuth film electrodes result from the formation of multi-component 'fusible' alloys [69], with a range of heavy metals that facilitate nucleation processes during the accumulation step [69]. Bismuth-modified electrodes can be fabricated by either *in situ* formation, where a bismuth salt was added to the solution under analysis, or where the bismuth(III) ions are electrochemically reduced to bismuth metal at facile (negative) potential. Alternatively, the bismuth-covered electrode can be prepared via previous *ex situ* electrolytic plating, which is then transferred into the solution under analysis. Other recent approaches have used screen-printed electrodes incorporated with bismuth oxide microparticles for a simplification of the *in situ* and *ex situ* formation.

In terms of clinical trials and the range of values that need to be measured in order to determine normal (10^{-7}–10^{-8} M) and exposed levels (10^{-4}–10^{-6} M) of metal poisoning, the sensor needs to be able to operate over a few orders of magnitude. The sensing of low levels in clinical samples such as blood is problematic, as highlighted earlier, due to the complexity and range of constituents in the blood sample. A feasible approach is the application of power ultrasound [71]. Bismuth film modified electrodes have been shown to withstand insonation at intensities of up to 130 W cm^{-2} and have shown that detection limits of $\sim 10^{-7}$ and 10^{-9} M are possible for zinc and cadmium ions respectively using a 60 second deposition time [71].

Work by Kruusma *et al.* [13] demonstrated for the first time that the quantification of lead in blood avoiding the use of mercury was possible, via an insonated *in situ* formed bismuth film electrode on a boron-doped diamond electrode. Kruusma demonstrated that 10^{-10} M detection limits were possible using a 300 second insonated deposition step. Application in diluted human blood samples revealed a significant reduction in the sensitivity of the process and it was found that a large overpotential was required (~ -3 V) to achieve the desired analytical performance. The beneficial utilisation of boron-doped diamond electrodes allows the high potential to be applied while insonation removed gas bubbles formed at the electrode surface, facilitating a highly reproducible signal yet also removing surface passivating species [13]. The protocol was shown to determine levels of lead in human blood at the clinical level for an unexposed person, and was shown to be in agreement with AAS, for which analysis requires a significant level of pre-treatment [13].

Other notable approaches have reported the sensing of lead in saliva which is quantitative in a manner similar to analytical measurements [72].

3.7 Biphasic Sonoelectroanalysis

The majority of the applications of sonoelectroanalysis are within aqueous media. In general terms, when quantification of species in non-aqueous media is needed, the use of sample degradation or separation is also required, which can lengthen the analytical protocol. Pioneering work by Hardcastle *et al.* [73] has shown in the case of the electroanalytical determination of vanillin that, rather than undertaking a sample preparative step, the use of power ultrasound permits quantification via biphasic sonoelectroanalysis. In this approach the vanillin, which is put into ethyl acetate, is added to an aqueous electrolyte where the application of ultrasound creates a sono-emulsion where the vanillin is extracted into the aqueous phase in order to be readily quantified [73–75]. Note that sono-emulsions have implications in electrosynthesis, see for example Wadhawan *et al.* [74]. In the study performed by Hardcastle, it is important to note that the sono-emulsion is created *in situ* in the 'face-on' mode and was shown to be in agreement with high-performance liquid chromatography [73].

The physical processes behind such analytical utilisation of biphasic sonoelectroanalysis has been shown to be due to thinning of the diffusion layer and release of the dissolved material from the suspended phase within the diffusion layer [75]. Additionally the biphasic system is also homogenised where the power ultrasound provides an intensive blending. This biphasic approach has been extended to the measurement of liquid/liquid kinetics in acoustically emulsified media [76] and measurement of the PZC [77]. In the former case the interfacial kinetics of copper ion extraction in a heterogeneous solution of ethyl acetate containing the ligand Acorga P50 (2-hydroxy-5-nonyl benzaldoxime) and aqueous copper solution under acoustic emulsification can be simply estimated by monitoring the bulk concentration [75]. It was further shown that sub-micron sized droplets are produced from acoustic emulsification providing fast rates of mass transport to the droplet interface and providing an extremely rapid rate of copper ion extraction.

Such benefits have been utilised in the application of a double liquid/liquid extraction in which lead ions were first extracted from an aqueous solution into an organic phase containing the chelating ligand dithizone and later stripped back from the organic phase into a clean 1 M HNO_3 aqueous solution which does not contain possible surface passivating species [78]. The aqueous lead(II) solution can then be quantified via mercury-free acoustically assisted anodic stripping voltammetry, providing a simple, novel biphasic approach to the analytical detection of lead ions [78]. This was extended to the quantitative analysis of cadmium or zinc ions removed via double sono-extraction from an aqueous media hostile to voltammetric analysis. This technique was shown to be viable for the quantification of cadmium ions from surfactant-containing media where direct electroanalytical techniques fail. The method was also applied to the analysis of zinc in commercial shampoo and toothpaste, giving results in close agreement with those obtained from independent analysis [79]. Biphasic sono-extraction, synergistically coupled with sono-SWASV, presents an attractive technique for target metal ions in the electrode-passivating media. This technique likely removes contaminants present in the test solution since these will 'prefer' to remain in the initial aqueous phase, or will transfer to the extracting organic phase, but are unlikely to be doubly transferred into the clean final aqueous phase [79].

Such approaches have been extended to the detection of copper and lead in lubricating oils, where the determination of metals in lubricating oils has been used as an important way to prevent component failures of vehicular engines [80, 81], to control the quantity of engines [82, 83] and to identify some adulteration. Additionally, these determinations can provide valuable environmental information [84]. Munoz *et al.* [84] have shown that an extraction step where the desired oil sample is subjected to 30 minutes insonation with a $1:1$ (v/v) of concentrated HCl and H_2O_2 achieves an extraction efficiency of $\sim$100% [84].

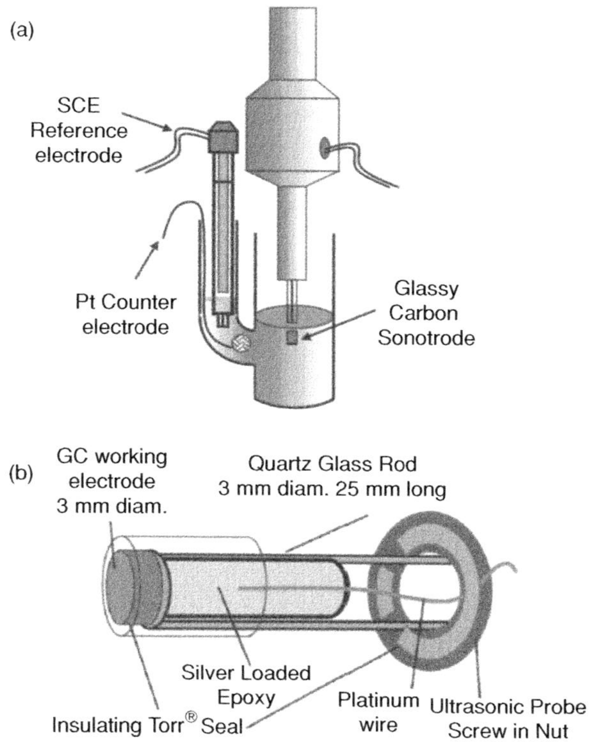

Figure 3.7 *Schematic representation of a typical sonotrode cell arrangement (a) and the tip design of the sonotrode (b) (Reprinted with permission from [92] Copyright (2002) Royal Society of Chemistry).*

In the extraction phase the metals were analysed using square-wave stripping voltammetry at gold macroelectrodes. The methodology was found to be equivalent to GF-AAS [84].

3.8 Applying Ultrasound into the Field: The *Sonotrode*

As we have noted above, the 'face-on' configuration, where the ultrasonic probe is placed directly in front of the electrode's substrate, is utilised. To simplify matters, Compton and co-workers reported the use of a new type of electrode, the 'sonotrode' [9]. With commercial and experimental constraints such as 'smaller, cheaper, faster, simpler' and environmental concerns, which necessitate analysis 'in the field', this geometry can be adapted to fulfil such criteria. The alternative geometry makes use of the fact that the tip of the ultrasonic probe can be made to act as the working electrode itself [85–87] or following the insertion of an isolated metal disk [88–91].

Figure 3.7 depicts a typical experimental arrangement of a sonotrode. Floate *et al.* [92] have reported on a sonotrode encompassing a glassy carbon electrode substrate which is constructed as shown in Figure 3.7. It was demonstrated that the sonotrode can act akin to that of the 'face-on' approach with the example of the electroanalysis of cobalt.

The sonotrode is shown to be suitable for quantifying copper(II) ions in surfactants in an otherwise passivating media and also for the sensing of copper ions in beer, which is known to be a highly electrode fouling media and which makes the analysis of copper ions in beer a challenging task [34]. The benefits of using a 'single probe' system over a traditional 'face-on' arrangement include: improved simplicity of the cell arrangement, the increased possibility for miniaturisation and also the more realistic situation of performing experiments in the field making the sonotrode an attractive and viable alternative to existing experimental set-ups [92].

We thus expect further utilisation of the sonotrode where other electrode substrates can be incorporated into the sonotrode allowing sonoelectroanalysis to become truly applicable for use 'in the field'.

3.9 Conclusions

In this updated overview we have demonstrated that the application of power ultrasound into electroanalytical processes is high beneficial in allowing the quantification of trace species in otherwise passivating media. The virtues of applying power ultrasound arise from the enhancement in mass transport from acoustic streaming, allowing highly sensitive analytical methodologies to be realised coupled with acoustic cavitation. This permits quantitative electroanalytical measurements to be performed in real samples where *in situ* cavitational cleaning allows measurements which would otherwise be beyond the realms of conventional electroanalysis. Finally, Table 3.1 shows a summary of a non-exhaustive list of various analytes that can be determined using sonoelectroanalysis.

References

(1) Banks, C.E. and Compton, R.G. (2004) Ultrasound: promoting electroanalysis in difficult real world media. *Analyst*, **129**, 678.

(2) Florence, T.M. (1970) Determination of iron by anodic stripping voltammetry. *Electroanal. Chem.*, **26**, 293.

(3) Gonzalez-Garcia, J., Saez, V., Esclapez, M.D., Bonete, P. and Vargas, Y. (2010) Nonlinear acoustics in ultrasound metrology and other selected applications. *Phys. Proc.*, **3**, 117.

(4) Banks, C.E. and Compton, R.G. (2003) Sonoelectroanalysis: A review. *Chem. Anal.*, **48**, 159.

(5) Sljukic, B.R., Banks, C.E. and Compton, R.G. (2009) Sonoelectroanalysis – application to lead determination. *Hemijska Industrija*, **63**, 529.

(6) Banks, C.E. and Compton, R.G. (2003) Voltammetric exploration and applications of ultrasonic cavitation. *Chem. Phys. Chem.*, **4**, 169.

(7) Del Campo, F.J., Melville, J., Hardcastle, J.L. and Compton, R.G. (2001) Differential pulse and chronoamperometric studies of insonated systems: Acoustic streaming and cavitational effects. *J. Phys. Chem. A.*, **105**, 666.

(8) Banks, C.E. and Compton, R.G. (2003) Ultrasonically enhanced voltammetric analysis and applications: An overview. *Electroanalysis*, **15**, 329.

(9) Eklund, J.C., Marken, F., Waller, D.N. and Compton, R.G. (1996) Voltammetry in the presence of ultrasound: A novel sono-electrode geometry. *Electrochim. Acta*, **41**, 1541.

(10) Moriguchi, N. (1934) The effect of supersonic waves on chemical phenomena, (III). The effect on the concentration polarization. *J. Chem. Soc. Jpn.*, 55.

(11) Marken, F., Akkermans, R.P. and Compton, R.G. (1996) Voltammetry in the presence of ultrasound: the limit of acoustic streaming induced diffusion layer thinning and the effect of solvent viscosity. *Electroanal. Chem.*, **415**, 55.
(12) Cooper, E.L. and Coury, L.A. (1998) Mass transport in sonovoltammetry with evidence of hydrodynamic modulation from ultrasound. *Electrochem. Soc.*, **145**, 1994.
(13) Kruusma, J., Banks, C.E. and Compton, R.G. (2004) Mercury-free sono-electroanalytical detection of lead in human blood by use of bismuth-film-modified boron-doped diamond electrodes. *Anal. Bioanal. Chem.*, **379**, 700.
(14) Ward-Jones, S., Simm, A.O., Banks, C.E. and Compton, R.G. (2006) Acoustically fabricated random microelectrode assemblies. *Ultrason. Sonochem.*, **13**, 261.
(15) Montenegro, M.I. (1994) Applications of microelectrodes in kinetics. *Res. Chem. Kinetics.*, **2**, 1.
(16) Aoki, K. (1993) Theory of ultramicroelectrodes. *Electroanal.*, **5**, 627.
(17) Valizadeh, S., George, J.M, Leisner, P. and Hultman, L. (2001) Electrochemical deposition of Co nanowire arrays; quantitative consideration of concentration profiles. *Electrochim. Acta*, **47**, 865.
(18) Madigan, N.A., Hagan, C.R.S. and Coury, L.A. Jr (1994) Preparation of microarray electrodes by sonochemical ablation of polymer films. *Electrochem. Soc.*, **141**, L23.
(19) Barton, A.C., Collyer, S.D., Davies, F., Gornall, D.D., Law, K.A., Lawrence, E.C.D., Mills, D.W., Myler, S., Pritchard, J.A., Thompson, M. and Higson, S.P. (2004) Sonochemically fabricated microelectrode arrays for biosensors offering widespread applicability: Part I. *Biosens. Bioelectron.*, **20**, 328.
(20) Menon, V.P. and Martin, C.R. (1995) Fabrication and evaluation of nanoelectrode ensembles. *Anal. Chem.*, **67**, 1920.
(21) Sapelnikova, S., Dock, E., Ruzgas, T. and Emneus, J. (2003) Amperometric sensors based on tyrosinase-modified screen-printed arrays. *Talanta*, **61**, 473.
(22) Fletcher, S. and Horne, M.D. (1999) Random assemblies of microelectrodes (RAM^{TM} electrodes) for electrochemical studies. *Electrochem. Commun.*, **1**, 502.
(23) Scholz, F. and Meyer, B. (1994) Electrochemical solid state analysis: state of the art. *Chem. Soc. Rev.*, **23**, 341.
(24) Scholz, F. and Lange, B. (1992) Abrasive stripping voltammetry — an electrochemical solid state spectroscopy of wide applicability. *Trends Anal. Chem.*, **11**, 359.
(25) Jayarama, R.S., Hermes, M. and Scholz, F. (1996) The application of abrasive stripping voltammetry for a simple and rapid screening of pesticides. *Electroanalysis*, **8**, 955.
(26) Clegg, A.D., Rees, N.V., Banks, C.E. and Compton, R.G. (2006) Ultrafast chronoamperometry of single impact events in acoustically agitated solid particulate suspensions. *Chem. Phys. Chem.*, **7**, 807.
(27) Omanovic, D., Peharec, Z., Magjer, T., Lovric, M. and Branica, M. (2005) Cyclic voltammetry of ferrocenecarboxylic acid monomolecular films and their reaction with glucose oxidase. *Electroanalysis*, **6**, 11.
(28) Ball, J.C. and Compton, R.G. (1997) Anodic stripping voltammetry at hydrodynamic mercury thin film electrodes. Numerical simulation of stripping peaks for reversible processes at uniformly accessible electrodes. *Electroanalysis*, **9**, 765.
(29) Banks, C.E., Simm, A.O., Bowler, R., Dawes, K. and Compton, R.G. (2005) Hydrodynamic electrochemistry: design for a high-speed rotating disk electrode. *Anal. Chem.*, **77**, 1928.
(30) Wang, J., Polksy, R., Tian, B. and Chatrathi, M.P. (2000) Voltammetry on microfluidic chip platforms. *Anal. Chem.*, **72**, 5285.
(31) Maisonhaute, E., Brookes, B.A. and Compton, R.G. (2002) Surface acoustic cavitation understood via nanosecond electrochemistry. 2. The motion of acoustic bubbles. *J. Phys. Chem. B*, **106**, 3166.

(32) Maisonhaute, E., White, P.C. and Compton, R.G. (2001) Surface acoustic cavitation understood via nanosecond electrochemistry. *J. Phys. Chem. B*, **105**, 12087.
(33) Maisonhaute, E., Prado, C., White, P.C. and Compton, R.G. (2002) Surface acoustic cavitation understood via nanosecond electrochemistry. Part III: shear stress in ultrasonic cleaning. *Ultrason. Sonochem.*, **9**, 297.
(34) Agra-Gutierrez, C., Hardcastle, J.L., Ball, J.C. and Compton, R.G. (1999) Anodic stripping voltammetry of copper at insonated glassy carbon-based electrodes: application to the determination of copper in beer. *Analyst*, **124**, 1053.
(35) Hu, Y.X., Yang, X.M., Wang, C., Zhao, J., Li, W.N. and Wang, Z. (2008) A sensitive determination method for carbendazim and thiabendazole in apples by solid-phase microextraction–high performance liquid chromatography with fluorescence detection. *Food Additives and Contaminants*, **25**, 314.
(36) Maldaner, L., Santana, C.C. and Jardim, I.C.S.F. (2008) HPLC determination of pesticides in soybeans using matrix solid phase dispersion. *J. Liquid Chromatogr. Relat. Technol.*, **31**, 972.
(37) Ravelo-Perez, L.M., Hernandez-Borges, J., Borges-Miguel, T.M. and Rodriguez-Delgado, M.A. (2008) Pesticide analysis in tomatoes by solid-phase microextraction and micellar electrokinetic chromatography. *J. Chromatogr. A*, **1185**, 151.
(38) Gustavo, S., Salazar-Banda, G.R. and Avaca, L.A. (2009) Sonovoltammetric determination of toxic compounds in vegetables and fruits using diamond electrodes. *Food Chem.*, **116**, 1029.
(39) Arribas, A.S., Bermejo, E., Chicharro, M. and Zapardiel, M.A. (2007) Application of matrix solid-phase dispersion to the propham and maleic hydrazide determination in potatoes by differential pulse voltammetry and HPLC. *Talanta*, **71**, 430.
(40) Ni, Y.N., Qiu, P. and Kokot, S. (2004) Study of the voltammetric behaviour of maleic hydrazide and its determination at a hanging mercury drop electrode. *Talanta*, **63**, 561.
(41) Garbellini, G.S., Salazar-Banda, G.R. and Avaca, L.A. (2007) Sonovoltammetric determination of 4-nitrophenol on diamond electrodes. *J. Brazilian Chem. Soc.*, **18**, 1095.
(42) Garbellini, G.S., Salazar-Banda, G.R. and Avaca, L.A. (2009) Sonovoltammetric determination of toxic compounds in vegetables and fruits using diamond electrodes. *Food Chem.*, **116**, 1029.
(43) Moorcroft, M.J., Davis, J. and Compton, R.G. (2001) Detection and determination of nitrate and nitrite: a review. *Talanta*, **54**, 785.
(44) Kozub, B.R., Rees, N.V. and Compton, R.G. (2010) Electrochemical determination of nitrite at a bare glassy carbon electrode; why chemically modify electrodes? *Sensors and Actuators*, **143**, 539.
(45) Nakata, K., Doi, Y., Kubota, S. and Shimazu, K. (2010) Reduction of nitrite on tin-modified noble metal electrodes. *Electroanal. Chem.*, **647**, 187.
(46) Pal, M. and Ganesan, V. (2010) Electrochemical determination of nitrite using silver nanoparticles modified electrode. *Analyst*, **135**, 2711.
(47) Khairy, M., Kadara, R.O. and Banks, C.E. (2010) Electroanalytical sensing of nitrite at shallow recessed screen printed microelectrode arrays. *Anal. Methods*, **2**, 851.
(48) Davis, J. and Compton, R.G. (2000) Sonoelectrochemically enhanced nitrite detection. *Anal. Chim. Acta*, **404**, 241.
(49) Omanovic, D., Kwokal, Z., Goodwin, A., Lawrence, A., Banks, C.E., Compton, R.G. and Komorsky-Lovric, S. (2006) Trace metal detection in Šibenik Bay, Croatia: Cadmium, lead and copper with anodic stripping voltammetry and manganese via sonoelectrochemistry. A case study. *J. Iranian Chem. Soc.*, **3**, 128.
(50) Saterlay, A.J., Foord, J.S. and Compton, R.G. (1999) Sono-cathodic stripping voltammetry of manganese at a polished boron-doped diamond electrode: application to the determination of manganese in instant tea. *Analyst*, **124**, 1791.

(51) Goodwin, A., Lawrence, A.L., Banks, C.E., Wantz, F., Omanovic, D., Komorsky-Lovric, S. and Compton, R.G. (2005) On-site monitoring of trace levels of free manganese in sea water via sonoelectroanalysis using a boron-doped diamond electrode. *Anal. Chim. Acta*, **533**, 141.

(52) Rodrigues, S., Munichandraiah, N. and Shukla, A. (1998) A cyclic voltammetric study on the kinetics and mechanism of electrodeposition of manganese dioxide. *J. Appl. Electrochem.*, **28**, 1235.

(53) Simm, A.O., Banks, C.E. and Compton, R.G. (2005) A cyclic voltammetric study on the kinetics and mechanism of electrodeposition of manganese dioxide. *Electroanalysis*, **17**, 335.

(54) Smith, A.H., Lingasm, E.O. and Rahman, N. (2000) Contamination of drinking-water by arsenic in Bangladesh: a public health emergency. *Bul. World Health Organization*, **78**, 1093.

(55) Feeney, R. and Kounaves, S.P. (2002) Voltammetric measurement of arsenic in natural waters. *Talanta*, **58**, 23.

(56) Alam, M.G.M., Allinson, G., Stagnitti, F., Tanaka, A. and Westbrooke, M. (2002) Arsenic contamination in Bangladesh groundwater: A major environmental and social disaster. *Int. J. Environ. Health Res.*, **12**, 235.

(57) Dai, Z. and Compton, R.G. (2006) Detection of As(III) via oxidation to As(V) using platinum nanoparticle modified glassy carbon electrodes: arsenic detection without interference from copper. *Analyst*, **131**, 516.

(58) King, J.R. (1959) *Qualitative Analysis and Electrolytic Solutions*, Harcourt, Brace, and World, New York.

(59) Wagtech, http://www.wagtech.co.uk/products/water-and-environmental/water-test-kits/arsenator%C2%AE-digital-arsenic-test-kit (accessed Nov. 2010).

(60) Forsberg, G., O'Laughlin, J.W. and Megargle, R.G. (1975) Determination of arsenic by anodic stripping voltammetry and differential pulse anodic stripping voltammetry. *Anal. Chem.*, **47**, 1586.

(61) Simm, A.O., Banks, C.E. and Compton, R.G. (2004) Sonically assisted electroanalytical detection of ultratrace arsenic. *Anal. Chem.*, **76**, 5051.

(62) Burtis, C.A. and Ashwood, E.R. (1996) *Tietz Texbook of Clinical Chemistry*, Saunders, Philadelphia.

(63) Wang, J. (1982) Anodic stripping voltammetry at graphite-epoxy microelectrodes for in vitro and in vivo measurements of trace metals. *Anal. Chem.*, **54**, 221.

(64) Hardcastle, J.L., West, C.E. and Compton, R.G. (2002) The membrane free sonoelectroanalytical determination of trace levels of lead and cadmium in human saliva. *Analyst*, **127**, 1495.

(65) Jagner, D., Josefon, M., Westerland, S. and Aren, K. (1981) Simultaneous determination of cadmium and lead in whole blood and in serum by computerized potentiometric stripping analysis. *Anal. Chem.*, **53**, 1406.

(66) Hoyer, B. and Florence, T.M. (1987) Application of polymer-coated glassy carbon electrodes to the direct determination of trace metals in body fluids by anodic stripping voltammetry. *Anal. Chem.*, **59**, 2839.

(67) Wang, J., Lu, J., Hocevar, S.B., Farias, P.A.M. and Ogorevc, B. (2000) Bismuth-coated carbon electrodes for anodic stripping voltammetry. *Anal. Chem.*, **72**, 3218.

(68) Wang, J., Lu, J., Hocevar, S.B. and Ogorevc, B. (2001) QC test for noninvasive glucose monitoring system. *Electroanalysis*, **13**, 12.

(69) Wang, J. (2005) Stripping analysis at bismuth electrodes: A review. *Electroanalysis*, **17**, 1341.

(70) Svancara, I., Prior, C., Hocevar, S.B. and Wang, J. (2010) A decade with bismuth-based electrodes in electroanalysis. *Electroanalysis*, **22**, 1405.

(71) Banks, C.E., Kruusma, J., Hydemm, M.E., Salimi, A. and Compton, R.G. (2004) Sonoelectroanalysis: investigation of bismuth-film-modified glassy carbon electrodes. *Anal. Bioanal. Chem.*, **379**, 277.

(72) West, C.E., Hardcastle, J.L. and Compton, R.G. (2002) Sono-electroanalytical determination of lead in saliva. *Electroanalysis*, **14**, 1470.
(73) Hardcastle, J.L., Paterson, C.J. and Compton, R.G. (2001) Biphasic sonoelectroanalysis: Simultaneous extraction from, and determination of vanillin in food flavoring. *Electroanalysis*, **13**, 899.
(74) Wadhawan, J.D., Markenm, F. and Compton, R.G. (2001) Biphasic sonoelectrosynthesis. A review. *Pure Appl. Chem.*, **73**, 1947.
(75) Banks, C.E., Rees, N.V. and Compton, R.G. (2002) Sonoelectrochemistry in acoustically emulsified media. *Electroanal. Chem.*, **535**, 41.
(76) Banks, C.E., Klymenko, O.V. and Compton, R.G. (2003) Voltammetric exploration and applications of ultrasonic cavitation. *Chem. Phys. Chem.*, **4**, 169.
(77) Banks, C.E., Rees, N.V. and Compton, R.G. (2002) Sonoelectrochemistry understood via nanosecond voltammetry: Sono-emulsions and the measurement of the potential of zero charge of a solid electrode. *J. Phys. Chem.*, **106**, 5810.
(78) Tomcik, P., Banks, C.E. and Compton, R.G. (2003) Sonoelectrochemistry in acoustically emulsified media: The detection of lead. *Electroanalysis*, **15**, 1661.
(79) Kruusma, J., Tomcik, P., Banks, C.E. and Compton, R.G. (2004) Sonoelectroanalysis in acoustically emulsified media: Zinc and cadmium. *Electroanalysis*, **16**, 852.
(80) Ekanem, E.J., Lori, J.A. and Thomas, S.A. (2000) The determination of wear metals in used lubricating oils by flame atomic absorption spectrometry using sulphanilic acid as ashing agent. *Talanta*, **44**, 2103.
(81) Zieba-Palus, J. (1998) Examination of used motor oils by flame AAS for criminalistic purposes: a diagnostic study. *Forensic Sci. Int.*, **91**, 171.
(82) Goncalves, I.M., Murillo, M. and Gonzalez, A.M. (1998) Determination of metals in used lubricating oils by AAS using emulsified samples. *Talanta*, **47**, 1033.
(83) Pouzar, M., Cernohorsky, T., Ciraolo, L. and Novelli, V. (2003) Polycyclic aromatic hydrocarbons in Brazilian vegetables and fruits. *Food Control*, **14**, 53.
(84) Munoz, R.A.A., Oliveira, P.V. and Angnes, L. (2006) Combination of ultrasonic extraction and stripping analysis: An effective and reliable way for the determination of Cu and Pb in lubricating oils. *Talanta*, **68**, 850.
(85) Reisse, J., Francois, H., Vandercammen, J., Fabre, O., Kirsch-de-Mesmaeker, A., Maerschalk, C. and Delpancke, J.L. (1994) Sonoelectrochemistry in aqueous electrolyte: A new type of sonoelectroreactor. *Electrochim. Acta*, **39**, 37.
(86) Durant, A., Francois, H., Reisse, J. and Kirsch-de-Mesmaeker, A. (1996) Sonoelectrochemistry: The effects of ultrasound on organic electrochemical reduction. *Electrochim. Acta*, **41**, 277.
(87) Compton, R.G., Akkermans, R.P., Coles, B.A. and Marken, F. (1997) Ultrasound in photoelectrochemistry: A new approach to the enhancement of the efficiency of semiconductor electrode processes. *Ultrason. Sonochem.*, **4**, 223.
(88) Compton, R.G., Eklund, J.C., Marken, F. and Waller, D.N. (1996) Electrode processes at the surfaces of sonotrodes. *Electrochim. Acta*, **41**, 315.
(89) Marken, F. and Compton, R.G. (1996) Electrochemistry in the presence of ultrasound: the need for bipotentiostatic control in sonovoltammetric experiments. *Ultrason. Sonochem.*, **3**, 131.
(90) Compton, R.G., Eklund, J.C., Marken, F., Rebbitt, T.O., Akkermans, R.P. and Waller, D.N. (1996) Dual activation: coupling ultrasound to electrochemistry – an overview. *Electrochim. Acta*, **42**, 2919.
(91) McMurray, H.N. (1998) Hydrogen evolution and oxygen reduction at a titanium sonotrode. *Chem. Commun.*, **8**, 887.

(92) Floate, S., Hardcastle, J.L., Cordemans, E. and Compton, R.G. (2002) A sonotrode for electroanalysis: the determination of copper in passivating media. *Analyst*, **127**, 1094.

(93) Matysik, F.-M., Matysik, S., Brett, A.M.O. and Brett, C.M.A. (1997) Ultrasound-enhanced anodic stripping voltammetry using perfluorosulfonated ionomer-coated mercury thin-film electrodes. *Anal. Chem.*, **69**, 1656.

(94) Akkermans, R.P., Ball, J.C., Marken, F. and Compton, R.G. (1998) The use of sonotrodes for electroanalysis: Sono-ASV detection of lead in aqueous solution. *Electroanalysis*, **10**, 26.

(95) Blythe, A.N., Akkermans, R.P. and Compton, R.G. (2000) Sono-electroanalysis: Application to the detection of lead in petrol. *Electroanalysis*, **12**, 16.

(96) Akkermans, R.P., Ball, J.C., Rebbit, T.O., Marken, F. and Compton, R.G. (1998) Sono-electroanalysis: Application to the detection of lead in wine. *Electrochim. Acta*, **43**, 3443.

(97) Simm, A.O., Banks, C.E. and Compton, R.G. (2005) Sonoelectroanalytical detection of ultra-trace arsenic. *Electroanalysis*, **17**, 335.

(98) Agra-Gutierrez, C. and Compton, R.G. (1988) Adsorptive stripping voltammetry in the presence of power ultrasound: The detection of vanadium in aqueous media. *Electroanalysis*, **10**, 204.

(99) Akkermans, R.P., Wu, M. and Compton, R.G. (1998) A comparison between pulsed sono-voltammetry and low power laser activated voltammetry for the electroanalysis of ascorbic acid in a commercial fruit drink. *Electroanalysis*, **10**, 814.

(100) Hardcastle, J.L. and Compton, R.G. (2001) Sonoelectroanalytical determination of heavy metals in fish gill mucous. *Electroanalysis*, **13**, 89.

(101) Davis, J., Moorcroft, M.J., Wilkins, S.J., Compton, R.G. and Cardosi, M.F. (2000) Electrochemical detection of nitrate at a copper modified electrode under the influence of ultrasound. *Electroanalysis*, **12**, 1363.

(102) Tsai, Y.C., Davis, J. and Compton, R.G. (2000) Sono-anodic stripping voltammetric determination of cadmium in the presence of surfactant. *Fresenius J. Anal. Chem.*, **368**, 415.

(103) Beckett, E.L., Lawerence, N.S., Evans, R.G., Davis, J. and Compton, R.G. (2001) Sonoelectrochemically enhanced determination of 5-aminosalicylic acid. *Talanta*, **54**, 871.

4

Sonoelectrochemistry in Environmental Applications

Pedro L. Bonete Ferrández, María Deseada Esclapez, Verónica Sáez Bernal and José González-García

4.1 Introduction

Among the large number of techniques found in the literature that are available for the degradation of recalcitrant compounds, a number of well-established technologies have been revisited and new technologies have emerged. For example, the electrochemical (ECT), the sonochemical (SCT) and especially the sonoelectrochemical (SECT) technologies are very promising [1, 2]. Several attractive features are shared among these different methods: (i) the procedures are performed at room temperature and atmospheric pressure; (ii) they involve the unique use of electrical energy as the 'reactant', allowing them to be considered as 'green' technologies; (iii) their energetic consumption is adjustable to the contaminant level; (iv) the treatment/disinfection rates are easily controlled by means of the electrical current together with the ultrasonic power level within the sonoelectrochemical reactor; (v) the processes are safe; (vi) no issues are observed with the resulting by-products (especially from aqueous wastes where the main by-products, if any, are hydrogen and oxygen of high purity – see Chapter 10); and (vii) the main reactant, that is, the electrical current, is freely available, without any market restrictions.

The application of sonoelectrochemistry to the degradation of pollutants, especially for contaminated soil, sludge and wastewater samples, looks very promising for those compounds which are readily degradable by electrochemistry. Although ultrasound usually enhances the effects induced by electrochemistry, further studies are required in order to draw better and stronger conclusions.

Power Ultrasound in Electrochemistry: From Versatile Laboratory Tool to Engineering Solution, First Edition. Edited by Bruno G. Pollet.

4.2 Sonoelectrochemical Degradation of Persistent Organic Pollutants

ECT is considered to be a 'clean' and valid alternative method to chemical depollution techniques for effluents containing organic pollutants. However, the ECT process possesses a few disadvantages such as: (i) low efficiencies especially at low pollutant concentrations, and (ii) electrode fouling frequently occurring whereby the electrode is passivated. Thus, the use of ultrasound in combination with electrochemical oxidation can solve these problems, in turn providing a powerful method for organic pollutant decontamination for toxic samples and effluents.

4.2.1 Sonoelectrochemical Applications

4.2.1.1 Sonoelectrochemical Degradation of Textile Dyes

Effluents containing dyes and pigments from the textile and printing industries are a major concern for the environment. Various dye concentrations and types are present in the effluents which makes them difficult to be removed easily by using conventional wastewater treatment methods. There are over 10^4 different synthetic dyes and pigments, and around 10^6 tons per year of these products are produced globally; it has been estimated that about 15% of these dyes are released into the environment during dyeing processes.

Over the last few years, several methods have been proposed and employed to 'decolorize' wastewater samples and effluents containing dyes and pigments. Filtration, coagulation, absorption on activated carbon, and other physical, chemical, and biological techniques are often inefficient. Some electrochemical technologies [3] and a number of advanced oxidation processes have been developed to 'decolorize' and/or degrade dyeing effluents for environmental protection, and, among these, sonoelectrochemical treatment (SECT) has been proposed to be a promising method. This is due to the fact that ultrasound can help with solving the main problems appearing in electrochemical processes, such as electrode fouling caused by electrode passivation or/and gas evolution (e.g. H_2 and O_2 bubbles blocking the electrode surface) and poor mass transport [4].

Sandolan Yellow was the first model dye used in dye effluent decolorization by ultrasonically assisted electrooxidation [5]. Solutions of this acidic dye were subjected to sonolysis (see Chapter 1), electrolysis, and sonoelectrolysis. It was observed that the decolorization process did not take place in the presence of ultrasound alone (20 and 40 kHz), but was achieved using electrooxidation. It was found that the decolorization process was dependent on the type of electrode used, the electrolyte concentration, and the current density. In this study, it was observed that only alkali metal chloride solutions promoted the decolorization of dye solutions due to the production of hypochlorite during the electrolysis. Furthermore, platinum electrodes led to better efficiencies than carbon electrodes when galvanostatic electrolysis was carried out. The electrooxidation of Sandolan Yellow using platinum electrodes was further enhanced under sonication (40 kHz), but an unexpected decrease in the decolorization process was observed when the ultrasonic power was increased. It was shown that this observation was mainly due to the degassing effect of ultrasound and the gas nature of chlorine, which acted as an intermediate product in the conversion of chloride to hypochlorite. Consequently, when a sealed sonoelectrochemical cell was used, an improvement in process efficiency was observed. Similar results were observed using a basic dye such as Maxilion Blue 5G in high-frequency (510 kHz) sonoelectrochemical oxidation [6]. In this work, the decolorization of the dye solutions was monitored by UV-Vis spectrophotometry. It is also noteworthy that changes in the absorbance of the solutions do not correspond to a decrease in toxicity!

Sonovoltammetric oxidation studies of Procion Blue on a boron-doped diamond (BDD) electrode was carried out in phosphate buffer solutions [7]. The dye concentration and pH were seen to have a significant influence on the oxidation process. Thus, in this case, the oxidation of Procion Blue was most easily achieved in acidic solutions and at low dye concentrations. In this study, it was also observed that electrode fouling occurred in more alkaline solutions and at very high dye concentrations. Upon increasing the dye concentration, a proportional increase in the limiting current with a distortion of the sigmoidally shaped voltammetric wave was also observed (see Chapter 1). It was concluded that a resistive layer was formed on the electrode surface as a result of the adsorption and desorption of the large dye molecules at the BDD electrode prior and after oxidation.

The advantages of SECT processes over conventional methods have been demonstrated for various reactive dyes. In fact, the negligible effect of ultrasound in the decolorization [8] of reactive dyes has been reported, and only basic dyes are degraded using high-frequency ultrasound alone [6]. Sonoelectrochemical treatment does not always need to be carried out in the presence of an alkali metal chloride electrolyte solution, but the addition of SO_4^{2-}, NO_3^-, and especially Cl^- ions into the solution significantly enhances the degradation process [9].

Lissamine Green B [10], Trupocor Red [10], Reactive Black 5 [10], Acid Black [10], Methyl Orange[10–12], Rhodamine B [12], Methylene Blue [12], Reactive Brilliant X-3B [12], and Reactive Blue 19 [13] have been used as reactive dyes for acoustic cavitation coupled with electrochemical treatment. Hydrodynamic cavitation has also been used to enhance the electrochemical treatment of Brilliant Red X-3B [9]. For all these studies, various electrode materials (Ti-IrO_2 [9], graphite [10], platinum [12], PbO_2 [13]) and three-dimensional electrodes [11] were employed and the effect of several parameters were studied, including the ultrasonic power/intensity, the cell design, pH, initial dye concentrations, and so on. It was observed that pseudo-first-order kinetics were obtained in most cases where di-azo dyes showed slow degradation rates compared with mono-azo dyes [10]. It is worth pointing out that the optimal reaction conditions for dye degradation cannot be deduced from all these studies, considering the wide variety of chemical structures and chromophore groups of the dyes and the various conditions employed. Nevertheless, it can be concluded that the use of electrooxidation combined with ultrasound increases the efficiency of both treatments, whereby a 'positive' synergic effect has been observed.

The decolorization and degradation of dye solutions are usually monitored by UV-Vis spectrophotometry and total organic carbon (TOC) and chemical oxygen demand (COD) are often measured. It was shown that mass spectrometry-gas chromatography (GC-MS) can also be used to identify the degradation intermediates. For example, it was found that the intermediate compounds for the Reactive Blue 19 dye [13] were acetic acid, benzoic acid, and products of low molecular weight. Once again, it should be emphasized that the decolorization of solutions does not correspond to a toxicity 'decrease'; however, the total mineralization is the final goal of the treatment.

4.2.1.2 Sonoelectrochemical Degradation of Chlorinated Pollutants

Chlorinated compounds are one of the most widespread pollutants found in any effluents in the environment and, because of this, a large number of techniques are being continuously developed to provide an efficient solution to this important issue.

For example, 2,4-dichlorophenoxyacetic acid is one of the most widely used herbicides worldwide. It is a poorly biodegradable pollutant with mild toxicity, which can be converted into highly toxic chlorinated compounds such as 2,4-dichlorophenol. The oxidative degradation of 2,4-dichlorophenol has been studied sonoelectrochemically [14] using platinum electrodes in an ultrasonic bath (44 kHz). Cyclic **sonovoltammograms** showed an increase in currents under insonation, leading to an overall overpotential decrease together with no deactivation of the electrode. Bulk electrolyses (1.0 M Na_2SO_4) were carried out galvanostatically at various current densities, and the degradation of 2,4-dichlorophenol was monitored by TOC, GC, and Cl^- concentration analyses. It was observed that high degradation rates were obtained when the current density was increased to 70 mA cm^{-2}, leading to a reduction of *circa* 78% in TOC. The mass balance based on chloride ions indicated the formation of unanalyzed low-molecular-weight chlorinated compounds.

Perchloroethylene (PCE) is a widely used solvent in many industrial sectors, and can be found in all types of effluents. Furthermore, it has been reported that PCE is a major intermediate compound in the degradation of most chlorinated systems. Sáez *et al.* studied the effect of ultrasound (20 kHz) at various ultrasonic powers upon the electrochemical degradation of perchloroethylene in aqueous solutions [15]. In this study, the experimental data obtained by SECT were compared with those previously reported electrochemically [16–18] and sonochemically [19, 20]. In their work, bulk electrolyses were carried out in galvanostatic mode (3.5 mA cm^{-2}) using a lead (Pb) cathode and a lead dioxide (PbO_2) anode in a well-characterized sonoelectrochemical cell [21, 22], which had previously been employed for sonochemical treatment (SCT) [19, 20]. Sáez *et al.* found that (i) no erosion occurred on the electrode surface and (ii) a current efficiency enhancement compared to silent conditions was observed for all output ultrasonic powers employed. Taking into account the low initial concentration of pollutant (60 ppm), reasonable current efficiency values were obtained during the first hour of treatment (Figure 4.1).

The concentrations of PCE and degradation intermediates (mainly trichloroethylene and dichloroethylene) were measured in the aqueous and gaseous phases using several analytical methods (HPLC, GC, and IC), and a total degradation of all these volatile compounds in the first 2.5 hours of the process was reported. No detectable differences were observed for most of the ultrasonic intensities employed, but it should be noted that the energy consumption with SECT was lower than that presented by SCT due to the fact that the treatment time was significantly reduced.

In a recent study, the scale-up of the SECT for aqueous solutions of trichloroacetic acid (TCAA) from voltammetric analyses to pre-pilot stages has been reported [23]. A titanium (Ti) rotating disc electrode (RDE) for the voltammetric studies and a mesh-plate titanium for the bulk electrolyses were used as cathode materials. Two different titanium ultrasonic horns were also used: a 24 kHz horn (200 W, 0.7 cm^2 emitter surface Ti tip) was used for the sonovoltammetry and batch electrolysis experiments, and a 20 kHz horn (1000 W, 12.5 cm^2 emitter surface Ti tip) was employed for the flow cell electrolysis experiments. For the

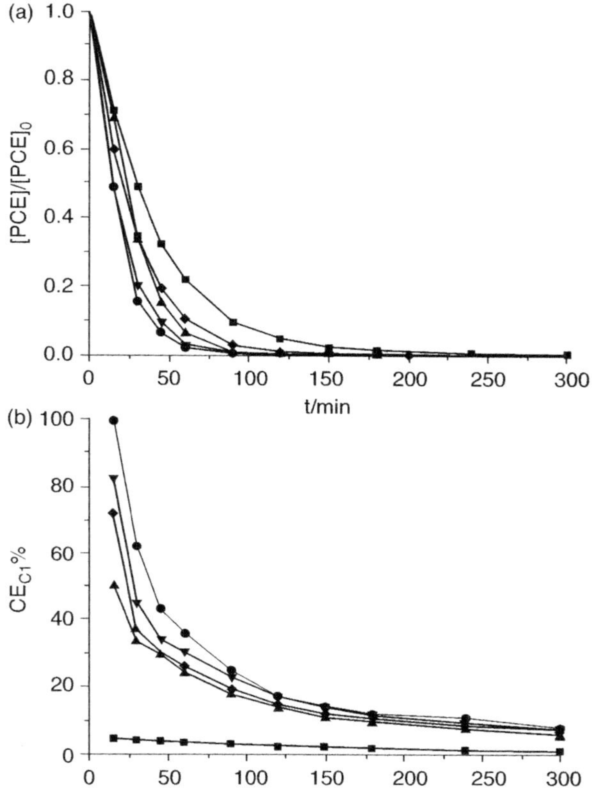

Figure 4.1 (A) The normalized concentration decay of PCE, and (B) the current efficiency for the ultrasonic intensity series: ■ 1.84, • 3.39, ▲ 5.09, ▼ 6.36, ♦ 7.64 W cm^{-2}. f = 20 kHz, 3.5 mA cm^{-2}, 20 °C (Reprinted with permission from [15] Copyright (2010) Elsevier Ltd).

voltammetric and sonovoltammetric studies, a typical sonoelectrochemical cell that allowed the 'face-on' configuration (see Chapter 3) was used, whereas for bulk electrolysis at batch scale an H-type electrochemical cell was employed, using a Nafion 450 cationic membrane when electrolyses were carried out (in divided mode to separate catholyte and anolyte chambers). A non-optimized flow sonoelectrochemical system was used for process scale-up (Figure 4.2).

In this work, the polypropylene sonoelectrochemical reactor was designed to be used in divided and undivided configurations, whereby the acoustic field in the reactor was calculated according to the finite-element simulation method [24, 25].

According to the voltammetric results, it was found that low current efficiencies in the galvanostatic electrochemical reduction of TCAA on titanium (10 mM Na_2SO_4) could be achieved. Here, the sonovoltammetric studies also provided the range of potentials and/or current densities to be employed (in views of minimizing the lowest current efficiency penalty). RDE and sonovoltammetry experiments suggested that the electrochemical

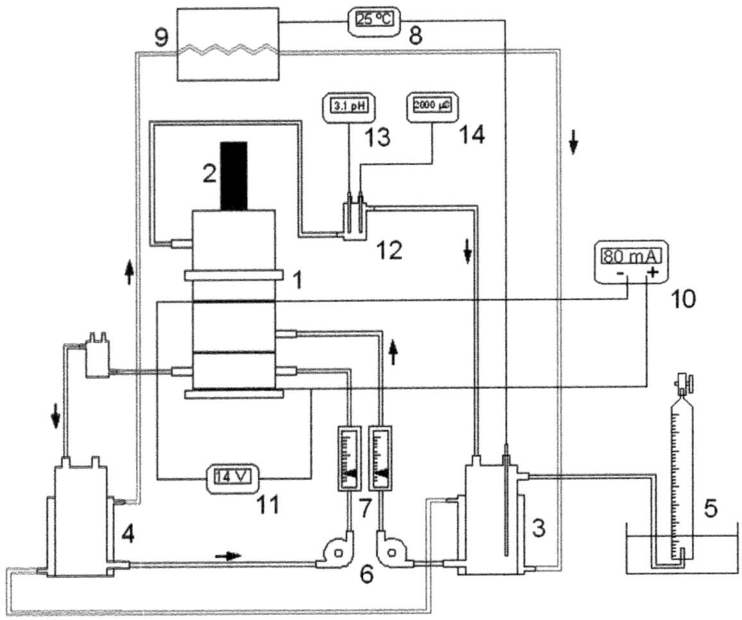

Figure 4.2 Flow reactor electrolyses experimental setup. (1) flow sonoelectrochemical reactor, (2) ultrasound transducer, (3) catholyte tank, (4) anolyte tank, (5) gas measurement system, (6) pumps, (7) temperature sensor, (8) cooling system, (9) power supply, (10) voltmeter, (11) sensor holder, (12) pH sensor, and (13) conductivity sensor (Reprinted with permission from [23] Copyright (2010) Elsevier Ltd).

reduction of TCAA on titanium is under mass-transport control, and that a significant current density enhancement can be achieved under insonation at potentials lower than 2.2 V *versus* Ag/AgCl. It was also observed that the degradation of TCCA in silent conditions yielded poor efficiencies, which were improved in the presence of ultrasound.

The reactor configuration, electrode/horn distance, and ultrasonic power values were determined using bulk electrolyses. A higher reduction rate was observed for the divided configuration (Figure 4.3), and, for this reaction configuration, the electrode/horn distance proved to be an insignificant parameter. Bulk electrolyses carried out at different ultrasound intensities showed that the kinetics of the electrochemical reaction are mainly under mixed control, but an increase of the mass transport conditions using a higher output power of ultrasound does not markedly increase the current efficiency of the process. In these experiments, dichloroacetic acid (DCAA) was the main product obtained.

Scale-up bulk electrolysis experiments, using the flow sonoelectrochemical reactor, allowed the study of the combined effects of ultrasound and fluid flow. It was observed that ultrasound led to better results, for example a fractional conversion of 97%, a degradation efficiency of 26%, a selectivity of 0.92, and a current efficiency of 8% were found at low volumetric flows (Figure 4.4). The undivided configuration for the flow sonoelectrochemical reactor yielded a drop in trichloroacetic acid degradation efficiencies.

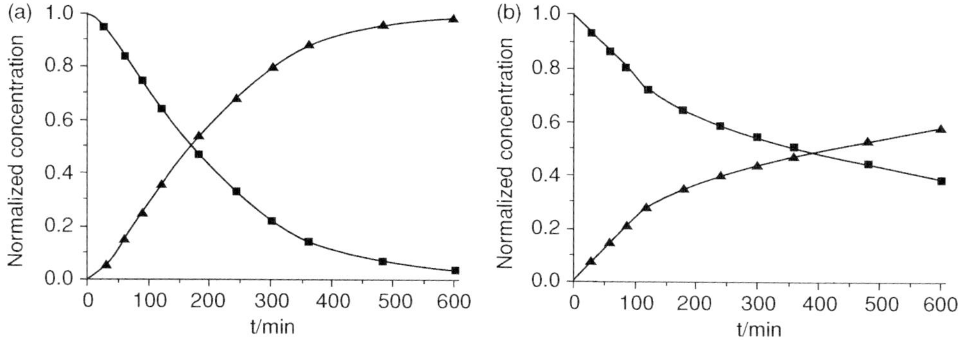

Figure 4.3 *Normalized concentrations related to the starting TCAA for TCAA (■) and DCAA (▲) in divided (A) and undivided (B) cell configurations for batch-scale sonoelectrolysis (Reprinted with permission from [23] Copyright (2010) Elsevier Ltd).*

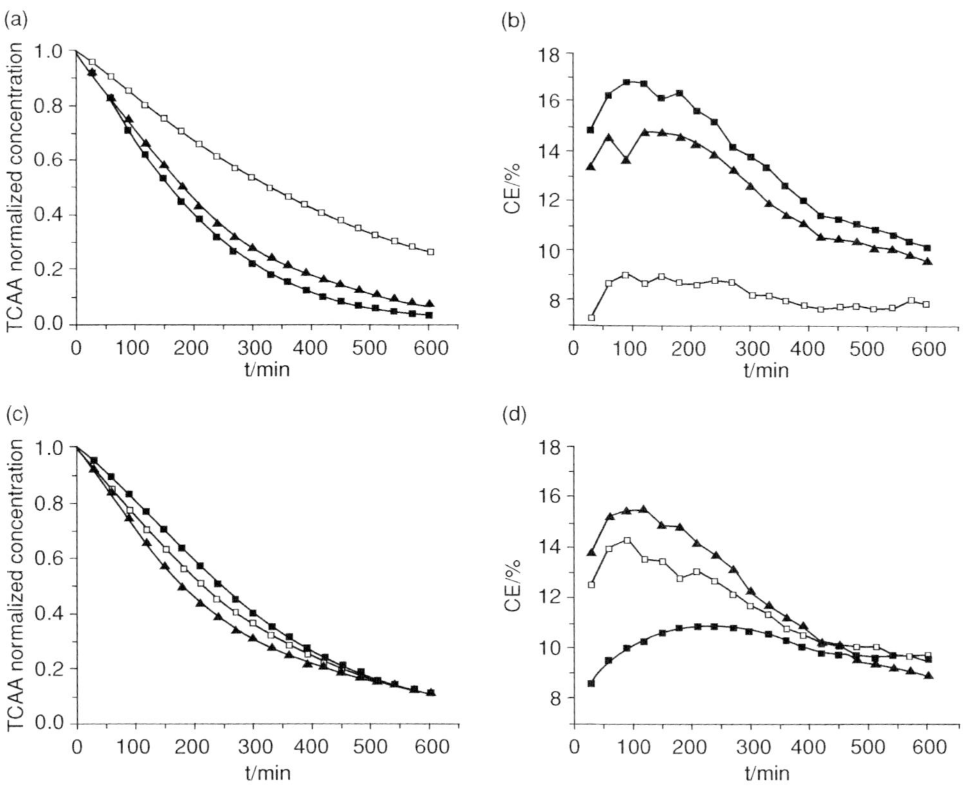

Figure 4.4 *Normalized concentration decay for TCAA at 100 L h^{-1} (A) and 300 L h^{-1} (C), and current efficiency for Cl^- at 100 L h^{-1} (B) and 300 L h^{-1} (D), in the flow sonoelectrochemical reactor at different ultrasound powers: 0 (□), 3.8 (■), and 9.5 W cm^{-2} (▲) (Reprinted with permission from [23] Copyright (2010) Elsevier Ltd).*

$$R-Ph(NO_2)_2 \xrightarrow[-H_2O]{4e^- + 4H^-} R-Ph(NO_2)(NHOH) \xrightarrow[-H_2O]{4e^- + 4H^-} R-Ph(NHOH)_2 \xrightarrow[-2H_2O]{4e^- + 4H^-} R-Ph(NH_2)_2$$
$$R = H, CH_3$$

Scheme 4.1 *Proposed reduction mechanism for nitro compounds.*

4.2.1.3 Sonoelectrochemical Degradation of Nitro Compounds

The removal of nitro aromatics from polluted effluents using SECT has been proposed and studied sonovoltammetrically [26] and galvanostatically [27]. Under both silent and ultrasonic conditions (20 kHz and up to 40 W cm^{-2}), nitrobenzene in aqueous solutions (pH = 13) was reduced on glassy carbon (GC) and gold electrodes [26] in a chemically reversible one-electron ($1e^-$) process followed by an irreversible three-electron ($3e^-$) reduction. At sufficiently negative potentials, an overall four-electron ($4e^-$) process occurred leading to the production of phenylhydroxylamine. In these experiments, the GC electrode surface was damaged for very short electrode/horn distances and the gold electrode led to more complicated mechanisms due to surface reaction pathways.

It was found that the reduction of 1,3-dinitrobenzene and 2,4-dinitrotoluene in acid media occurs in three stages, according to Scheme 4.1. Galvanostatic reduction on titanium [27] under silent conditions takes place slowly, and low current efficiencies are observed in an unoptimized cell. It was observed that the use of ultrasound (titanium sonotrode) enhances the electrochemical reduction rate, but the overall rate of the process is still slow.

4.2.1.4 Sonoelectrochemical Degradation of Aromatic and Phenolic Derivatives

Phenol is one of the most common pollutants generated by many industries, giving an unpleasant odor and high toxicity to effluents, even when present at low concentrations. The high degree of toxicity and low degradation makes it difficult for a single treatment to meet legal requirements. For this reason, the degradation of phenolic compounds continues to be an active research area, where some advanced oxidation processes have been employed for treating phenolic polluted wastewater samples and effluents, including sonoelectrochemical processes [28].

Trabelsi *et al.* [29] reported the electrochemical remediation of aqueous phenol solutions under low- and high-frequency ultrasound irradiation. In their study, the authors stressed the importance of the reactor and emitter geometries. Due to the multiple reflections occurring on the interfaces, the acoustic energy is distributed heterogeneously in the sonoelectrochemical reactor, and measurements [30] either as calculus simulations [21, 23–25], or as visualization [31, 32] should be performed in order to (i) obtain information about the acoustic field in the reactor and (ii) optimize the process. The determination of the active zones is of particular interest, because it allows the location of the electrodes where the maximum mass transfer occurs. In the low-frequency sonoelectrochemical reactor used by Trabelsi *et al.*, the maximum cavitational activity was found to be very near the ultrasonic transducer, and to be comparatively higher in the central plane just below it. Thus, the authors recommended that the electrodes were located in the central cylindrical portion, which is the zone of high cavitational energy (Figure 4.5A). In the high-frequency sonoelectrochemical reactor, the maximum cavitational activity was found to be in the intermediate zone between the axis and the reactor wall and near the air/liquid interface (Figure 4.5B).

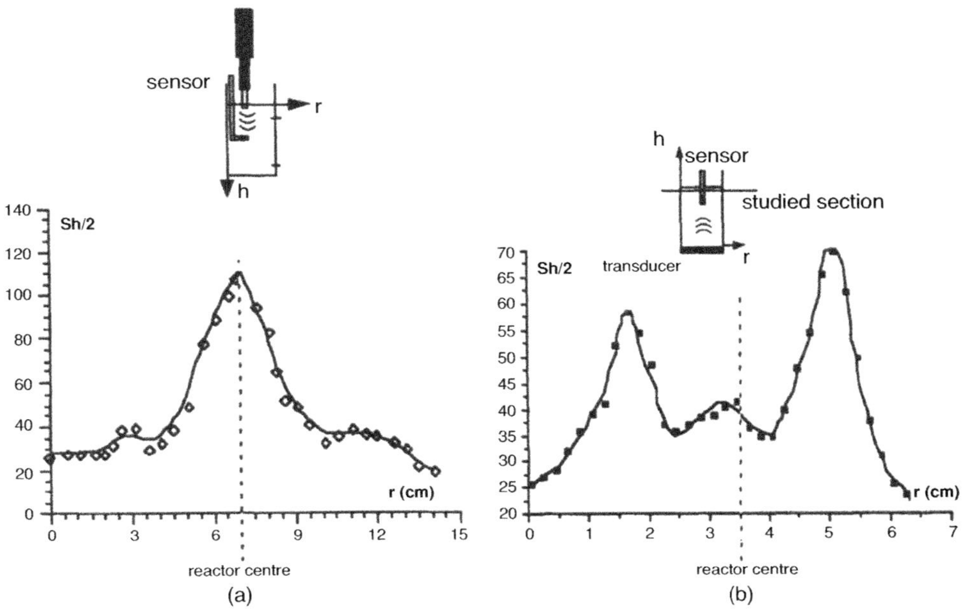

Figure 4.5 *Radial distribution of Sherwood number for (A) low- and (B) high-frequency sonoelectroreactors (Reprinted with permission from [23] Copyright (1996) Elsevier Ltd).*

The SECT of phenol solutions (0.1 g L^{-1} with NaCl 0.5 g L^{-1} as supporting electrolyte) was performed in galvanostatic mode (6.8 mA cm^{-2}) using a cylindrical nickel foam cathode and a cylindrical expanded platinized titanium anode, according to the experimental setup depicted in Figure 4.6.

Mass transport was shown to be significantly enhanced at 500 kHz (70-fold increase in the diffusional mass-transfer rate), and yet higher effect can be obtained at 20 kHz (120-fold increase). This observation can be attributed to the enhanced physical effects in terms of the ultrasonic turbulence generated under low-frequency irradiation. Although higher fractional conversions were obtained at both frequencies, the use of high-frequency irradiation was more beneficial in terms of chemical effects (enhanced generation of free radicals – see Chapter 1), and resulted in complete degradation of phenol after 20 min, with no production of toxic aromatic intermediates (Table 4.1).

From a vast variety of materials used as anodes for pollutant degradation, the BDD electrode has received much attention due to its excellent properties such as its wide electrochemical window, high oxygen overpotential, low background currents, and high anodic stability [33, 34], and BDD has therefore been used for the efficient degradation of phenolic derivatives in ECT experiments [35–37]. BDD electrode performance has been compared with platinum for the electrochemical oxidation of phenol under silent and ultrasonic conditions (30 kHz, 50 W ultrasonic horn), from the viewpoints of both fundamental and applied aspects [38]. Although both anode materials (BBD and Pt) are highly fouled, a much stronger electrode passivation occurs on Pt. It was found that the

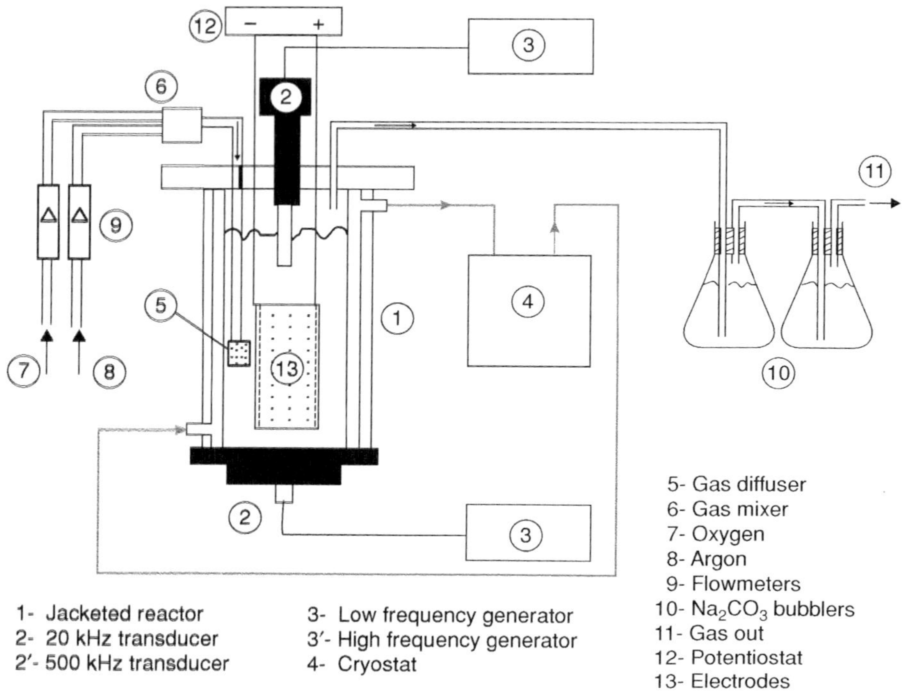

Figure 4.6 *Schematic diagram of the sonoelectrochemical reactors (Reprinted with permission from [23] Copyright (1996) Elsevier Ltd).*

Table 4.1 *SECT of phenol at 20 kHz and 540 kHz (Reprinted with permission from [29] Copyright (1996) Elsevier Ltd).*

Time (min)	Phenol (mg L^{-1})	*p*-Quinone (mg L^{-1})	Acetic acid (mg L^{-1})	Phenol conversion (%)
		20 kHz		
0	100	0	0	0
5	54	7.9	2.4	46
10	24	12.6	3.1	76
20	12.8	11.1	4.6	87.2
45	7.8	7.6	4.8	92.2
60	3	5.5	6.2	97
		540 kHz		
0	100	—	0	0
5	26.4	—	2.7	73.6
10	5	—	3.5	95
45	—	—	11.5	100
60	—	—	7.4	100

ultrasonic irradiation improves the phenol degradation, leading to high current efficiencies on both electrodes (although much higher when using BDD electrodes). Mass transport and adsorption phenomena, as well as electrode reactions, can explain the different behaviors of these electrode materials. It was observed that for the BDD electrode, (i) reaction rates and current efficiencies were increased by 300 and 100%, respectively under sonication and (ii) lower yields of intermediates compared to silent conditions. Although in the presence of ultrasound the variety of intermediates does not change for both electrodes used, it was found that the production and degradation rates of intermediates are especially enhanced by ultrasound on BDD electrodes (Figure 4.7). In addition, the energy consumption of the electrochemical oxidation of phenol decreases when ultrasound is used, due to the fact that the treatment time is significantly reduced.

The potentiostatic SECT oxidation of 2,4-dihydroxybenzoic acid (2,4-DHBA) (0.1–0.3 g L^{-1}) was also performed using both 20 and 500 kHz ultrasonic irradiations [39]. The geometry and position of the electrodes were optimized in order to maximize the acoustic effects. The platinized titanium grid anode (cylinder or disc at high-frequency or low-frequency ultrasonic experiments respectively) was located either in the region of highest ultrasonic intensity (situated near the air/liquid interface) or at the transducer surface for low-frequency and high-frequency ultrasonic experiments respectively (Figure 4.8).

In these experiments, organic pollutants dissolved in water underwent degradation through reactions involving highly powerful oxidizing agents such as hydroxyl radicals generated mainly by high-frequency ultrasound (via sonolysis – see Chapter 1); however, the ECT degradation of 2,4-DHBA was hardly accelerated by the ultrasonic field. Nevertheless, by using low-frequency ultrasound, large increases in 2,4-DHBA degradation rates giving lower final TOC values were achieved (Figure 4.9).

In this work, current densities of 10 and 30 mA cm^{-2} were used for both ultrasonic frequencies (20 and 500 kHz). It was observed that for complete degradation of the 2,4-DHBA pollutant to occur, the required coulombic charge was lower when using lower current densities, mainly due to energy loss caused by the oxygen generation at high current densities. It was seen that a dramatic enhancement of the degradation process occurred at low-frequency ultrasound at the two current densities employed, but the synergistic effect was more pronounced when low-frequency ultrasound was used at low current densities, as shown in Table 4.2.

It was also observed that compounds generated by the sonoelectrochemical oxidation of 2,4-DHBA were the same as for electrooxidation alone (silent conditions) under

Table 4.2 *Charge (Q) required for complete sonoelectrochemical conversion of 2,4-DHBA (Reprinted with permission from [39] Copyright (2002) Elsevier Ltd).*

	Q (A h)	
	10 mA cm^{-2}	30 mA cm^{-2}
Electrooxidation (*silent*)	>1.38	>4.14
H-F sonoelectrooxidation	1.15	2.76
L-F sonoelectrooxidation	0.34	0.78

$T = 30\ ^\circ C$, $C_o = 100$ mg L.

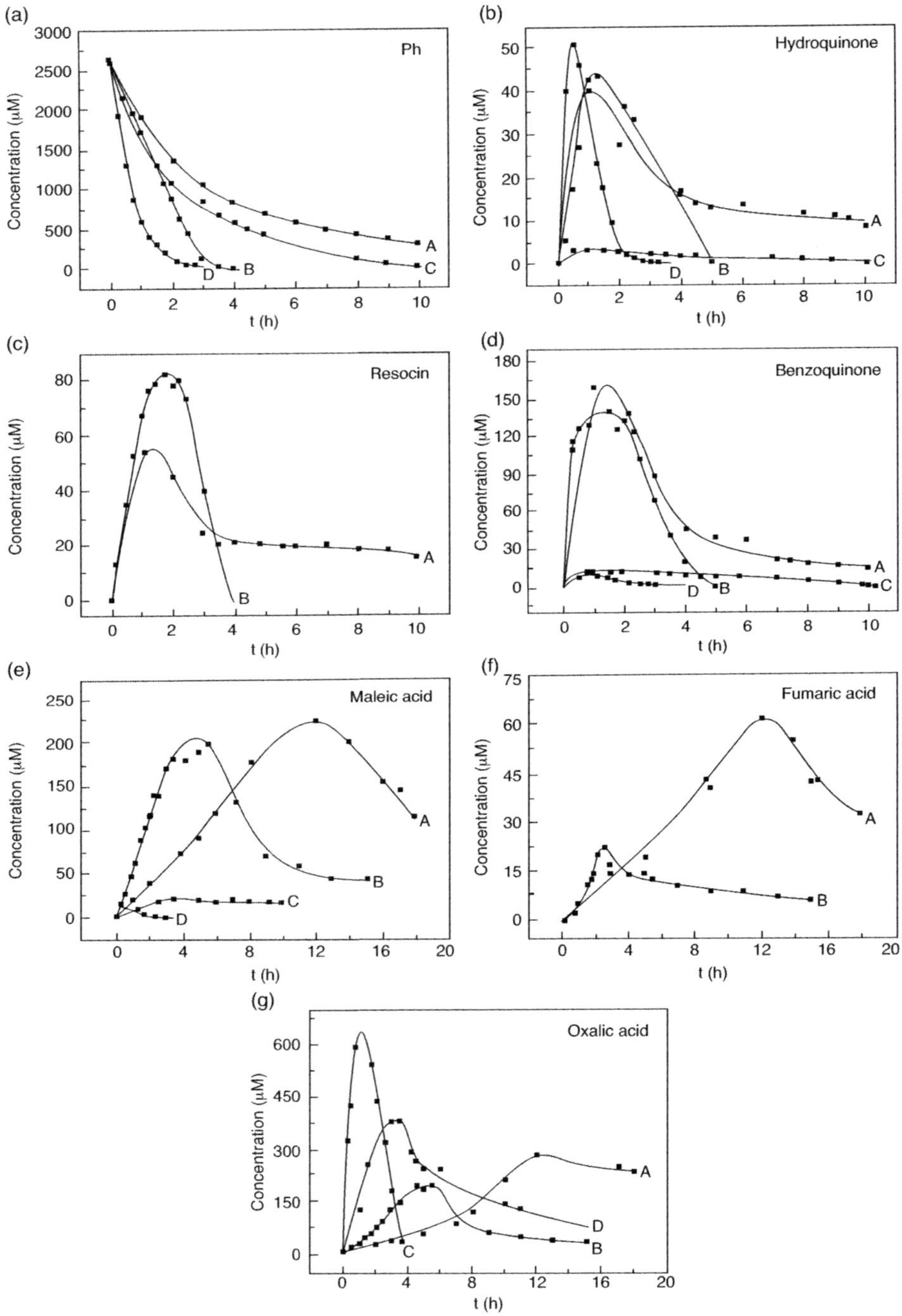

Figure 4.7 *Change of the concentration of phenol and intermediates during the degradation process: (A) Pt, ECT; (B) Pt, SECT; (C) BDD, ECT; (D) BDD, SECT (Reprinted with permission from [38] Copyright (2008) Elsevier Ltd).*

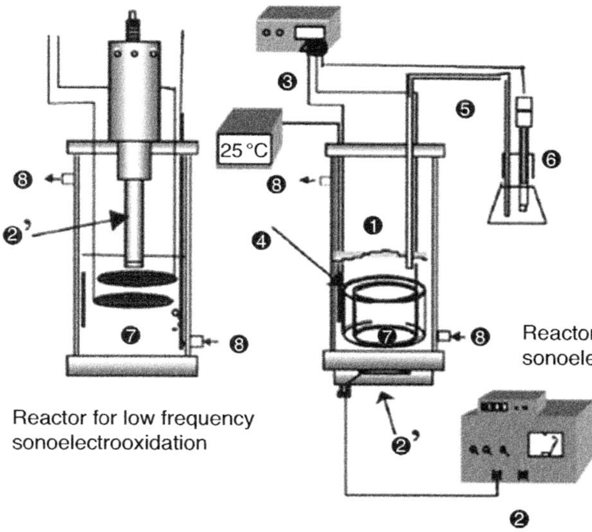

Figure 4.8 *Schematic diagram of the sonoelectrochemical reactors: (1) reactor, (2) ultrasound generator, (2′) transducer, (3) potentiostat/galvanostat, (4) thermometer, (5) agar-agar bridge, (6) reference electrode, (7) electrodes, (8) cooling fluid (Reprinted with permission from [39] Copyright (2002) Elsevier Ltd).*

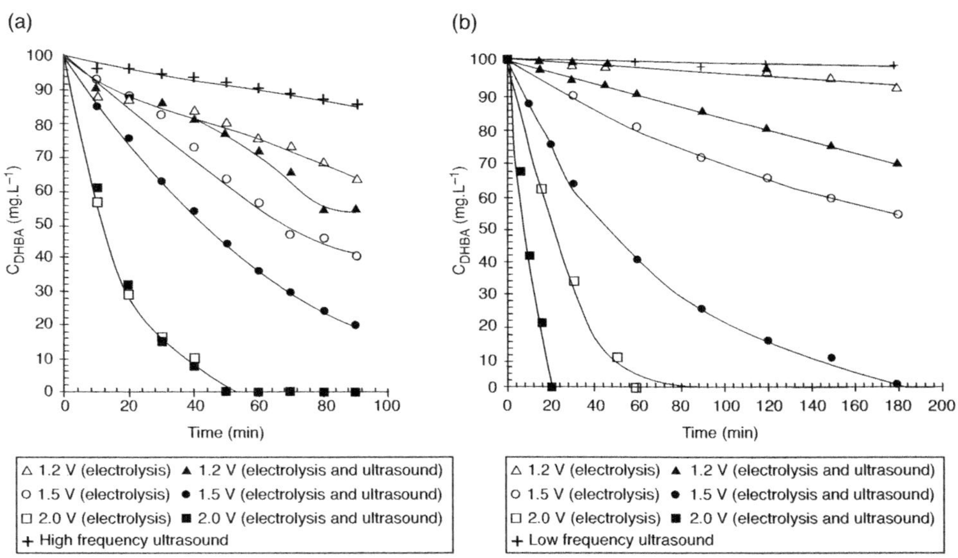

Figure 4.9 *2,4-DHBA concentration profiles for potentiostatic electrolyses, with and without (A) high-frequency (500 kHz) and (B) low-frequency (20 kHz) ultrasound (Reprinted with permission from [39] Copyright (2002) Elsevier Ltd).*

pseudo-first-order kinetics; however, fewer intermediate aromatic compounds were produced at low-frequency irradiation.

It should be emphasized at this point that ultrasonic irradiation does not always enhance the electrochemical oxidation of pollutants at the anode. For example, Zhao *et al.* recently observed that the electrochemical oxidation of phthalic acid on sonicated BDD electrodes led to poor efficiencies compared to silent conditions due to enhanced adsorption of materials at the electrode surface under insonation [40]. Electrochemical oxidation processes at the electrode surface are complex, and may undoubtedly involve the following steps: (i) mass transport – the diffusion of pollutants from the bulk solution to the electrode surface; (ii) adsorption or/and desorption – the adsorption of pollutants near the electrode to the electrode surface and desorption of intermediates from the electrode surface; and (iii) electrochemical reaction, that is, electron transfer. In this case, ultrasound may have an overall positive or negative effect depending on the electrochemical properties of the electrode and pollutant. The enhancement of degradation efficiencies on various electrodes and pollutants is directly related to the influence of ultrasound on the three steps described above.

4.2.1.5 *Sonoelectrochemical Treatment of Wastewater with High Organic Content*

Many industrial processes generate wastewater with a high level of salinity and TOC; for example, the food, the fish, olive oil, and leather processing industries. The treatment of saline wastewater is a challenging task as it also involves biological treatment. High operating/running costs resulting from high energy requirements are one of the many disadvantages of ECT; however, it is thought that ultrasound may reduce dramatically energy consumption in these processes.

In order to study the effect of SECT on pollution attenuation, the electrooxidation of a model protein (bovine serum albumin, 2.5 g L^{-1}) in synthetic hypersaline wastewater (NaCl, 40 g L^{-1}) was carried out [41]. SECT experiments were performed galvanostatically at 10 mA cm^{-2} using concentrically placed cylinder electrodes (200 cm^2 titanium electrodes coated with lead, manganese, and tantalum oxides) and a 20 kHz horn-type sonication device. The attenuation of COD and total Kjeldahl nitrogen (TKN) in the absence and presence of ultrasound during electrooxidation was studied. It was found that the introduction of ultrasound significantly enhances pollution attenuation during the early reaction stages of the process. The attenuation of COD and TKN values suggested that pollution attenuation is directly proportional to the ultrasonic power. Total current efficiencies and COD removal rates were also found to be proportional to the ultrasonic power (up to 100 W). However, increasing the ultrasonic power above 100 W did not render a significant increase in the overall process. The energy consumption for this electrooxidation was found to be inversely proportional to the ultrasonic power up to 100 W. Although the total energy consumption is relatively high for commercial applications, ultrasound may be introduced initially in the process as a pre-treatment technique where important benefits could be observed.

The capability of the two types of pre-pilot-scale electrochemical systems [42] (treatment capacities of 4 and 0.5 m^3 h^{-1}) developed for effective organic pollutant oxidations in various wastewater (effluents from an anaerobic digester of cattle wastewater, supernatants from primary sedimentation in a sewage plant, and domestic wastewater) was evaluated. For both the anaerobic digester and the sewage plant, the 4 m^3 h^{-1} system was found to remove: 87–91% of total phosphorus, 74–96% of total nitrogen, 70–94% of NH_4-N,

88–91% of TOC, and 75–87% of COD. Similarly, for wastewater, the 0.5 $m^3\ h^{-1}$ system was found to remove 62–90% of total phosphorus, 83–92% of total nitrogen, 90–100% of NH_4-N, 75–83% of TOC, and 80–100% of COD. In these experiments, it was observed that the high-voltage pulses and ultrasonic treatment decomposed NH_4-N and TOC effectively due to the formation of radical species (hydroxyl radicals) and hypochlorite (here, total phosphorus was removed by electrocoagulation).

Although ultrasound is currently used in the food industry (as an effective 'emulsifier'), the separation of oil-in-water emulsions may have broad applications in manufacturing and wastewater treatments. However, the application of ultrasound to coagulate liquids effectively presents some drawbacks which may be overcome by the use of electrolysis. Stack *et al.* [43] proposed the utilization of standing ultrasonic waves to flocculate micelles, and simultaneous application of electrolysis to coalesce the micelles, in view of overcoming the electrostatic barrier between micelles and by destroying the structure of the proteins on the surface of micelles. They also studied important parameters such as pH, temperature, conductivity, ultrasonic intensity, electrolysis voltage and current, and duration, and their contribution to the sonoelectrocoagulation process. In this study, they selected a stable oil-in-water emulsion, for example, the wastewater from washing raw wool. Sonoelectrochemical experiments were performed in an ultrasonic bath (40 kHz and up to 22 $W\ L^{-1}$) using two disks of copper screen (located in the antinodes of the sound wave) as electrodes which were mounted in the tank perpendicular to the direction of the wave propagation. They observed that the wastewater conductivity was found to be the most significant parameter influencing the effectiveness of the treatment. For example, increasing the conductivity through the dissolution of salts directly influences the breaking of emulsions by compressing the electric double layer surrounding the soap-encased grease micelle, thus overcoming the electrostatic repulsion that prevents micelles from coalescing (identified as the most important factor in the stability of wool-scouring emulsions). According to this study, the developed procedure is capable of removing up to 100% of the grease from wastewater.

4.2.2 Hybrid Sonoelectrochemical Techniques Applications

The use of indirect electrochemical degradation of organic pollutants in industry has increased significantly, and often involves electrogeneration of strong oxidizing or reducing agents that improve the quality of the resulting solution. The *in situ* electrochemical production of hypochlorite and hydrogen peroxide (H_2O_2) has been employed extensively for this purpose. Furthermore, its combination with ultrasonic irradiation and the use of catalysts usually enhances the pollution attenuation. For example, the addition of Fe^{2+} to a solution containing hydrogen peroxide is known to 'catalyze' the organic compound degradation through the generation of the highly reactive hydroxyl radical ($OH^\bullet$). The combination of ultrasonic waves, electrochemistry, and Fenton's reagents has been employed as a novel hybrid technique for the degradation of organic pollutants in water [44–46]. Powerful oxidizing $OH^\bullet$ radicals are produced according to a Fenton-type mechanism, as described in Scheme 4.2.

$$Fe^{2+} + H_2O_2 \longrightarrow Fe^{3+} + OH^- + OH^\bullet$$

Scheme 4.2 *Fenton's reaction mechanism.*

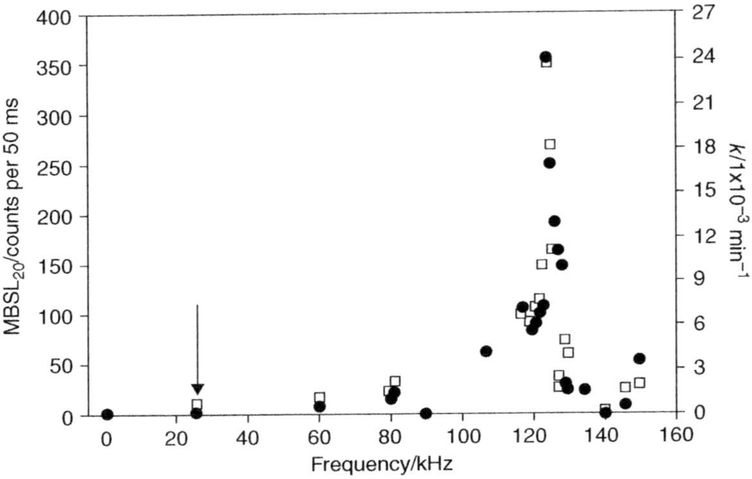

Figure 4.10 *Plot showing the variation in the rate constant (•) for degradation of Meldola Blue as a function of ultrasonic frequency, compared to the light output (□) from multibubble sonoluminiscence (MBSL) over the same frequency range. The cell contained 100 cm*^{3} *of an aqueous solution comprising 50 mmol L*$^{-1}$ Na_2SO_4*, 10 mmol L*$^{-1}$ H_2SO_4 *(pH 2), and 0.5 mmol L*$^{-1}$ $FeSO_4$*. In the degradation cases the solution also contained 0.5 mmol L*$^{-1}$ *Meldola Blue. The arrow indicates the resonance frequency of the transducer (Reprinted with permission from [44] Copyright (2002) Royal Society of Chemistry).*

Using this sonoelectro-Fenton (SEF) treatment procedure, the degradation of an organic dye molecule, Meldola Blue, has been studied [44]. Electrochemical generation of hydrogen peroxide was accomplished potentiostatically using a reticulated vitreous carbon cathode under acidic conditions (50 mM Na_2SO_4, 10 mM H_2SO_4, pH = 2) and $FeSO_4$ (0.5 mM) was added for SEF experiments. For the optimal performance of the sonoelectrochemical reactor, the pollutant degradation was measured as a function of ultrasonic frequency in the range 20 to 150 kHz (Figure 4.10). The degradation of the organic dye was monitored by UV-Vis and COD measurements, and it was shown to be significantly enhanced by both the presence of Fe^{2+}, electrochemically generated hydrogen peroxide, and the presence of ultrasound (Table 4.3). Others reaction parameters were also optimized.

The SEF procedure was also applied for the degradation of the common herbicide 2,4-dichlorophenoxyacetic acid (2,4-D) and its derivative 2,4-dichlorophenol, but using nickel electrodes in an unoptimized sonoelectrochemical reactor [45]. Electrolysis experiments were conducted galvanostatically (10–100 mA cm^{-2}, 0.5 g L^{-1} Na_2SO_4), and the 20 kHz ultrasonic field was generated by a horn emitter (75 W). A dramatic enhancement in the degradation rate for both compounds occurred using the SEF procedure, but only for low pollutant concentrations (Figure 4.11). The degradation intermediates, identified by MS-GC, were similar to those obtained by UV-photolysis (Scheme 4.3).

The production of the highly toxic intermediate 2,4-dichlorophenol on the oxidation of 2,4-D is an important drawback of the presented degradation procedure. To avoid this problem, Yasman *et al.* [47] studied the use of palladium-based catalysts in the sonoelectrochemical degradation of 2,4-D. Suspensions of finely divided Pd and Fe-Pd

Table 4.3 Variation in the rate constant for the degradation of Meldola Blue as a function of the physical parameters of the degradation treatment employed (Reprinted with permission from [44] Copyright (2002) Royal Society of Chemistry).

Treatment	E/V	$FeSO_4$ (mmol·L^{-1})	f (kHz)	(k/min^{-1})/1 × 10^{-3}
Electrochemical	−0.7	0	[a]	[b]
Sonochemical	[a]	0	124	4.4 ± 0.3
Sonoelectrochemical	−0.7	0	124	12.6 ± 0.4
Sono-Fenton	[a]	0.5	124	8.1 ± 0.7
Electro-Fenton	−0.7	0.5	[a]	14.2 ± 0.5
Sonoelectro-Fenton	−0.7	0.5	124	23.7 ± 0.4

50 mmol L Na_2SO_4, 10 mmol L H_2SO_4 and 0.1 mmol L Meldola Blue solutions.
[a]Absence of electrochemical generation of hydrogen peroxide or ultrasonic irradiation.
[b]Not measurable.

powders were used in the sonoelectrochemical setup previously employed for SEF treatment. The authors state that coupling ultrasound and electrocatalytic degradation of 2,4-D results in complete mineralization of the pollutant, and a greatly enhanced reaction rate was observed compared with traditional electrocatalytic processes (Scheme 4.4). Nevertheless, further experimental details and investigations of the mechanisms involved in this sonoelectrocatalytic procedure are needed.

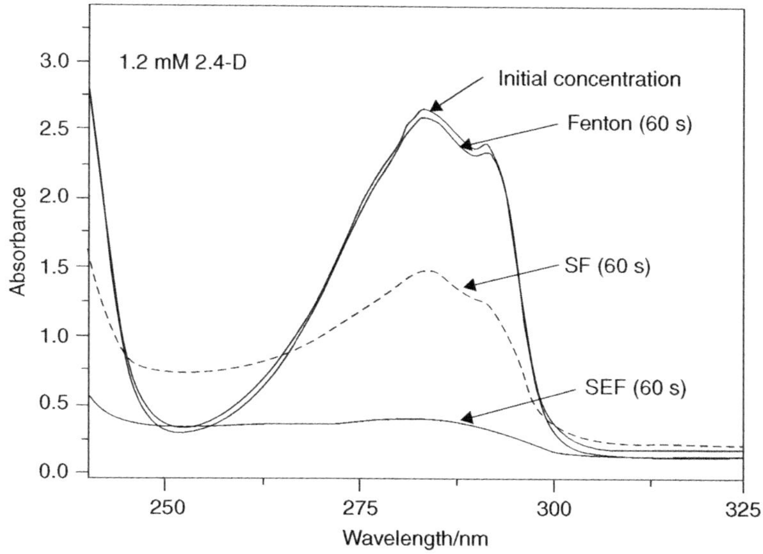

Figure 4.11 UV absorbance spectra of as-prepared 2,4-D (1.2 mM); and following 60 s of Fenton, sono-Fenton (SF), and SEF processes. [Fe^{2+}] = 3.0 mM, [H_2O_2] = 3.0 mM (Reprinted with permission from [45] Copyright (2004) Elsevier Ltd).

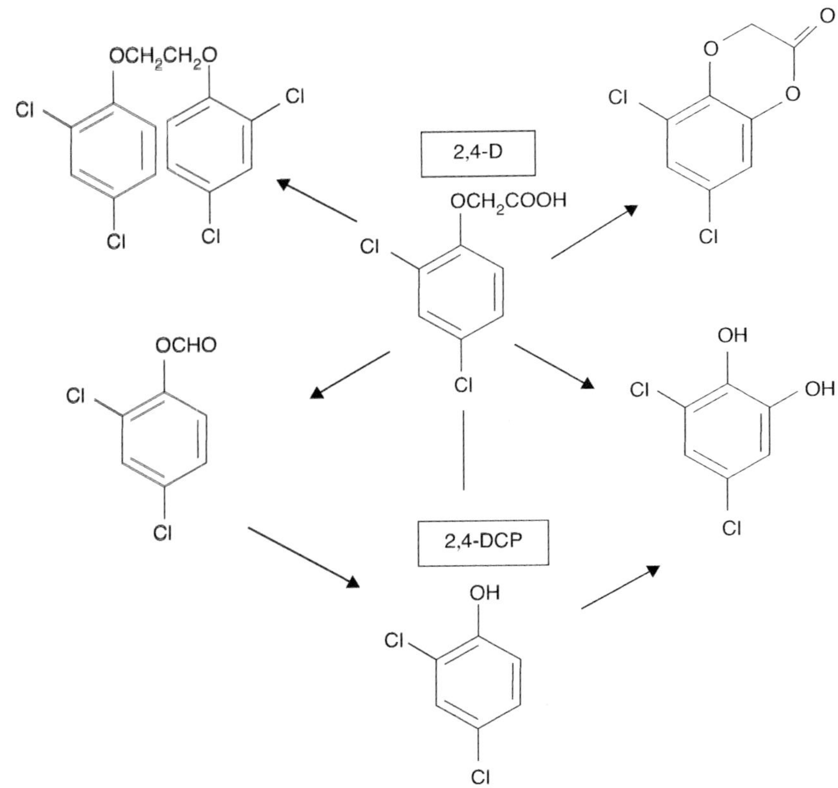

Scheme 4.3 *Schematic presentation of the proposed first chemical stages in the SEF decomposition of 2,4-D (Reprinted with permission from [45] Copyright (2004) Elsevier Ltd).*

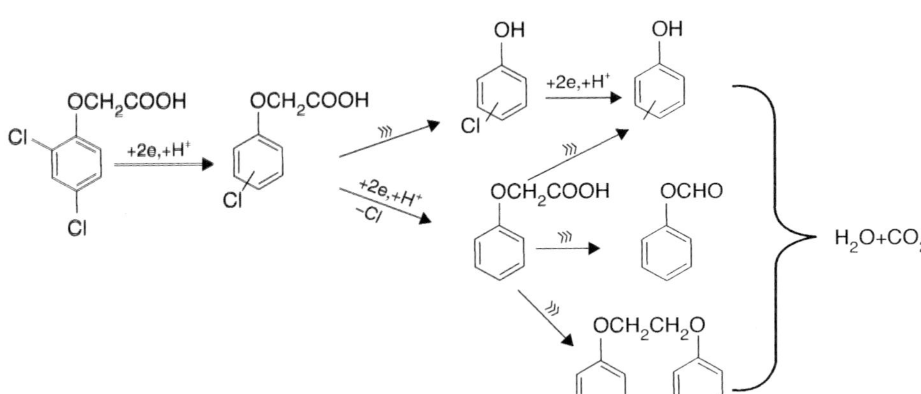

Scheme 4.4 *Proposed mechanism for the sonoelectrocatalytic degradation of 2,4-D on black Pd, based on intermediate determination (Reprinted with permission from [45] Copyright (2004) Elsevier Ltd).*

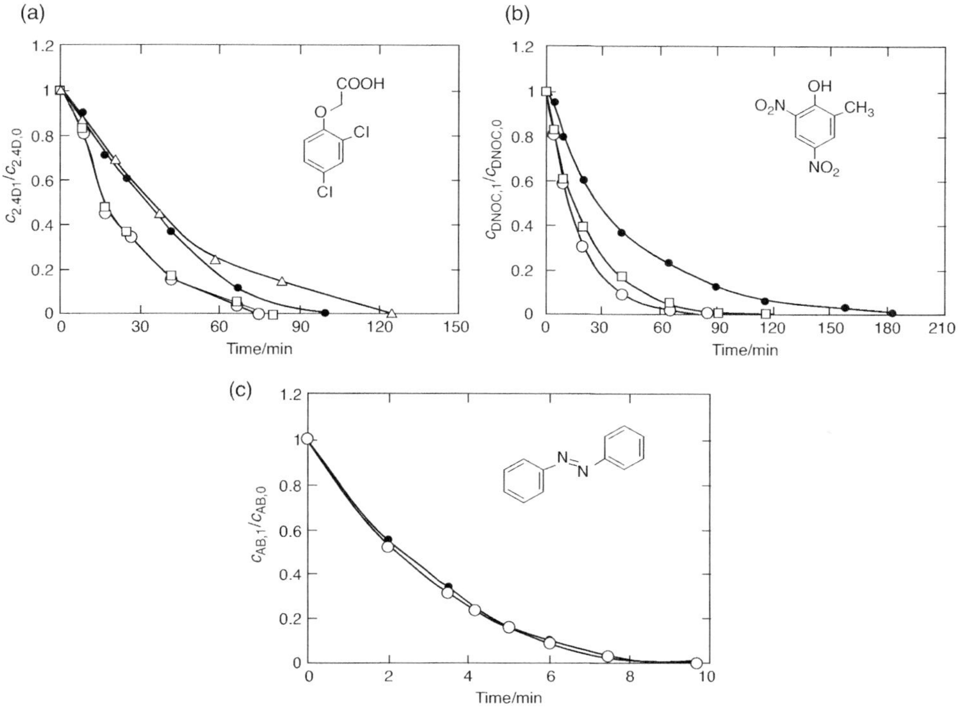

Figure 4.12 Decay of pollutant concentration during the treatment of 250 mL of aqueous solutions containing (A) 1 mM of 2,4-D; (B) 0.5 mM of 4,6-dinitrocresol; (C) 0.025 mM of dye AB and 50 mM Na_2SO_4 and 0.1 mM Fe^{3+} as catalyst, at 200 mA, pH 3.0, and room temperature, using a Pt anode and a carbon-felt cathode at different ultrasound powers: (♦) 0, (○) 20, (□) 60 and (Δ) 80 W, at a frequency of 28 kHz (Reprinted with permission from [46] Copyright (2008) Elsevier Ltd).

Oturan *et al.* [46] also reported the results of the degradation of 2,4-D and other pollutants such as the herbicide 4,6-dinitrocresol and the azo-dye AB by SEF treatment (Figure 4.12). Hydrogen peroxide was produced galvanostatically (50 mM Na_2SO_4) at pH 3, using a carbon felt cathode and continuous air bubbling to ensure oxygen saturation of the solution. Ultrasound irradiation at 28 kHz and different ultrasonic output powers were used for this study.

SCT degradation (28 kHz and 460 kHz frequencies), electro-Fenton, and SEF treatment processes were compared, and the SEF treatment was noted to be the most efficient (Table 4.4). Pseudo-first-order apparent rate constants for the SEF process could be measured at different ultrasonic powers. It was concluded, however, that increasing the output power greatly decreases the degradation efficiency because higher values hamper the dissolved oxygen concentration, and, consequently, this affects the cathodic hydrogen peroxide electrogeneration required for Fenton's reaction. Ultrasound did not improve the results obtained by the electro-Fenton process for the degradation of the readily oxidizable azo-dye AB.

Table 4.4 Apparent rate constant (k_{app}) values obtained for the kinetic analysis of the concentration decays assuming pseudo-first-order reaction kinetics (Reprinted with permission from [46] Copyright (2008) Elsevier Ltd).

	k_{app} (min^{-1})			
Pollutant	Electro-Fenton	SEF (20 W)	SEF (60 W)	SEF (80 W)
2,4-D	0.025	0.046	0.043	0.023
4,6-dinitrocresol	0.023	0.065	0.051	–
Dye AB	0.36	0.38	–	–

Another hybrid sonoelectrochemical process reported in the literature is the simultaneous application of ultrasound and ozonation [27] to the electrochemical degradation of 1,3-dinitrobenzene and 2,4-dinitrotoluene in acid media, which allows virtually complete consumption of these pollutants in a shorter time than in the absence of ozone. The increase in the reaction rate of the degradation of nitroaromatics is attributed by the authors to the formation of intermediates that react readily with ozone.

Among the new advanced oxidation methods, photocatalytic technology has been applied to decolorize aqueous solutions of azo-dyes with TiO_2 as photocatalyst. Unfortunately, the low photocatalytic efficiency has limited the application of this technology in practical water treatment. It has been reported that the photocatalytic efficiency is enhanced when combined with other techniques, resulting in electro-assisted photocatalysis, and ultrasonic-assisted photocatalysis. The simultaneous use of all these techniques, a sonophotoelectrocatalytic process[48], has been studied for the degradation of the azo-dye Methyl Orange in aqueous solution, and this has been compared with photoelectrocatalytic, photocatalytic, and sonophotocatalytic processes using a TiO_2 nanotube electrode as the photocatalyst (Figure 4.13).

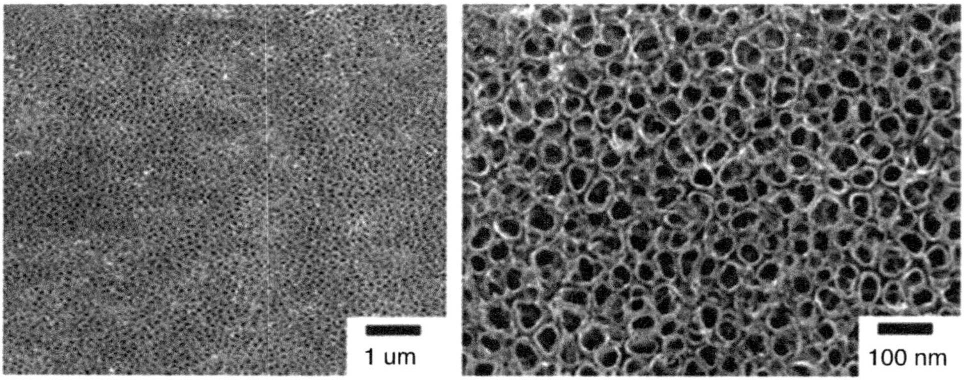

(a) (b)

Figure 4.13 FE-SEM images of TiO_2 nanotubes: (a) low magnification of top view, and (b) high magnification of top view (Reprinted with permission from [48] Copyright (2008) Elsevier Ltd).

Table 4.5 Rate constants (k) and regression coefficients (R^2) of Methyl Orange degradation with different photocatalytic processes (Reprinted with permission from [48] Copyright (2008) Elsevier Ltd).

Process	k (min^{-1})	R^2
Sonophotoelectrocatalytic	0.0732	0.9858
Photoelectrocatalytic	0.0523	0.9951
Sonophotocatalytic	0.0073	0.9874
Photocatalytic	0.0035	0.9936

The sonophotoelectrocatalytic process was carried out on a quartz photoreactor immersed in an ultrasonic cleaning bath (40 kHz) using an 11-W UV lamp (253.7 nm) and a Pt counter electrode. The photocurrents registered on linear sweep photovoltammograms were potential dependent, and showed a significant increase when the solution was irradiated with ultrasound. A remarkable synergistic effect exists in the ultrasonic, electro-assisted, and photocatalytic processes, according to the measured degradation rate constants (Table 4.5).

Moreover, the degradation efficiency of the dye could be increased significantly by increasing the ultrasonic power or decreasing the initial dye concentration (Figure 4.14). Good stability of the electrode was reported, but longer reaction times should be tested in order to validate the method for practical applications.

4.3 Recovery of Metals and Treatment of Toxic Inorganic Compounds

Metal-containing wastewaters are produced as by-products of manufacturing processes. These include spent electroplating or electroless baths, pickling acids, acids used for cleaning or activation of metal surfaces, residues from electrowinning, and used photographic

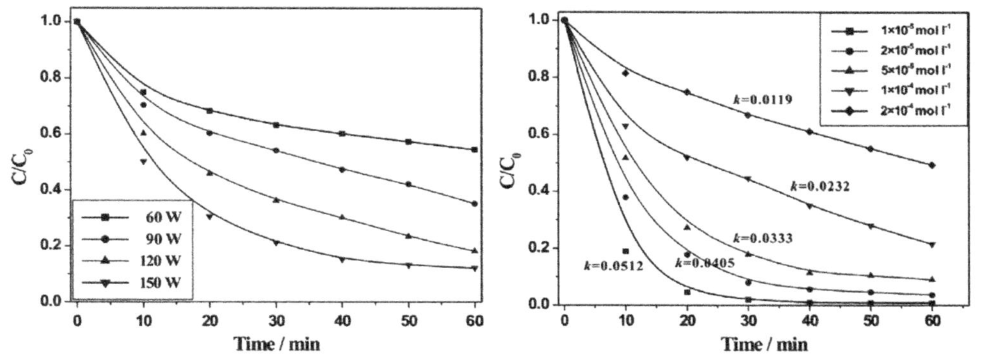

Figure 4.14 *Sonophotoelectrocatalytic degradation of Methyl Orange (A) with different ultrasonic powers ($C_0 = 5 \times 10^{-5}$ M), and (B) with different initial concentrations (ultrasound power, 150 W) (Reprinted with permission from [48] Copyright (2008) Elsevier Ltd).*

solutions. The recovery of the dissolved metals from these electrolytic solutions in a valuable form may be possible, and ion exchange, membrane separation, electrolysis, precipitation, and solvent-extracting techniques have been applied for this purpose. Currently, the most widely employed metal recovery method is electrolysis. The low metal concentration of the wastewater is one of the most important problems of electrodeposition, because the mass transport of cations to the cathode, often the rate-controlling step, can be very slow. Obviously, the nature of the cation also has a great influence on the electrodeposition process.

The deposition of metals under the influence of an ultrasonic field has received significant attention [49], particularly in the applied literature. It is claimed that ultrasound not only confers various benefits (hardness, brightness, grain size, film thickness, adhesion) over conventional silent electrodeposition, but also over electrochemical factors such as current efficiency and energy consumption [50, 51]. Nevertheless, this work has been focused on the electrodeposits rather than on the recovery process [52]. Much of this work has been carried out from an industrial standpoint, particularly on metals that are important in electroplating.

Electrochemical copper deposition has long been studied, and copper recovery has been observed to be more efficient in the presence of ultrasound. Ultrasonic agitation has been probed and found to be useful in increasing the rate of deposition of metal from a dilute solution. The degassing effect and the cavitation enhance the mass transport of cations to the cathode surface, producing an effective thinning of the diffusion layer and reducing concentration polarization. All these effects improve the electrolytic recovery of copper. Ultrasound fields of different frequencies generated by cleaning baths have been used to study the electrodeposition current efficiency and the metal concentration remaining in the solution. Experiments were carried out with controlled currents. Unfortunately, the specific values of the ultrasonic power or the cathodic material have not always been reported for all experiments. The results of various studies [52–54] dealing with copper removal conclude that an improvement in the deposition rate is observed with an ultrasonically agitated bath. Farooq *et al.* [53] used electrodeposition and ultrasound at 35 kHz to remove copper in wastewater. They found that the electroreduction of copper followed a pseudo-first-order kinetic model, with a reaction constant of 2.19×10^{-3} min^{-1} (ultrasound) or 6.75×10^{-4} min^{-1} (silent). The copper removal efficiency was enhanced from 55.1 to 94.6%. The improvement produced by the ultrasound field in the cathodic current efficiency was more pronounced when the concentration of metal ions was decreased. However, similar values were observed with concentrated copper solutions except when high current densities were applied. In those cases, the diffusion of copper ions to the cathode was insufficient to maintain the current, and some hydrogen evolution was observed, giving rise to a lower efficiency value [52]. It is in diluted solutions that concentration polarization is high and the application of ultrasound is particularly beneficial. The effects of ultrasonic [54] frequency on both the mass transport process and the diffusion layer were also investigated during electrochemical deposition. By increasing frequencies from 40 to 100 kHz, a high value for the mass transport coefficient was observed. Although the effect of cathode geometry on the ultrasonically agitated electrodeposition process has been studied, the electrocatalysis of the cathodic material has not been taken into account. A mesh titanium cathode has been proved to be better than a plate stainless steel cathode for the removal of copper. Finally it can be concluded that the use of ultrasound with electrochemical deposition can provide a valuable means for the efficient removal of metallic ions from industrial wastewater, reducing the energy consumption and deposition time.

The removal of zinc [52] and lead [54] from industrial wastewater has also been studied. There are no significant differences in conclusion: ultrasound was found to have a significant effect on metal recovery under diffusion control, while little, if any, effect was observed on systems in which the deposition was under charge-transfer control. It is also noteworthy that the effect of applying power ultrasound to various metal systems while electrodepositing on the cathode was studied using cyclic voltammetry [55]. The response to ultrasound was found to be dependent on ion concentration (Figure 4.15). With solutions of high metal concentration, where deposition is under charge control, little alteration was observed in the deposition parts of the cyclic voltammograms. Solutions with low metal concentrations, in which deposition is under diffusion control, showed a decreased overpotential deposition and large increases in the total deposition current. According to more applied articles, Hyde

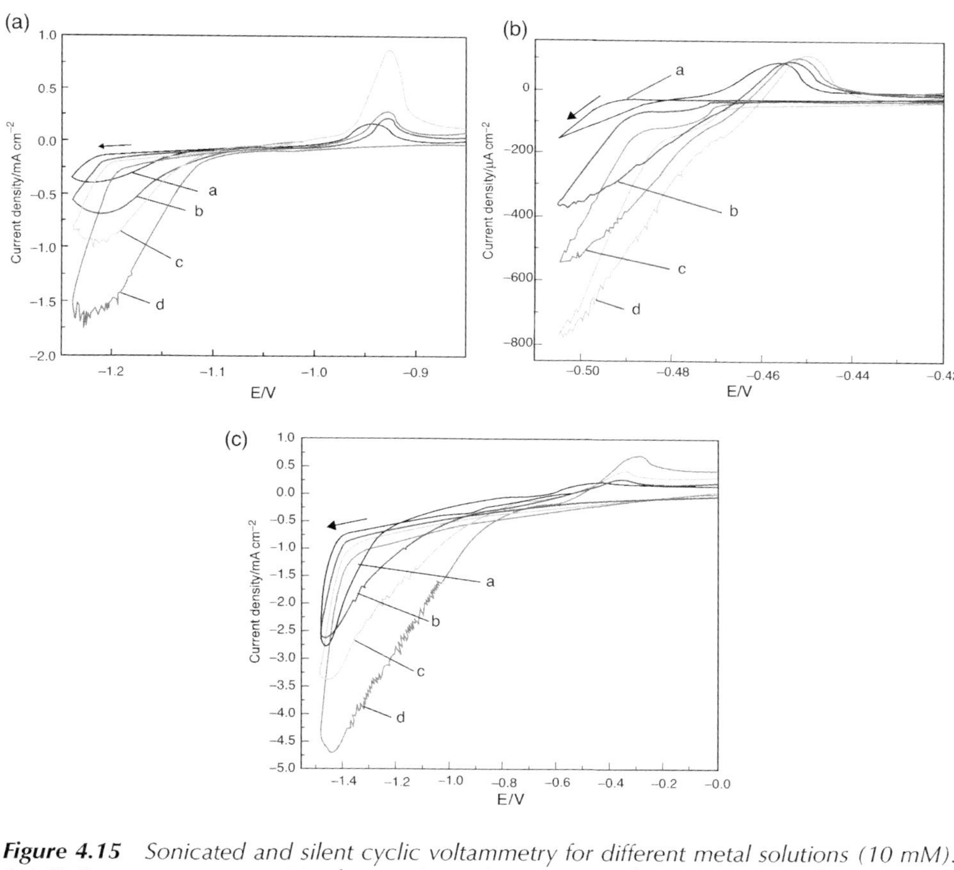

Figure 4.15 Sonicated and silent cyclic voltammetry for different metal solutions (10 mM). (A) $ZnBr_2$, scan rate 50 mV s^{-1}, (a) silent, (b) 37 W cm^{-2}, (c) 78 W cm^{-2}, (d) 238 W cm^{-2}. (B) $Pb(NO_3)_2$, 10 mV s^{-1}, (a) silent, (b) 37 W cm^{-2}, (c) 78 W cm^{-2}, (d) 238 W cm^{-2}. (C) $CoSO_4$, 100 mV s^{-1}, (a) silent, (b) 37 W cm^{-2}, (c) 64 W cm^{-2}, (d) 238 W cm^{-2} (Reprinted with permission from [55] Copyright (2002) Elsevier Ltd).

et al. [55] concluded that in electrodeposition under the influence of ultrasound, the critical effect is the increase in mass transport, which may be high enough to change a diffusion-controlled system into a charge-controlled system. Ultrasound also ablates material from the electrode surface, but has no effect on growth through charge transfer from the electrode to the metal ion. All these conclusions demonstrate that ultrasound can dramatically improve the efficiency of treating heavy-metal wastewater.

Industrial wastewaters usually contain metal complexes, for example, copper-ethylenediaminetetraacetic acid (EDTA), or silver-thiosulfate complexes appearing in photographic processing solutions. For this type of wastewater, other removal techniques [56] can be applied, but electrolysis is widely employed. During the electrodeposition, organic pollutants can be degraded at the anode while heavy metals are reduced at the cathode. Nevertheless, a very low current efficiency for treating wastewater containing chelated heavy metals is observed, reducing the economical viability of the electrodeposition technique; for example, a 97.7% recovery rate but a 3.1% current efficiency for EDTA-Cu system was obtained for the Cu electrodeposition process [57]. For this reason, the influence of ultrasound on electrodeposition applied to EDTA-Cu solutions has been studied [58]. In this case an ultrasonic transducer with a round tip was placed in the middle of the reactor to produce the ultrasonic wave, instead of using an ultrasonic bath. The study was carried out using a graphite anode and a copper plate cathode, under controlled voltage gradient mode. A possible reaction mechanism for the EDTA-Cu decomposition is depicted in Figure 4.16.

The pH always plays a major role in an electrodeposition process due to the migration velocity of H^+ and OH^- ions, the hydrogen evolution produced at the cathode at low pH values, and the oxygen evolution produced at the anode at high pH values, which influence the current efficiency. In addition, in this system, the pH influences the concentrations of the complex species and copper ions due to the weakly acidic character of EDTA. An increase in efficiency for copper removal with decreasing pH has been observed, which can be attributed to the EDTA and copper behavior at different pH values. A predominance of H_4EDTA and H_3EDTA^- species occurs at low pH values, resulting in more free copper ions which favor

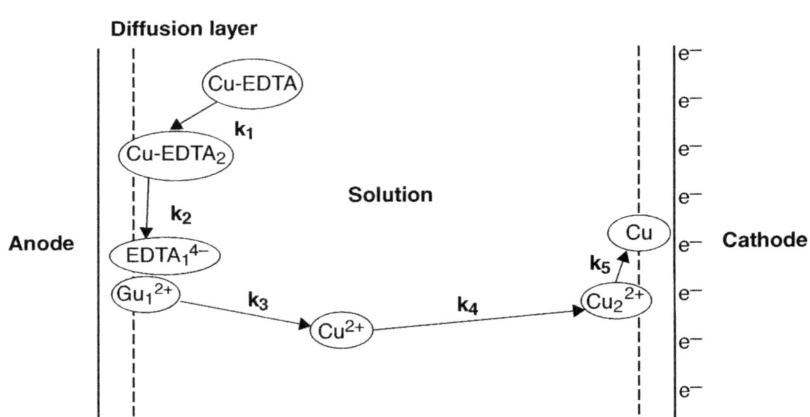

Figure 4.16 *Schematic presentation of the reaction mechanism for Cu-EDTA electrodeposition (Reprinted with permission from [58] Copyright (2009) Elsevier Ltd).*

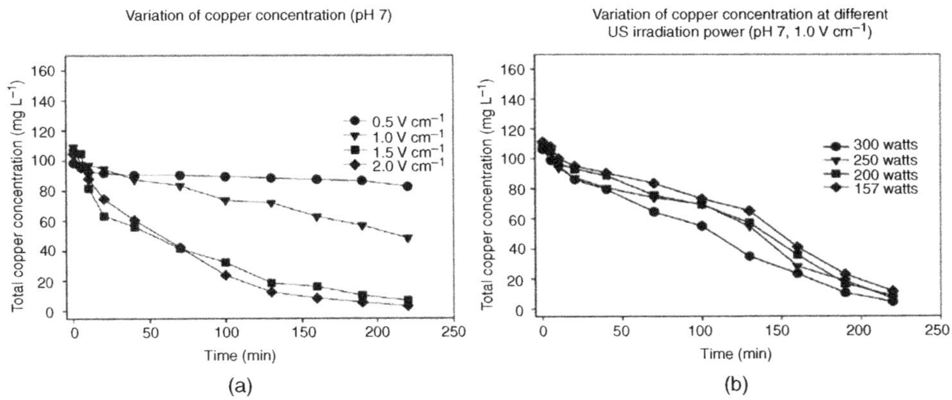

Figure 4.17 *Variations of copper concentration with electrodeposition: (A) under different voltage gradients (silent conditions); (B) under different ultrasonic powers (Reprinted with permission from [58] Copyright (2009) Elsevier Ltd).*

the electrodeposition of copper at the cathode. However, the highest current efficiency was obtained at pH 7, which can be attributed to the low hydrogen evolution at the cathode at neutral pH. The observed pH effect was not significant when a high voltage gradient was applied (Figure 4.17A). The efficiency of copper removal is enhanced when electrolysis is combined with increasing ultrasonic irradiation power (Figure 4. 17B), which also increases the current efficiency of the process. The apparent rate constant of copper removal fits to a pseudo-first-order kinetic model for silent conditions, and to a zero-order model in the presence of ultrasound. The different kinetics imply that the mass transport mechanism changes to a charge-controlled system under ultrasonic conditions. It was found that EDTA is also oxidized by the combined technique, and an 84% COD removal was also observed.

There is a vast amount of research dealing with the sonoelectrochemical removal of silver from 'fix' photographic processing solutions, which consist mainly of sodium thiosulfate, sodium bisulfite, and silver halides. Currently, the most widely employed method for silver removal is electrolysis. However, vigorous agitation is required in order to overcome electrode fouling. The reduction of silver cations, Ag^+, was studied by Pollet *et al.* [56, 59–62] at platinum, stainless steel, and carbon stationary or rotating electrodes in the absence and presence of ultrasound. All the experiments at low ultrasonic frequencies were performed using either a typical single-compartment voltammetric cell, or using a cell similar in design to that of Compton *et al.* [63] except that the ultrasound source was placed at the bottom of the electrochemical cell. When high-frequency ultrasound was used, the experiments were performed using a three-compartment cell [56]. For bulk electrolysis a specially designed cell, allowing the use of three ultrasonic probes, was employed [60].

Rotating cylinder electrode and RDE electrochemical experiments show that Ag^+ (4 g L^{-1}) reduction is under mass-transport control. However, at a sonicated (20 kHz) rotating cylinder electrode the variation in limiting current with rotation speed is not uniform, and the effect of ultrasound depends upon the rotation speed of the cylinder electrode [59]. This effect confirms the importance of configurational parameters in sonoelectrochemical systems. Linear sonovoltammograms reveal an onset shift to more positive potentials on the Ag^+ cathodic discharge potential that increases with ultrasonic power [59]. The effect of ultrasonic frequency upon kinetic parameters for the $Ag(S_2O_3)_2^{3-}/Ag$ redox couple [56] has also been studied. The results indicate that the Ag^+ cathodic discharge potential does not vary significantly with ultrasonic frequency, but rather with ultrasonic intensity. A bulk temperature of 400 K is needed to produce a similar shift to that observed using a maximum ultrasonic intensity of 43 W cm^{-2} (20 kHz) at 298 K. Current measurements at various potentials and different rotation speeds or ultrasonic intensities allowed Pollet *et al.* [62] to calculate the standard heterogeneous rate constant k_0 for silver thiosulfate, which was found to increase approximately 10-fold in the presence of ultrasound.

Although ultrasonic intensity appears to be one of the most important parameters in the improvement of the limiting current of an electrochemical process, the effect of positioning the ultrasonic probe with respect to the electrode is extremely important in the determination of mass-transport limited currents. Based on the silver electrodeposition process for the $Ag(S_2O_3)_2^{3-}/Ag$ redox couple, Pollet *et al.* [61] attempted to study the influence of both the 'face-on' and 'angular' geometrical configurations (45° between the electrode and the ultrasonic probe) on the limiting currents obtained directly from sonicated cylinder electrode voltammograms. Calculations were made of the hydrodynamics, diffusion layer thickness, and diffusion coefficients, but a more rigorous treatment should be used in order to obtain more accurate results. A linear relationship between limiting current and the square root of the emitted ultrasonic intensity for both the 'face-on' and 'angular' geometries was observed, but for an identical ultrasonic intensity transmitted to the total volume, the limiting current obtained for the 'angular' geometry was much lower than that for the 'face-on' geometry (Figure 4.18). The difference in limiting current between the two geometries may be attributed to a greater thinning of the diffusion layer in the case of the 'face-on' geometry, where the electroactive species uniformly reaches all points of the electrode. In the 'angular' geometry, the diffusion layer is non-uniform, and the effective sonicated areas for the two geometries are different.

More applied results for silver recovery from 'fix' photographic processing solutions were obtained using a combination of ultrasonic irradiation at 20 kHz and bulk potential-controlled electrolysis. For this study, specially designed sonoelectrochemical cells were used, which allowed the use of three ultrasonic probes in 'face-on' (probe 2) and 'side-on' (probes 1 and 3) geometries [60] (Figure 4.19).

Potentiostatic silver removal was noted to occur at an optimum electrode potential of −500 mV versus SCE using a stainless steel cathode [60]. The rate of deposition of silver increased with increasing ultrasonic intensity when 20 kHz sonication was used. The 'face-on' position of the ultrasonic probe led to higher rate constants compared to the 'side-on' geometry or silent conditions, showing 25-fold and 18-fold increases, respectively. For the 'face-on' geometry, 99.9% copper removal was achieved after 90 min of electrolysis in the presence of 107 W cm^{-2} ultrasound power; approximately 6 ppm of silver still remained in the photographic solution, compared with 440 ppm of silver (89% removal) at a maximum rotation speed of 8000 rpm in silent conditions or 3090 ppm (23% removal) at 0 rpm. These ultrasound conditions involve an effective rotation speed of 44 000 rpm. These results confirm that the cell geometry is an important parameter in the removal of silver and for most sonoelectrochemical processes. The specially designed cell used also allowed different ultrasound probe combinations. Operating one side probe ('side-on') and the bottom ultrasonic probes ('face-on') simultaneously (i.e. probes 1 and 2) led to a higher

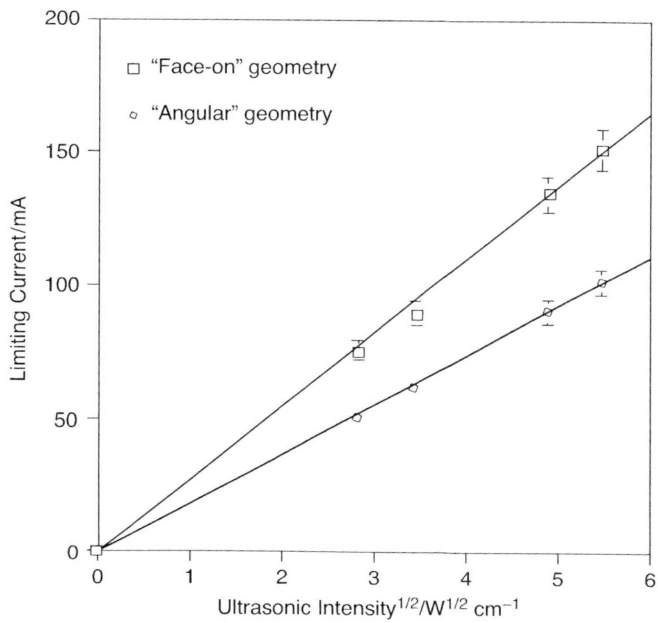

Figure 4.18 *Limiting currents plotted against the square root of ultrasonic intensity (20 kHz) for the reduction of silver (4 g L^{-1}) on a sonicated cylinder electrode at 298 K for the 'face-on' and 'angular' geometries (Reprinted with permission from [61] Copyright (2003) Elsevier Ltd).*

rate constant for a given ultrasonic power. Other probe combinations (probes 2 + 3 or 1 + 2 + 3) were less effective, and the use of the three ultrasonic probes operating at the same power did not significantly improve the rate constant of silver removal compared with that obtained for the 'face-on' geometry. Although silver recovery rates are very efficient using the combination of electrolysis and ultrasonic agitation, the current efficiency data have not been reported. However, this parameter does not seem too important given the high economic value of the recovered product.

The anodic reaction for silver removal reaction from 'fixed' photographic processing solutions has also been studied [64]. Biological treatment of recovered silver solution effluents is difficult due to the stability and antiseptic properties of sodium thiosulfate, which prevents the growth of microorganisms. The sonoelectrochemical treatment used in silver removal enhances not only the silver recovery rates, but also the thiosulfate oxidation process. A 50-fold increase compared with silent conditions was observed in the degradation of thiosulfate on platinum. Under the conditions studied, the thiosulfate degradation increased as the ultrasonic power was increased, suggesting that the oxidation of thiosulfate was also limited by diffusion.

Cyanide is a highly toxic species, mainly found in industrial effluents related to metallurgical industry activities such as metal surface treatment or mining. These effluents always contain cyanide, which must be treated before disposal in water environments. Cyanide can be electrochemically degraded by direct oxidation, first giving cyanate which is further

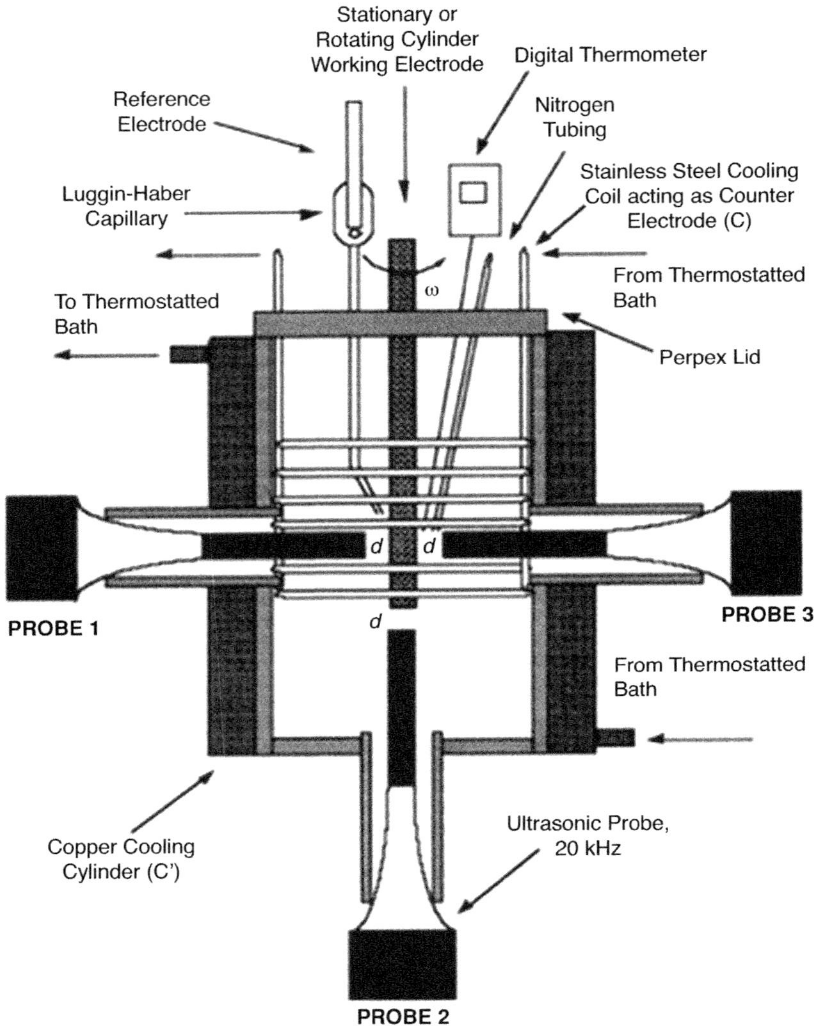

Figure 4.19 *Sonoelectrochemical cell for bulk electrolysis fitted with three ultrasonic probes (Reprinted with permission from [60] Copyright (2000) Elsevier Ltd).*

oxidized, or by indirect oxidation using chloride (forming ClO^-) as an oxygen carrier [65]. Both electrochemical procedures have been used in combination with ultrasound [66]. It has been observed that the efficiency of cyanide oxidation under galvanostatic conditions (0.3 mA cm^{-2}) in the presence and in the absence of ultrasound (37 kHz) depends upon the anode type used (titanium, nickel, copper, graphite, aluminum, and stainless steel). The highest degradation efficiency was obtained with the aluminum electrode. The enhancement of aluminum efficiency has been explained by the authors [66] as the result of continuous activation of the electrode surface, by a cleaning and abrading effect upon the passivating aluminum oxide coating on the electrode surface induced by the ultrasound irradiation. Cyanide indirect electrochemical degradation under sonication proved to give better results. Although up to 78% cyanide degradation efficiency after 60 min was reported, no data about current efficiency, energy consumption, and anode stability were given.

4.4 Disinfection of Water by Hypochlorite Generation

Common disinfection techniques used in water treatment include chlorination, ozonation, and ultraviolet irradiation. However, electrochemical water disinfection is rarely used, but is highly efficient. The disinfecting species, mainly hypochlorous acid and hypochlorite, can be produced electrochemically from the chloride ions in the electrolyzed water when an appropriate potential is applied to the electrodes introduced to the water stream to be disinfected. The widespread use of this method is mainly prevented by problems caused by the deposition of calcareous deposits on the cathode surface. Unfortunately, some species of bacteria also have the ability to mutate under the adverse conditions of chlorination. This results in the production of strains that are more tolerant to normal chlorine treatment levels. To overcome this problem, it is possible to use higher chlorine levels, but such treatment can result in unpleasant flavors and odors (due to the formation of chlorophenols and other halocarbons). The combination of ultrasound and electrochemistry has been used to solve all these problems.

Ultrasound is known to be an efficient tool for cleaning surfaces, and the possibility of using a horn which acts as cathode (sonotrode) has been probed. Kraft *et al.* [67] studied the electrochemical production of disinfecting species, measured as active chlorine, and the use of a sonoelectrode in order to avoid the deposition of calcareous deposits on the cathode surface. They concluded that the sonotrode efficiently removes calcium carbonate scales from the cathode surface. Thus, electrochemical water disinfection in potable water can be performed on a long timescale without the necessity for cathode cleaning through the use of acids or the polarity reversal method. The active chlorine production rate on the anode (ruthenium oxide-coated titanium expanded mesh, 22 mA cm^{-2}) is not significantly changed by the ultrasonic action (24 kHz, 50–250 W).

The use of ultrasound has been evaluated for the disinfection of water not only by electrochemical production of active chlorine, but also by measuring colony-forming units (CFU). Joyce *et al.* studied the electrolytic production of hypochlorite in conjunction with power ultrasound for the disinfection of *Klebsiella pneumonia* [68] and *Escherichia coli* [69] saline suspensions. Although sonication alone can provide powerful disinfection, a large amount of energy is required. Moreover, sonication at 40 kHz produces initial CFU values greater than 100%, which represent the disruption of bacterial clumps to produce a larger number of colony-forming units. It has been also observed that the biocide effect of electrolysis on bacterial suspensions is minimal, and adequate stirring is required. The electrolytic treatment is dependent upon the electrode material, and is improved by ultrasound. Carbon felt was proposed by the authors as the most suitable material for the anode as it provides an efficient kill and does not appear to suffer from pitting or erosion. The combined action of electrolysis and ultrasound results in a reduction of the treatment time and cell potential, and an increase in the biocide effect.

4.5 Soil Remediation

Soil contamination is an important issue for public health. Among the main soil pollutants, heavy metals, organic polychlorinated compounds and hydrocarbons, particularly polychlorinated biphenyls (PCBs) and polycyclic aromatic hydrocarbons (PAHs), create serious problems for the environment. These harmful pollutants are released into the environment accidentally or by industrial activities such as mining, electroplating, or distillation processes. Hazardous and persistent contaminants like these are of particular concern due to their accumulation in the food web, their long life, and their toxicity. Moreover, persistent organic pollutants (POPs) are strongly adsorbed on the soil due to the hydrophobic character of these organic compounds, especially in low permeable clay-contaminated soils [70].

Several clean-up technologies have been developed for contaminated soils, but they present important drawbacks. Electrochemical treatment has emerged recently as a promising, innovative, and cost-effective *in situ* soil remediation technology that employs electromigration, electroosmosis, and electrophoresis phenomena under electric fields for the transport, extraction, and separation of contaminants in cohesive soils.

Ionic compounds are removed from the soil mainly by electromigration, and the hydrophobic organic pollutants are transported primarily by electroosmosis. The driving mechanisms for species transport are ion migration by electrical gradients, pore fluid advection by prevailing electroosmotic flow, pore fluid flow due to any externally applied or internally generated hydraulic potential difference, and diffusion due to generated chemical gradients, as shown in Figure 4.20 [71]. As a result, cations are accumulated at the cathode and anions at the anode, while there is a continuous transfer of hydrogen and hydroxyl ions across the medium. Finally, the pollutants accumulate by the electrode or are transported into the water, which would need secondary treatment.

Although several studies have demonstrated the feasibility of the electrokinetic process for removing heavy metals, other cationic species, and POPs from contaminated soils, the process might not be effective unless the contaminants are soluble in pore fluid. Therefore, the desorption and solubility of PHPs should be enhanced to improve the mobility of these hydrophobic compounds. The acoustic waves promoted by ultrasound have been used for this enhancement in combination with electrolysis. The cavitation lowers the capillary forces on porous media, and then increases the porosity and permeability of the soil, promoting desorption and mobilization of hydrophobic compounds. The effects of acoustic waves on the porous grain framework of the soil and pore fluid are summarized in Figures 4.21 and 4.22.

In order to evaluate the coupling of electrokinetic and ultrasonic effects on the remediation of a soil contaminated with heavy metals, Chung *et al.* conducted laboratory-scale ECT and SECT experiments using natural clay spiked with Pb [72] or Cd [73] using the experimental setup described in Figure 4.23. Remediation tests were run using graphite electrodes and a constant voltage gradient of 1.0 V cm^{-1} or 50 mA constant current. The ultrasonic processor had a maximum power output of 200 W with a 30 kHz excitation frequency. The electroosmotic flow and electromigration by electric power and acoustic field made the pore water and contaminant flow and migrate from the inlet (anode side) to the outlet (cathode side), and the effluent was collected. Analysis of the effluent and soil results in a metal removal efficiency that increases with time. The removal rates of Pb and Cd average

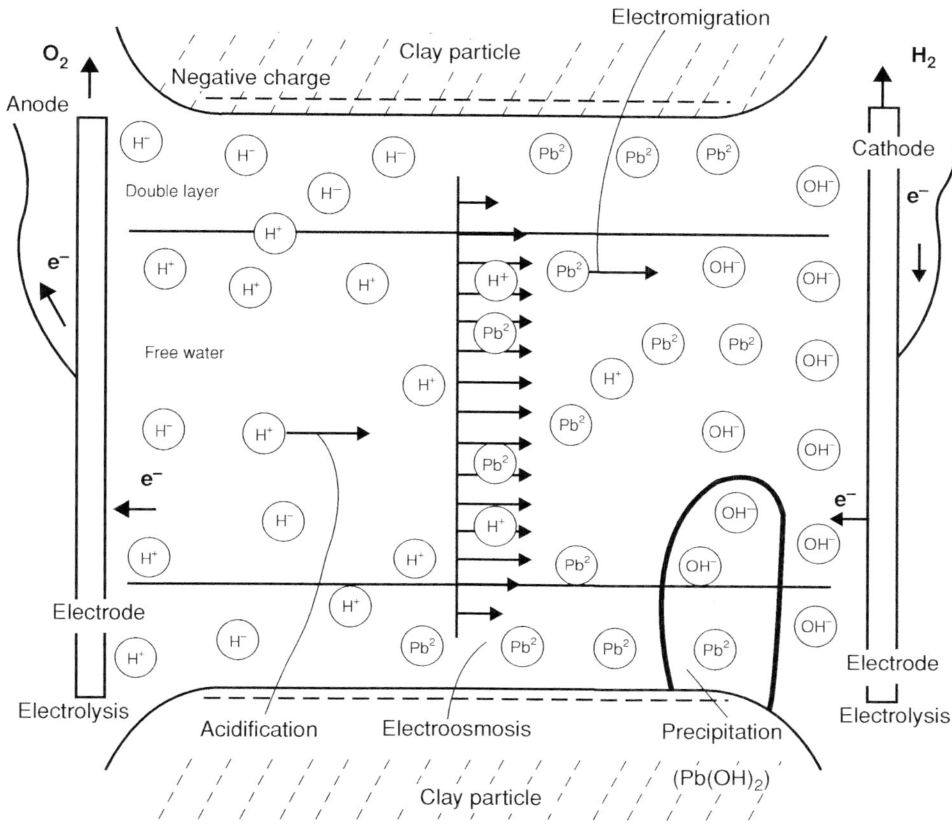

Figure 4.20 *The effects of electrokinetic phenomena on porous soil media (Reprinted with permission from [71] Copyright (1999) Elsevier Ltd).*

88 and 76% for ECT and 91 and 83% for SECT, respectively, thus the removal efficiencies are increased by about 3.4 and 9.2%. A more interesting parameter is the removal velocity of contaminant, which shows a high level in the case of SECT through the coupled action of electrokinetic and ultrasound phenomena.

When SECT was applied to remediate PAH-contaminated soils, natural clay and kaolin were used as low-permeable soils, and phenanthrene [72, 74, 75], diesel fuel [73], hexachlorobenzene [74, 75], fluoranthene [74], or chrysene [76] were used as PAH model compounds. Chung *et al.* [72, 73] carried out the experiments using graphite electrodes and a constant voltage gradient of 1.0 V cm^{-1} or 50 mA constant current. The ultrasonic processor had a maximum power output of 200 W with a 30 kHz excitation frequency. They observed that phenanthrene is migrated and transported toward the cathode zone, and removed from the soil specimen. The removal efficiency of phenanthrene by electrokinetic and ultrasonic techniques is increased by about 5.9% compared with the electrokinetic process. The removal efficiency of phenanthrene averaged 85% for electrokinetic tests, and 90% for electrokinetic and ultrasonic tests.

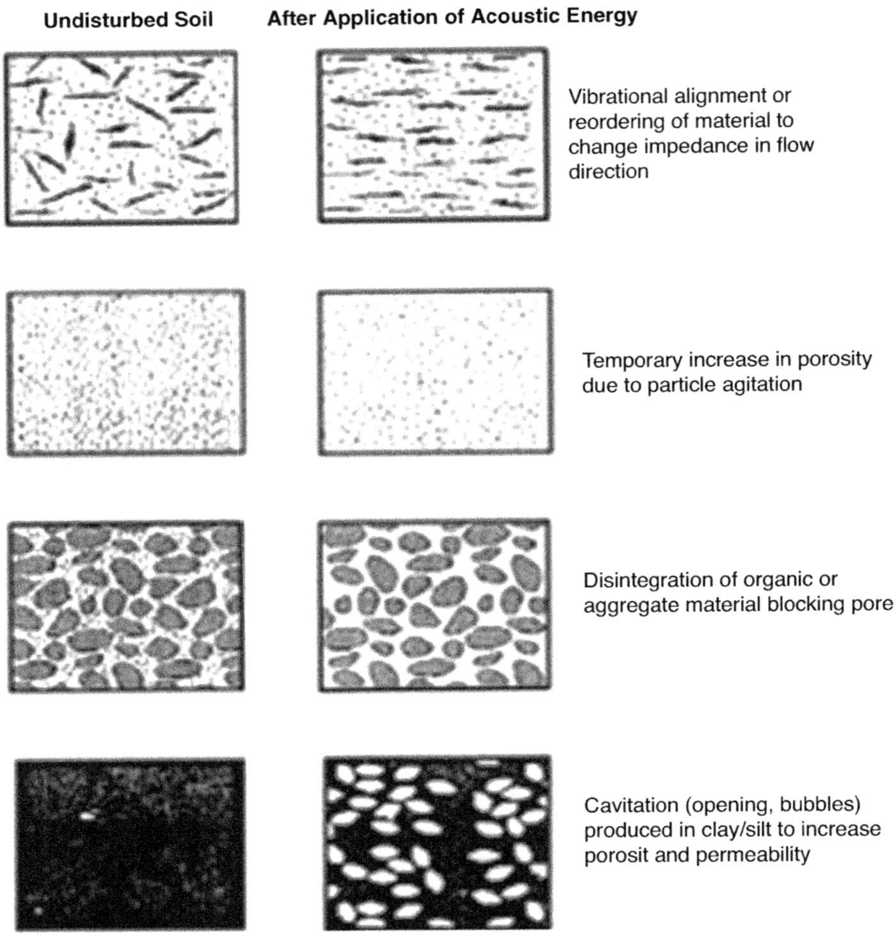

Figure 4.21 *The effects of ultrasonic phenomena on porous soil media (Reprinted with permission from [72] Copyight (2005) Elsevier Ltd).*

The use of kaolin as model clay soil in order to evaluate SECT soil remediation was studied recently by Sillanpää *et al.* [74–76]. Kaolin was artificially contaminated with phenanthrene, hexachlorobenzene, fluoranthene, or chrysene (100–500 ppm) and soaked homogeneously. The study was carried out using an electrochemical reactor with separation of compartments (three compartments separated by membranes of polypropylene) and without separation. Titanium or graphite was used as the cathode, with a titanium, graphite, or iron anode, and the voltage was kept constant at 30 V (1.5 V cm^{-1} voltage gradient). A 30 kHz and 100 W, or 20 kHz and 200 W, ultrasonic device was used for continuous or pulsed ultrasonic field generation. For the SECT a sonication period of only 1 h day^{-1} was used. SCT, ECT, and surfactant-assisted ECT were conducted and compared with the SECT treatment, and the results indicated that sonication increased the kaolin porosity and permeability, as well as increasing the desorption of these poorly soluble POPs. Along the kaolin

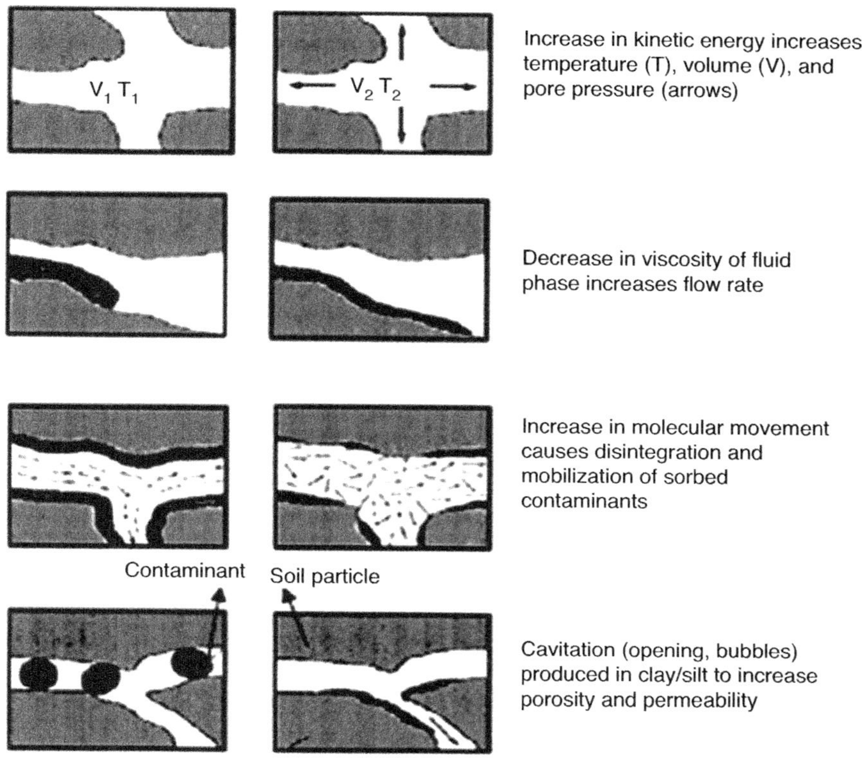

Figure 4.22 *The effects of ultrasonic phenomena on pore fluids (Reprinted with permission from [72] Copyright (2005) Elsevier Ltd).*

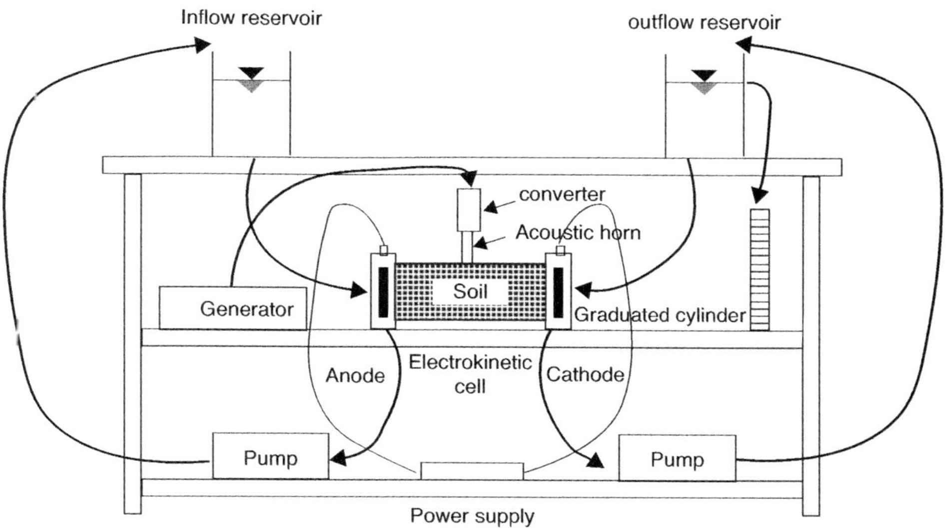

Figure 4.23 *Test setup for electrokinetic and ultrasonic experiments (Reprinted with permission from [72] Copyright (2005) Elsevier Ltd).*

profile, residual concentrations tended to be lower near the electrode zones, and to accumulate in the central part. This can be seen more clearly in the ECT test alone than in the SECT test, because of the physical water–kaolin mixing effect of ultrasound. Among the PHPs used, the concentration of hexachlorobenzene remains the highest in kaolin. Therefore, hexachlorobenzene is more difficult to remove, probably because it has a very stable chemical structure and low water solubility. When a 2-hydroxypropyl-β-cyclodextrin surfactant was used to provide surfactant assistance for the ECT, the hexachlorobenzene removal was slightly better, while SECT demonstrated better phenanthrene removal than surfactant-assisted ECT. Using SECT, an average removal efficiency higher than 80% was obtained for PAHs. Results from the experiments show that, generally, SECT tests have higher electroosmotic flow, higher current, and better performance than tests using ECT alone. However, ultrasonic enhancement can increase the removal rate by only up to about 10%.

4.6 Conclusions

It is clear that studies undertaken by the scientific community in sonoelectrochemistry applied to environmental remediation has been mainly focused on the analysis of the technical, the environmental, and the economic viability. Clear benefits have been pointed out and their explanations are normally based on the general mechanisms provided by the previous literature. Although scientists and engineers continue to study the application of sonoelectrochemistry on pollutant degradation, two main areas need to be further developed: (i) on the one hand, and despite of the inherent difficulty of the various systems studied, efforts should focus on detailed mechanisms of simple target molecules degradation in order to obtain fundamental information of sonoelectrochemical reactions and (ii) on the other hand, a strong work from a multidisciplinary point of view (chemistry, engineering, physics . . .) in order to develop optimized multipurpose sonoelectrochemical reactors which fit to the economical requirements.

List of Symbols and Abbreviations

2,4-D	2,4-dichlorophenol
2,4-DHBA	2,4-dihydroxybenzoic acid
BDD	boron-doped diamond
CFU	colony-forming unit
COD	chemical oxygen demand
ECT	electrochemical treatment
EDTA	ethylenediaminetetraacetic acid
FE-SEM	field emission scanning electron microscopy
GC	gas chromatography
H-F	high frequency
HPLC	high-performance liquid chromatography
IC	ionic chromatography
L-F	low frequency
MBSL	multibubble sonoluminescence

MS-GC	mass spectrum gas chromatography
PAHs	polyaromatic hydrocarbons
PCBs	polychlorinated biphenyls
POPs	persistent organic pollutants
RDE	rotating disk electrode
SCT	sonochemical treatment
SECT	sonoelectrochemical treatment
SEF	sonoelectro-Fenton
SF	sono-Fenton
TKN	Kjeldahl nitrogen
TOC	total organic carbon
UV-Vis	ultraviolet-visible

References

(1) González-García, J., Esclapez, M.D., Bonete, P., Vargas-Hernández, Y., Gaete-Garretón, L. and Sáez, V. (2010) Current topics on sonoelectrochemistry. *Ultrasonics*, **50**, 318–322.

(2) González-García, J., Sáez, V., Esclapez, M.D., Bonete, P., Vargas, Y. and Gaete, L. (2010) Relevant developments and new insights on Sonoelectrochemistry. *Physics Procedia*, **3**, 117–124.

(3) Martínez-Huitle, C.A. and Brillas, E. (2009) Decontamination of wastewaters containing synthetic organic dyes by electrochemical methods: A general review. *Applied Catalysis B: Environmental*, **87**, 105–145.

(4) Mason, T.J., Lorimer, J.P. and Walton, D.J. (1990) Sonoelectrochemistry. *Ultrasonics*, **28**, 333–337.

(5) Lorimer, J.P., Mason, T.J., Plattes, M. and Phull, S.S. (2000) Dye effluent decolourisation using ultrasonically assisted electro-oxidation. *Ultrasonics Sonochemistry*, **7**, 237–242.

(6) Lorimer, J.P., Mason, T.J., Plattes, M., Phull, S.S. and Walton, D.J. (2001) Degradation of dye effluent. *Pure and Applied Chemistry*, **73**, 1957–1968.

(7) Foord, J.S., Holt, K.B., Compton, R.G., Marken, F. and Kim, D.H. (2003) Mechanistic aspects of the sonoelectrochemical degradation of the reactive dye Procion Blue at boron-doped diamond electrodes. *Diamond and Related Materials*, **10**, 662–666.

(8) Ince, N.H. and Tezcanlí, G. (2001) Reactive dyestuff degradation by combined sonolysis and ozonation. *Dyes and Pigments*, **49**, 145–153.

(9) Wang, X., Jia, J. and Wang, Y. (2010) Electrochemical degradation of reactive dye in the presence of water jet cavitation. *Ultrasonics Sonochemistry*, **17**, 515–520.

(10) Rivera, M., Pazos, M. and Sanromán, M.A. (2009) Improvement of dye electrochemical treatment by combination with ultrasound technique. *Journal of Chemical Technology & Biotechnology*, **84**, 1118–1124.

(11) He, P., Wang, L., Xue, J. and Cao, Z. (2010) Electrolytic treatment of methyl orange in aqueous solution using three-dimensional electrode reactor coupling ultrasonics. *Environmental Technology*, **31**, 417–422.

(12) Ai, Z., Li, J., Zhang, L. and Lee, S. (2010) Rapid decolorization of azo dyes in aqueous solution by an ultrasound-assisted electrocatalytic oxidation process. *Ultrasonics Sonochemistry*, **17**, 370–375.

(13) Siddique, M., Farooq, R., Khan, Z.M., Khan, Z. and Shaukat, S.F. (2011) Enhanced decomposition of reactive blue 19 dye in ultrasound assisted electrochemical reactor. *Ultrasonics Sonochemistry*, **18**, 190–196.

(14) Padilla, K.R., Blanco, L.M., Orozco, S. and Jáuregui, F. (2004) Degradación oxidativa del 2,4-diclorofenol por vía sonoelectroquímica. *Ciencia UANL*, **VII**, 51–59.
(15) Sáez, V., Esclapez, M.D., Tudela, I., Bonete, P., Louisnard, O. and González-García, J. (2010) 20 kHz sonoelectrochemical degradation of perchloroethylene in sodium sulfate aqueous media: Influence of the operational variables in batch mode. *Journal of Hazardous Materials*, **183**, 648–654.
(16) Sáez, V., Esclapez, M.D., Frías-Ferrer, A., Bonete, P. and González-García, J. (2008) Electrochemical reduction of perchloroethylene in aqueous media: Influence of the electrode material. *Journal of New Materials for Electrochemical Systems*, **11**, 287–295.
(17) Sáez, V., Esclapez, M.D., Tudela, I., Bonete, P. and González-García, J. (2010) Electrochemical degradation of perchloroethylene in aqueous media: influence of the electrochemical operational variables in the viability of the process. *Industrial & Engineering Chemistry Research*, **49**, 4123–4131.
(18) Sáez, V., Esclapez Vicente, M.D., Frías-Ferrer, Á.J., Bonete, P. and González-García, J. (2009) Electrochemical degradation of perchloroethylene in aqueous media: An approach to different strategies. *Water Research*, **43**, 2169–2178.
(19) González-García, J., Sáez, V., Esclapez, M.D., Bonete, P., Walton, D.J., Rehorek, A. and Louisnard, O. (2010) Sonochemical degradation of perchloroethylene. *Physics Procedia*, **3**, 981–986.
(20) Sáez, V., Esclapez, M.D., Bonete, P., Walton, D.J., Rehorek, A., Louisnard, O. and González-García, J. (2011) Sonochemical degradation of perchloroethylene: The influence of ultrasonic variables, and the identification of products. *Ultrasonics Sonochemistry*, **18**, 104–113.
(21) Sáez, V., Frías-Ferrer, A., Iniesta, J., González-García, J., Aldaz, A. and Riera, E. (2005) Chacterization of a 20 kHz sonoreactor. Part I: Analysis of mechanical effects by classical and numerical methods. *Ultrasonics Sonochemistry*, **12**, 59–65.
(22) Sáez, V., Frías-Ferrer, A., Iniesta, J., González-García, J., Aldaz, A. and Riera, E. (2005) Characterization of a 20 kHz sonoreactor. Part II: analysis of chemical effects by classical and electrochemical methods. *Ultrasonics Sonochemistry*, **12**, 67–72.
(23) Esclapez, M.D., Sáez, V., Milán-Yáñez, D., Tudela, I., Louisnard, O. and González-García, J. (2010) Sonoelectrochemical treatment of water polluted with trichloroacetic acid: From sonovoltammetry to pre-pilot plant scale. *Ultrasonics Sonochemistry*, **17**, 1010–1020.
(24) Klíma, J., Frias-Ferrer, A., González-García, J., Ludvík, J., Sáez, V. and Iniesta, J. (2007) Optimisation of 20 kHz sonoreactor geometry on the basis of numerical simulation of local ultrasonic intensity and qualitative comparison with experimental results. *Ultrasonics Sonochemistry*, **14**, 19–28.
(25) Louisnard, O., González-García, J., Tudela, I., Klíma, J., Sáez, V. and Vargas-Hernández, Y. (2009) FEM simulation of a sono-reactor accounting for vibrations of the boundaries. *Ultrasonics Sonochemistry*, **16**, 250–259.
(26) Marken, F., Kumbhat, S., Sanders, G.H.W. and Compton, R.G. (1996) Voltammetry in the presence of ultrasound: Surface and solution processes in the sonovoltammetric reduction of nitrobenzene at glassy carbon and gold electrodes. *Journal of Electroanalytical Chemistry*, **414**, 95–105.
(27) Abramov, V.O., Abramov, O.V., Gekhman, A.E., Kuznetsov, V.M. and Price, G.J. (2006) Ultrasonic intensification of ozone and electrochemical destruction of 1,3-dinitrobenzene and 2,4-dinitrotoluene. *Ultrasonics Sonochemistry*, **13**, 303–307.
(28) Gogate, P.R. (2008) Treatment of wastewater streams containing phenolic compounds using hybrid techniques based on cavitation: A review of the current status and the way forward. *Ultrasonics Sonochemistry*, **15**, 1–15.
(29) Trabelsi, F., Ait-lyazidi, H., Ratsimba, B., Wilhelm, A.M., Delmas, H., Fabre, P.L. and Berlan, J. (1996) Oxidation of phenol in wastewater by sonoelectrochemistry. *Chemical Engineering Science*, **51**, 1857–1865.

(30) Trabelsi, F., Ait-lyazidi, H., Berlan, J., Fabre, P.L., Delmas, H. and Wilhelm, A.M. (1996) Electrochemical determination of the active zones in a high-frequency ultrasonic reactor. *Ultrasonics Sonochemistry*, **3**.

(31) Viennet, R., Ligier, V., Hihn, J.Y., Bereziat, D., Nika, P. and Doche, M.L. (2004) Visualisation and electrochemical determination of the actives zones in an ultrasonic reactor using 20 and 500 kHz frequencies. *Ultrasonics Sonochemistry*, **11**, 125–129.

(32) Mandroyan, A., Doche, M.L., Hihn, J.Y., Viennet, R., Bailly, Y. and Simonin, L. (2009) Modification of the ultrasound induced activity by the presence of an electrode in a sono-reactor working at two low frequencies (20 and 40 kHz). Part II: Mapping flow velocities by particle image velocimetry (PIV). *Ultrasonics Sonochemistry*, **16**, 97–104.

(33) Fujishima, A., Einaga, Y., Rao, T.N., Tryk, D.A., Akira, F., Yasuaki, E. and Donald, A.T. (2005) Fundamental and applied aspects of diamond electrodes, in *Diamond Electrochemistry* (eds A. Fujishima, Y. Einaga, T.N. Rao and D.A. Tryk), Elsevier Science, Amsterdam, pp. 556–570.

(34) Martin, H.B., Eaton, S.C., Landau, U., Angus, J.C., Akira, F., Yasuaki, E., Rao, T.N. and Donald, A.T. (2005) Electrochemical effects on diamond surfaces: wide potential window, reactivity, spectroscopy, doping levels and surface conductivity, in *Diamond Electrochemistry* (eds A. Fujishima, Y. Einaga, T.N. Rao, and D.A. Tryk), Elsevier Science, Amsterdam, pp. 28–34.

(35) Brillas, E., Boye, B., Sirés, I., Garrido, J.A., Rodríguez, R.M., Arias, C., Cabot, P.L. and Comninellis, C. (2004) Electrochemical destruction of chlorophenoxy herbicides by anodic oxidation and electro-Fenton using a boron-doped diamond electrode. *Electrochimica Acta*, **49**, 4487–4496.

(36) Boye, B., Brillas, E., Marselli, B., Michaud, P.A., Comninellis, C., Farnia, G. and Sandonà, G. (2006) Electrochemical incineration of chloromethylphenoxy herbicides in acid medium by anodic oxidation with boron-doped diamond electrode. *Electrochimica Acta*, **51**, 2872–2880.

(37) Zhu, X., Shi, S., Wei, J., Lv, F., Zhao, H., Kong, J., He, Q. and Ni, J. (2007) Electrochemical oxidation characteristics of p-substituted phenols using a boron-doped diamond electrode. *Environmental Science & Technology*, **41**, 6541–6546.

(38) Zhao, G., Shen, S., Li, M., Wu, M., Cao, T. and Li, D. (2008) The mechanism and kinetics of ultrasound-enhanced electrochemical oxidation of phenol on boron-doped diamond and Pt electrodes. *Chemosphere*, **73**, 1407–1413.

(39) Lima Leite, R.H., Cognet, P., Wilhelm, A.-M. and Delmas, H. (2002) Anodic oxidation of 2,4-dihydroxybenzoic acid for wastewater treatment: study of ultrasound activation. *Chemical Engineering Science*, **57**, 767–778.

(40) Zhao, G., Gao, J., Shen, S., Liu, M., Li, D., Wu, M. and Lei, Y. (2009) Ultrasound enhanced electrochemical oxidation of phenol and phthalic acid on boron-doped diamond electrode. *Journal of Hazardous Materials*, **172**, 1076–1081.

(41) Sundarapandiyan, S., Sivakumar, V., Rajeswari, G. and Saravanan, P. (2009) Effect of sonication on electro-oxidation of organic pollutants in saline wastewater. *Water Environment Research*, **81**, 319–324.

(42) Feng, C., Sugiura, N. and Maekawa, T. (2005) Performance of two new electrochemical treatment systems for wastewaters. *Journal of Environmental Science and Health, Part A: Toxic/Hazardous Substances and Environmental Engineering*, **39**, 2533–2543.

(43) Stack, L.J., Carney, P.A., Malone, H.B. and Wessels, T.K. (2005) Factors influencing the ultrasonic separation of oil-in-water emulsions. *Ultrasonics Sonochemistry*, **12**, 153–160.

(44) Abdelsalam, M.E. and Birkin, P.R. (2002) A study investigating the sonoelectrochemical degradation of an organic compound employing Fenton's reagent. *Physical Chemistry Chemical Physics*, **4**, 5340–5345.

(45) Yasman, Y., Bulatov, V., Gridin, V.V., Agur, S., Galil, N., Armon, R. and Schechter, I. (2004) A new sono-electrochemical method for enhanced detoxification of hydrophilic chloroorganic pollutants in water. *Ultrasonics Sonochemistry*, **11**, 365–372.

(46) Oturan, M.A., Sirés, I., Oturan, N., Pérocheau, S., Laborde, J.L. and Trévin, S. (2008) Sonoelectro-Fenton process: A novel hybrid technique for the destruction of organic pollutants in water. *Journal of Electroanalytical Chemistry*, **624**, 329–332.

(47) Yasman, Y., Bulatov, V., Rabin, I., Binetti, M. and Schechter, I. (2006) Enhanced electro-catalytic degradation of chloroorganic compounds in the presence of ultrasound. *Ultrasonics Sonochemistry*, **13**, 271–277.

(48) Zhang, Z., Yuan, Y., Liang, L., Fang, Y., Cheng, Y., Ding, H., Shi, G. and Jin, L. (2008) Sonophotoelectrocatalytic degradation of azo dye on TiO2 nanotube electrode. *Ultrasonics Sonochemistry*, **15**, 370–375.

(49) González-García, J. (2011) Sonoelectrochemical synthesis of materials, in *Theoretical and Experimental Sonochemistry Involving Inorganic Systems* (eds Pankaj and M. Ashokkumar), Springer, Berlin/Heidelberg.

(50) Walker, R. (1990) Ultrasound and electroplating. *Chemistry in Britain*, **26**, 251–254.

(51) Lorimer, P. and Mason, T.J. (1999) The applications of ultrasound in electroplating. *Electrochemistry*, **67**, 924–930.

(52) Walker, R. (1997) Ultrasound improves electrolytic recovery of metals. *Ultrasonics Sonochemistry*, **4**, 39–43.

(53) Farooq, R., Wang, Y., Lin, F., Shaukat, S.F., Donaldson, J. and Chouhdary, A.J. (2002) Effect of ultrasound on the removal of copper from the model solutions for copper electrolysis process. *Water Research*, **36**, 3165–3169.

(54) Yaqub, A., Ajab, H., Khan, S., Khan, S. and Farooq, R. (2009) Electrochemical removal of copper and lead from industrial wastewater: Mass transport enhancement. *Water Quality Research Journal of Canada*, **44**, 183–188.

(55) Hyde, M.E. and Compton, R.G. (2002) How ultrasound influences the electrodeposition of metals. *Journal of Electroanalytical Chemistry*, **531**, 19–24.

(56) Pollet, B., Lorimer, J.P., Phull, S.S., Mason, T.J., Walton, D.J., Hihn, J.Y., Ligier, V. and Wéry, M. (1999) The effect of ultrasonic frequency and intensity upon electrode kinetic parameters for the $Ag(S_2O_3)_{2^{3-}}$/Ag redox couple. *Journal of Applied Electrochemistry*, **29**, 1359–1366.

(57) Allen, H.E. and Chen, P.H. (1993) Remediation of metal contaminated soil by EDTA incorporating electrochemical recovery of metal and EDTA. *Environmental Progress*, **12**, 284–293.

(58) Chang, J.H., Ellis, A.V., Yan, C.T. and Tung, C.H. (2009) The electrochemical phenomena and kinetics of EDTA-copper wastewater reclamation by electrodeposition and ultrasound. *Separation and Purification Technology*, **68**, 216–221.

(59) Lorimer, J.P., Pollet, B., Phull, S.S., Mason, T.J. and Walton, D.J. (1998) The effect upon limiting currents and potentials of coupling a rotating disc and cylindrical electrode with ultrasound. *Electrochimica Acta*, **43**, 449–455.

(60) Pollet, B., Lorimer, J.P., Phull, S.S. and Hihn, J.Y. (2000) Sonoelectrochemical recovery of silver from photographic processing solutions. *Ultrasonics Sonochemistry*, **7**, 69–76.

(61) Pollet, B.G., Lorimer, J.P., Phull, S.S., Mason, T.J. and Hihn, J.-Y. (2003) A novel angular geometry for the sonochemical silver recovery process at cylinder electrodes. *Ultrasonics Sonochemistry*, **10**, 217–222.

(62) Pollet, B., Lorimer, J.P., Hihn, J.Y., Touyeras, F., Mason, T.J. and Walton, D.J. (2005) Electrochemical study of silver thiosulphate reduction in the absence and presence of ultrasound. *Ultrasonics Sonochemistry*, **12**, 7–11.

(63) Compton, R.G., Eklund, J.C., Marken, F. and Waller, D.N. (1996) Electrode processes at the surfaces of sonotrodes. *Electrochimica Acta*, **41**, 315–320.

(64) Pollet, B.G., Lorimer, J.P., Hihn, J.-Y., Phull, S.S., Mason, T.J. and Walton, D.J. (2002) The effect of ultrasound upon the oxidation of thiosulphate on stainless steel and platinum electrodes. *Ultrasonics Sonochemistry*, **9**, 267–274.

(65) Perret, A., Haenni, W., Skinner, N., Tang, X.M., Gandini, D., Comninellis, C., Correa, B. and Foti, G. (1999) Electrochemical behavior of synthetic diamond thin film electrodes. *Diamond and Related Materials*, **8**, 820–823.

(66) Iordache, I., Nechita, M.T., Rosca, I. and Aelenei, N. (2004) Ultrasound assisted electrochemical degradation of cyanides: Influence of electrode type. *Turkiss Journal of Engineering and Environmental Sciences*, **28**, 377–380.

(67) Kraft, A., Blaschke, M. and Kreysig, D. (2002) Electrochemical water disinfection Part III: Hypochlorite production from potable water with ultrasound assisted cathode cleaning. *Journal of Applied Electrochemistry*, **32**, 597–601.

(68) Joyce, E., Mason, T.J., Phull, S.S. and Lorimer, J.P. (2003) The development and evaluation of electrolysis in conjunction with power ultrasound for the disinfection of bacterial suspensions. *Ultrasonics Sonochemistry*, **10**, 231–234.

(69) Joyce, E.M., Mason, T.J. and Lorimer, J.P. (2006) Application of UV radiation or electrochemistry in conjunction with power ultrasound for the disinfection of water. *International Journal of Environment and Pollution*, **27**, 222–230.

(70) Krauss, M., Wilcke, W. and Zech, W. (2000) Availability of polycyclic aromatic hydrocarbons (PAHs) and polychlorinated biphenyls (PCBs) to earthworms in urban soils. *Environmental Science & Technology*, **34**, 4335–4340.

(71) Chung, H.I. and Kang, B.H. (1999) Lead removal from contaminated marine clay by electrokinetic soil decontamination. *Engineering Geology*, **53**, 139–150.

(72) Chung, H.I. and Kamon, M. (2005) Ultrasonically enhanced electrokinetic remediation for removal of Pb and phenanthrene in contaminated soils. *Engineering Geology*, **77**, 233–242.

(73) Chung, H.I. (2007) Treatment of contaminated groundwater in sandy layer under river bank by electrokinetic and ultrasonic technology. *Water Science & Technology*, **55**, 329–338.

(74) Pham, T.D., Shrestha, R.A., Virkutyte, J. and Sillanpää, M. (2009) Combined ultrasonication and electrokinetic remediation for persistent organic removal from contaminated kaolin. *Electrochimica Acta*, **54**, 1403–1407.

(75) Pham, T.D., Shrestha, R.A. and Sillanpää, M. (2010) Removal of hexachlorobenzene and phenanthrene from clayey soil by surfactant- and ultrasound-assisted electrokinetics. *Journal of Environmental Engineering*, **136**, 739–742.

(76) Shrestha, R., Pham, T. and Sillanpää, M. (2010) Electro ultrasonic remediation of polycyclic aromatic hydrocarbons from contaminated soil. *Journal of Applied Electrochemistry*, **40**, 1407–1413.

5
Organic Sonoelectrosynthesis

David J. Walton

5.1 Introduction

Electrochemical techniques have a respected tradition in chemical synthesis, and the electrooxidation of aqueous acetate (ethanoate) solutions to give ethane as a key product is one of the oldest known organic reactions: the Kolbé reaction [1]. Expanding the range of chemical systems amenable to the methodology awaited the development and greater availability of high-dielectric organic solvents in the mid-twentieth century, with organic-soluble electrolyte salts to allow the necessary current densities, as well as the development of electrode materials suitable for scaling-up to provide sufficient quantities of products for synthetic purposes [2–6].

However, it is fair to say that the exploitation of electroorganic synthesis has rather lagged behind expectations [3], and it is the aim of workers in the field to explore methodologies that may help to redress this situation. In particular the use of an additional perturbation to modify electrochemical phenomena has attracted attention, and is one explanation for the increasing interest in sonoelectroorganic synthesis.

In other chapters of this book the key effects of insonation upon various electrochemical applications have been described. In particular the vast increase in hydrodynamics and mass transport is of importance in synthesis. This is because, as shown below, simple electron transfers to neutral organic molecules can lead directly to ion-radicals, which may then undergo chemical reactions via the charged sites, or else react via unpaired electrons, giving immediate opportunity for mechanistic distinction. The early-formed intermediates may then undergo subsequent electron transfers and then further chemical steps may occur. This leads to a range of possible products, discriminated by kinetic regimes that are greatly perturbed by insonation, and hence ultrasound can substantially modify the products, yield ratios and current efficiencies in electroorganic synthetic systems.

Power Ultrasound in Electrochemistry: From Versatile Laboratory Tool to Engineering Solution, First Edition. Edited by Bruno G. Pollet.

Depending upon system parameters, the following benefits of ultrasound, which have been elucidated in other chapters of this book, may variously be observed in electrosynthesis [8–11]:

(i) enhancement of mass transport phenomena
(ii) alteration of adsorption and surface effects at the electrodes
(iii) diminution of diffusion layers near the electrode, allowing higher current densities, lessened cell power requirements and reduced concentrations of necessary electrolyte salts
(iv) reduction of electrode fouling and removal of electrode coatings
(v) modification of electro-deposited materials
(vi) induction of new reactions by sonochemical enhancement of follow-up reactions after electron transfer
(vii) sonochemical generation of electro-active species in systems that are electro-inactive when silent
(viii) greater general reproducibility towards new and/or unusual products
(ix) effects of ultrasonic mixing in multi-phase systems, for example to produce organic substrates in aqueous emulsions to replace expensive and/or toxic organic solvents
(x) ultrasonic degassing influences gas-evolving processes
(xi) effects upon the production and applications of nanomaterials.

5.2 Scale-Up Considerations

In respect of other chapters in this book, comparisons between electroanalytical and electrosynthetic systems are not always easy to draw, because of the different conditions employed. For synthesis electrolytes may be concentrated, with high current densities used, and precipitation or other phase-separation phenomena may take place. Since cavitation and other effects of coupling ultrasonic irradiation into a medium are dependent upon parameters such as viscosity, density, presence of dissolved gases or solid particles and so on, and since larger-area electrodes are employed with dimensions greater than the wavelength of sound then the effects of insonation on a preparative electrolysis may not be easy to elucidate, and rigorous sonoelectrochemical analysis of the system becomes complicated. However, it is clear that for any individual electroorganic synthetic system, the use of ultrasound provides an extra parameter to provide a perturbation that can give important mechanistic information.

For example, the relatively reversible and mechanistically simplistic evolution of hydrogen and chlorine, well-behaved in analytical studies at platinised electrodes, are both beset (chlorine evolution most severely) in large-scale chlor-alkali cells by adherence of gas bubbles to the electrode surface. This problem can be ameliorated by ultrasound [12], and furthermore the parasitic irreversible oxygen evolution reaction, problematic in the commercial application, is differently affected by ultrasound, offering a practical benefit for mechanistic discrimination by use of sonoelectrochemistry. Overall, the combination of the usual electrochemical parameters with the additional ultrasonic parameters of power, frequency, sound-source, and the geometric and physical properties of the system provides a rich vein for manipulation by the experimentalist.

5.3 Early History of Organic Sonoelectrochemistry

Truly pioneering experiments in many fields are often found to have taken place even longer ago than might be expected, but a benchmark widely quoted report concerns the effects of ultrasound upon water electrolysis by Moriguchi in 1934 [13]. Later a review of acoustoelectrokinetics and similar physicochemical electroanalytical phenomena by Yeager in 1953 contained 105 references [14], but there had been almost no references on electrosynthetic systems until the early 1980s. Xanthate electrooxidation is used to assist froth flotation in the mineral-ore industry [8], which might not necessarily be considered as electrosynthesis. The mechanisms are complex, and involve dixanthogen and other species. A series of papers by Chanturiya, Malakhov *et al.* investigated the effects of insonation. Among other effects, ultrasound modified the product mixture, favouring dixanthogen, and gave a more uniform emulsion, with 80% of droplets below 5 μm diameter, improving the recovery of galena, sphalerite and chalcopyrite, as shown in Equation (5.1) [15, 16]:

$$\begin{aligned} \text{RO-C(=S)S}^- \rightarrow \text{e-/H}_2\text{O} \rightarrow\ & \text{ROC(=S)-S-S-(S=)COR} \\ & + \text{ROC(=O)-S-S-(O=)COR} \end{aligned} \tag{5.1}$$

A study of butylxanthate sonoelectrooxidation showed butylmonothiocarbonate disulfide and bis(butylxanthogen) among the products [16], and a more recent account of this system shows that similar sonoelectrochemical effects pertain from insonation at 20 kHz (ultrasound) and 50 Hz (audible sound) [17]. Insonation powers are not easy to compare from the report, but this result has great interest regarding the significance of cavitational phenomena in the system, and underlines the importance of ultrasonic system parameters in mechanistic elucidation. It is worth mentioning here that the effects of ultrasonic frequency in electrosynthetic systems remain a matter for debate. In the work in Reference [17] there were parallels between 20 kHz ultrasound and infrasound (audible sound). Described elsewhere in this chapter are several systems where comparison between higher ultrasonic frequencies (often around 800 kHz) and the more commonly used 20–40 kHz frequency region shows that similar effects seem to pertain, certainly in regard to switches in product distributions (which are most important to synthetic chemists). However, effects upon electrode surface erosion, electrodeposited materials, electrodissolution and similar phenomena are more affected by frequency, as to be expected.

Xanthate electrooxidations show benefits of insonation in a multiphase system. Another early set of studies in multiphase systems concerned electrolytic dehalogenation for the destruction of environmental pollutants [18]. Simple halobiphenyls were reduced at mercury-pool cathodes, presumably at low effective ultrasonic powers from immersion in a cleaning bath, while in more forcing conditions DDT itself [1,1-bis(4-chlorophenyl)-2,2,2-trichloroethane] was degraded at carbon-felt cathodes using insonation from a horn source. Both aromatic and aliphatic chlorine atoms were removed [19]. Rusling *et al.* also used cobalt bipyridines, various phthalocyanines and other species as redox mediators [19, 20]. In these studies surfactants such as didodecyldimethyl ammonium bromide (DDAB) were present, and the field of electrocatalysis in microemulsions, micelles and systems of controlled microstructure has been reviewed by these workers [21–23].

The range of pollutants extends from a wide range of halo-compounds, phenols and aromatics, and particularly azo- and other textile-dyes. The sonoelectrochemical degradation of pollutants is the subject of a separate chapter of this book, and though not strictly organic electrosynthetic systems, these do involve the effects of ultrasound upon electroorganic mechanisms. However, the organic component is present in very low concentrations, and identification of all products remains a problematic aspect. For example, dyes such as Orange II, Sandolan Yellow and Indigo Carmine are amenable to degradation in silent conditions by the electrogeneration of hypochlorite from chloride-containing media (see Chapter 4). Not only is this notably enhanced by ultrasound, but interestingly now there need be no chloride present and instead degradation occurs by sonoelectrochemical reaction in the presence of either nitrate or sulfate ions [24]. The advantage of this is that chloride might not be appreciably present in say a polluted agricultural water stream, but nitrate may well be there. The likely mechanism involves electrogeneration of peroxy species, but in this work the degree of dye degradation was measured by simple loss of colouration in the analyte, without product identification.

There is an issue that pollutant degradation could lead to the formation of still-toxic species, and a recent comparative study of perchloroethene degradation by electrolysis, ultrasound alone, and sonoelectrolysis (discussed elsewhere in this book), shows not only a surprising range of different product mixtures and reaction efficiencies, but also addresses a known mass loss in this system [25]. It transpires that the missing mass is due to the formation of chloroacetate species, themselves toxic, so here pollutant degradation by these routes might have alternative environmental consequences.

There have been numerous recent studies on the degradation of dyes, phenols and other organic pollutants, including the use of modern electrode materials such as boron-doped diamond, with recent reviews and the reader is referred to Chapter 4 of this book.

Recent developments concern the use of 'Advanced Fenton' methodology, using ultrasound [26]. A strategy used by Bremner *et al.* is that the ultrasound need not be produced by an electronic transducer, as is most often the case in laboratory sonoelectrochemical studies. Instead the solution can be pumped through the orifice of a 'whistle reactor', which also produces cavitation [27] (the condiment 'salad cream' is made as a stable emulsion on a large commercial scale by this method). Although the above applications fall better under the heading of 'pollutant degradation' there are clear overlaps with synthetic studies in methodologies for systems that operate by sonoelectrogeneration of hydrogen peroxide, or another powerful oxidant. These can be used on a production scale, and the use of electrochemistry is environmentally friendly since it obviates the need for toxic metal oxidants, or similar. Thus Compton *et al.* have sonoelectrochemically generated hydrogen peroxide for the 'green' oxidation of benzonitrile to benzamide, appropriate for a synthetic scale [28].

5.4 Electroorganic Syntheses

We now turn to systems that are designed for organic electrosynthesis.

5.4.1 Electroreductions

An early report concerned improvement of a persistently problematic system in electrosynthesis, the electroreduction of a nitroarene to the amino derivative. This is a complicated system with competing reactions such as hydrodimerisation to hydrazo compounds [2].

Gigi *et al.* performed the sonoelectroreduction of 4-nitrophenol to 4-aminophenol on a 50-g scale in 30% aqueous H_2SO_4 containing 0.5% $SnCl_2$ and 0.01% of a polyethyleneglycol-based surfactant, under galvanostatic control at 50 mA cm^{-2} to give c. 70% yield [29].

Ultrasound is applied at 2.375 kW power (full cell dimensions are not given, so precluding assessment of power density), electrolysis is continued for 16–30 hours and the cell temperature rises to 80–90 °C. A simple work-up yields some 70% yield of product. The analyte in the divided cell contains only a more dilute 8% solution of H_2SO_4. Despite the multipathway nature of nitroelectroreductions, insonation gives a preponderance of one major product.

Figure 5.1 shows a series of feasible reaction pathways from a neutral (i.e. uncharged) organic substrate starting material, in this case the reduction of a carbonyl compound. The commonest first chemical follow-up reaction once electron addition has produced a radical anion is protonation, if labile protons are present in the system. Protonation of the radical anion leads to an uncharged protonated radical species that may dimerise in an overall one-electron per molecule route from the starting material (this dimerisation step is the origin of the dimeric hydrazo species often found in electroreductions of aromatic nitro compounds). The electrosynthetic mechanistic shorthand *EC* is used to show one-electron transfer followed by (any number of) chemical step(s). Hydrodimers are commonly found

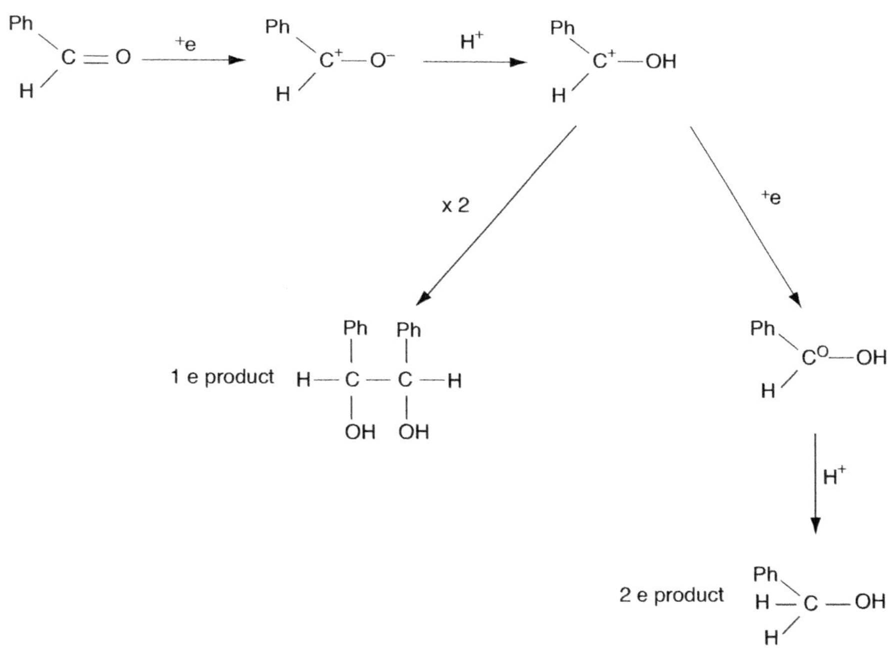

Figure 5.1 *Representative electroreduction pathways for a neutral reactant molecule in the presence of a proton donor, giving one-electron per molecule versus two-electron opportunities, here exemplified for the carbonyl compound benzaldehyde (NB: alternative cleavage reaction mechanisms are not shown).*

in electroreductions of carbonyl compounds, and also of activated alkenes, such as the commercially important monomer adiponitrile used in the manufacture of nylon-6,6, made by electrohydrodimerisation of acrylonitrile [7].

However, the protonated radical may undergo a second electron transfer to give an anion that can react with another proton or any other electrophile. This gives a two-electron per molecule *ECE* mechanism. This may be a desired reaction (e.g. to produce an alcohol from two-electron carbonyl electroreduction) or give an undesired side-product (e.g. propionitrile contaminant in adiponitrile production from acrylonitrile). There are other feasible reactions, for example if there is a facile anionic leaving group from the first-formed radical anion then loss of X^- could give a different uncharged fragment radical, which would dimerise to give a different one-electron *EC* product (not shown here). Such a radical could undergo a second electron addition to give a neutral anion that can react with a proton or another electrophile to give a different two-electron *ECE* product. This shows some of the varied inherent opportunities available in apparently simple electroreductions, and control of conditions to maximise any desired product is a key aim of workers in the field.

Discrimination between the various reaction mechanisms in Figure 5.1 depends of course upon the chemical identities of the reactants, and also on the solvents, electrolytes, other system components and reaction parameters chosen. What has attracted attention is the well-described effect of ultrasound upon hydrodynamics, which can clearly modify the build-up of key intermediates and so ought to modify the reaction mechanisms, all other parameters remaining the same in any system.

A consistent body of work in this field was performed by Nonaka and Atobe *et al.* at the Tokyo Institute of Technology. Their series of papers 'Ultrasonic Effects in Electroorganic Processes' had reached Part 24 in by the year 2003 [30], and output has continued from the laboratory. In early studies they first examined a range of electroreductions that offered products from both one-electron or two-electron pathways. Insonation switched to the lower electron-value route for carbonyl electroreductions (i.e. pinacol hydrodimers from benzaldehyde and derivatives rather than the simple alcohol products). Thus the electroreduction of benzaldehyde can lead to either the hydrodimer in a one-electron substrate molecule process, or to the benzyl alcohol in a two-electron process, as indicated in Figure 5.1.

Using a lead cathode, in dilute methanolic sulfuric acid, at constant current of 20 mA cm^{-2}, the benzyl alcohol was the major product from an unstirred solution, while mechanical stirring reversed the position to favour the hydrodimer, but ultrasonic irradiation from a cleaning bath (100 W, 36 kHz) so strongly favoured the hydrodimer that the alcohol was barely evident [31]. The effect varied somewhat with the position of the cell in the ultrasonic bath, and increased in magnitude with ultrasonic power; and throughout all electrolyses there was no appreciable change in the stereochemistry (dl/meso-ratio) of the benzoin hydrodimer.

For benzaldehyde, there was always some 30% yield of unidentified products, whether the reaction was sonicated or not. These side-products are thought to be hydroquinoid dimers; and *p*-methylbenzaldehyde, which also shows the substantial sonoelectrochemical switch to the hydrodimer, has a blocked para-position and is less affected by side-reactions. Similar effects were obtained by ultrasonic irradiation from a horn probe (20 kHz); and by employing a range of different silent and sonicated conditions (powers, sources and cell geometries) and comparing product ratios and limiting current densities from voltammetric data, the authors then used their previously reported method for estimation of mass-transport coefficients [32] to conclude that ultrasound was acting simply by agitation and not by cavitation. They further pursue the thesis that simple enhancement of mass transport will produce the same product switch in a more recent paper [33], but also describe work in which an increase in bulk reaction temperature from 0 to 60 °C produces a similar shift in products [34]. However, here the authors do not discuss parallels between ultrasonic and thermal effects.

These authors also noted still significant switches towards the one-electron products in other sonoelectrochemical reductions [35] including dimethylmaleate at a lead cathode in an aqueous mixed phosphate buffer, and benzyl bromide at a lead cathode in methanolic tetraethylammonium bromide solution. In this last case the ratio of one-electron to two-electron products was 0.3 when mechanically stirred, and was 0.9 with insonation from a horn source.

The reduction of benzoic acid at a lead cathode in aqueous/sulfuric/citric acids does not in any case give a one-electron hydrodimer, but instead yields the two-electron products benzaldehyde and a four-electron product benzyl alcohol. Here ultrasound produces some switch from four-electron towards the two-electron products; in all the cases that they studied the authors found that ultrasound favoured the process involving the smaller number of electrons per starting molecule. However, this is not the case in other reactions (see below) and it is clear that the nature of the electrochemical system is an important consideration before generalising about sonoelectrochemical phenomena in electroorganic synthetic systems.

The exact practical procedure is of importance. So, for example, the Tokyo group have used the titanium alloy tip of a sonic horn as a sonoelectrode for electroreductions of various benzaldehyde derivatives, after insulating the metal of the submerged part of the horn barrel with heat-shrink plastic [30]. They also found an improvement in current efficiency with insonation, but in addition noted some change in product selectivity towards one electron per molecule products. However, when both benzaldehyde and benzoquinone were electroreduced by a group in Brussels using the titanium tip of a 20 kHz sonic horn as the electrode ('sonoelectrode') no significant switch of mechanism from two-electron to one electron per molecule was observed, compared to a silent system with the same electrode [36]. However, titanium is not the best electrode material for electrosynthesis in view of its poor conductivity and its reactivity, which can lead to surface coatings, depending on the history of the surface. Thus the voltammetry of benzaldehyde given in this paper shows broad peaks substantially shifted from the values found at platinum, which complicates interpretation of the sonoelectrochemical results. These authors did, however, observe an improvement in product yield under ultrasound, which is a commonly noted benefit in many different systems.

Studies at Coventry also showed the same switch from a two-electron to a one-electron pathway in carbonyl electroreduction (also for *p*-substituents methyl-, methoxy- and cyano-). Figure 5.2 shows raw GC-MS data for a 'silent' but stirred reaction (top trace), compared to (bottom trace) an insonated system (38 kHz and 800 kHz give very similar results). Also studied was the effect of insonating the electrode separately from applying the electrolysis potential (middle trace). This showed that no electrode 'pretreatment' was occurring that might modify a silent electrolysis. Benzaldehyde reduction at 38 kHz and 800 kHz gives almost entirely the pinacol, but a clear benefit of the higher frequency is

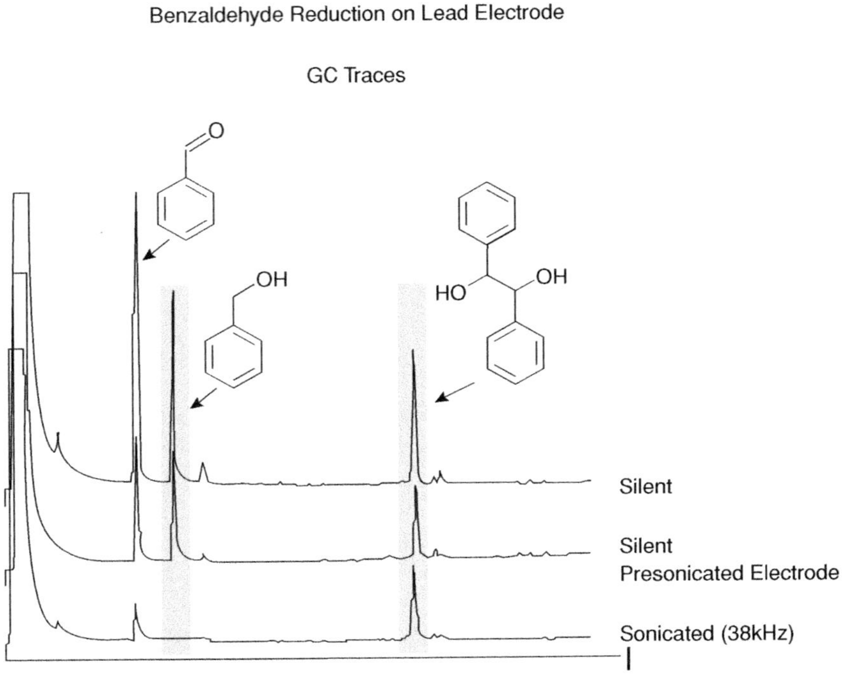

Figure 5.2 *Experimental GC analysis data from electroreductions of benzaldehyde at a lead electrode in the presence of a proton donor (reactions not run to completion). Triple comparison between 'silent' but stirred solution (i.e. 'normal' conditions), then with an electrode that was exposed to ultrasound without an applied potential ('pre-sonicated') before a 'silent' electrolysis stirred without ultrasound, then with a system insonated throughout electrolysis (results using 38 and 800 kHz baths very similar).*

that there is little weight loss of the lead cathode. This soft metal is significantly abraded away by insonation in the 20–40 kHz range but remains substantially intact at the higher frequency, an important procedural advantage.

The electroreduction of methyl halides at a reactive tin cathode has also been examined by the group at Tokyo [37]. This is a different type of electrochemical system in which the cathode is a reactive metal and a surface intermediate is formed on the electrode which reacts to give either a three-electron dimeric distannane or a four-electron tetramethyl stannane, as shown in Equation (5.2):

$$CH_3I\,(+e^-\text{at Sn Cathode}) \rightarrow (CH_3)_3Sn\text{-}Sn(CH_3)_3 + (CH_3)_4Sn \qquad (5.2)$$

For reduction of CH_3I in DMF/Bu_4NClO_4 at room temperature galvanostatically at 10 mA cm^{-2}, an unstirred reaction gave a 1 : 10 ratio of di-stannane to stannane, slightly improved by mechanical stirring, while ultrasound from a 20 kHz probe increased the dimer to give a 1 : 5 ratio. A similar trend was observed for methyl bromide in the same solvent system, and also in an aqueous $MeCN/Et_4NBr$ solvent system, but here there was also a small amount of a solid precipitate formed. Interestingly the amount of this precipitate was significantly increased under ultrasound, but since this material was not characterised, the role of ultrasound in its formation could not be stated.

In their work on carbonyl electroreductions these authors quote enhanced mass transfer across the electrode interface as the origin of the sonoelectrochemical trend towards products from the lesser amount of electrons per substrate molecule, but here the involvement of surface species at the reactive electrode provides a complication.

The group at Wesleyan University, Middletown, Connecticut, have examined reductive silylation of geminal dihalides [38]. Here ultrasound was originally introduced only because the design of cell was too constricted to be stirred magnetically [38]. The reaction employed a stainless steel cathode, a sacrificial magnesium anode and occurred in two steps with stereochemical implications, though there are significant competitive pathways [Equation (5.3)]:

$$R_1R_2CCl_2 + e^-/Me_3SiCl \rightarrow R_1R_2C(Cl)SiMe_3 + e^-/Me_3SiCl \rightarrow R_1R_2C(SiMe_3)_2 \quad (5.3)$$

Ultrasound substantially enhanced current efficiency and stereoselectivity over silent electrolyses, but results were somewhat erratic, attributed to the need for care in positioning the electrolysis cell in the Branson ultrasonic cleaning bath. Again the observation refers to the need to address geometric considerations when dealing with sound waves. The situation became complicated when a divided cell was employed. Although less suitable for effective preparative electrolyses because of resistance losses, this control experiment yielded an unexpected reduced product, an α-halosilane, from the anode (oxidising) compartment containing the sacrificial magnesium electrode. This suggested the involvement of a sonochemically induced reaction, without electrolysis; and therefore complicated the interpretation of the sonoelectrochemical events in the normal undivided cell.

The possibility that ultrasound can be involved chemically in an intended electrochemical system is not without precedent. In earlier work at the Weslyan University involving the electroreduction of α,α'-dibromo ketones at a mercury cathode, ultrasound was also employed just for stirring [39, 40]. It was then realised that although the dihaloketone was stable to mercury (without electrolysis) over weeks in silent conditions, the ultrasonic irradiation, which tended to produce a range of finely divided mercury droplets above the pool, induced sonochemical reduction on the metal [41]. The authors later deduced experimental conditions using either electrochemistry or chemical reaction on ultrasonically dispersed mercury that could select from the range of possible products [42].

There have been recent developments that could be useful in these and other electroreductions since a procedure has been reported for producing ultrasonically stable mercury cathodes, as thin film coatings onto platinum and other solid electrode materials [43].

5.4.2 Organochalcogenides

One of the first series of reports about ultrasonically enhanced electrosynthesis came in the mid 1980s from a French group, who used the technique to explore routes to organoselenium and organotellurium derivatives. Thus electroreductive synthesis of Se_2^{2-} and Se^{2-} di-anions was enhanced by simultaneous irradiation from an ultrasonic cleaning bath [44]. Instead of employing a sacrificial cathode of elemental selenium, this procedure allowed the direct use of selenium powder with carbon cloth as cathode. A further benefit was that this method also allowed production of the corresponding tellurium anions. These species could be employed *in situ* in aprotic solvents such as DMF, THF and MeCN for the synthesis of selenides and tellurides by nucleophilic displacement from haloalkanes [Equation (5.4)].

$$E + e^- \rightarrow E^{2-},\ E_2^{2-} + RX \rightarrow \text{R-}E\text{-R} + \text{R-}E\text{-}E\text{-R} \qquad (5.4)$$

(E = Se, Te; X = Cl, Br)

This system was extended to the electroreductive synthesis of a cyclopentadienyl titanium pentaselenide $[(C_5H_4R)_2TiSe_5]$ from selenium powder and $[(C_5H_4R)_2TiCl_2]$ and when R was Me the optimised process gave a 70% yield [45].

The same group also addressed the formation of unsymmetrical (di)aryl chalcogenides, by electroreduction of the symmetrical diaryl chalcogenide (e.g. PhSeSePh) at a carbon cloth cathode to give for example the $PhSe^-$ anion. Bromobenzonitrile was then added and electroreduction under ultrasound was continued, leading to 57% yield of $C_6H_5SeC_6H_4CN$ and a 42% yield of $C_6H_5TeC_6H_4CN$ with some of the corresponding symmetrical functionalised dichalcogenide as side product [46]. This SRN_1 mechanism was further exploited to produce a number of unsymmetrical phenylselenobenzophenones and their tellurium analogues [47], again by a two-step electroreductive process, both steps being performed under ultrasound. In the first procedure, the potential was made more negative after formation of the $PhSe^-$ anion at which point the haloketone was then added. Thus $PhCOC_6H_4SePh$ could be made in 49% yield, but production of the tellurium analogue was less efficient, giving only 17% isolated, with symmetrical species such as $PhCOC_6H_4TeC_6H_4COPh$ as by-products. These are complex reaction systems with a number of side reactions, and it was found that addition of weakly acidic species such as malononitrile and fluorene suppressed competitive solvent deprotonation and side reaction pathways so initiated. In a further sophistication a redox mediator such as azobenzene was added. Now the cathodic potential in the second step of the procedure could be substantially reduced, until only the mediator remained electroactive. Thus the electroreductive generation of RSe^- or RTe^- comprised the first step; then the haloketone, the acidic species such as fluorene and the azobenzene redox mediator were all added and the potential reduced for the second step. By this means isolated yields of 86% $PhCOC_6H_4SePh$ (before crystallisation) and 45% for the tellurium analogue were found. The role of ultrasound in such complex systems is not easy to ascertain, and interestingly in this paper the isolated yield for the tellurium derivative under mechanical stirring is given as 44% showing little difference from the sonoelectrochemical yield.

These reactions are of course different from those in Reference [40] where activation of the solid elemental chalcogenide is involved and where particle-size breakdown and increased mass transport of solids to the electrode surface will themselves be useful benefits of ultrasonic irradiation.

The French group also reported the anodic cleavage of REER or RER (where E = chalcogenide) in MeCN, in contrast to the previous electroreductive systems, and a number of products were prepared where RE^+ had been trapped by appropriate nucleophiles [48].

In respect of organosilicon and organogermanium, Japanese effort in this area involved the production of organosilanes and organogermanes, as small molecules or polymeric systems, by electroreduction of the appropriate halo species at a reactive metal cathode in aprotic media. The same metal, Mg, Al or sometimes Cu, is used as a sacrificial anode, and the cell is undivided. Thus with lithium perchlorate as preferred electrolyte salt in

THF solvent, a dichlorosilane such as $PhMeSiCl_2$ gives a polysilane of $M_n \sim 3000$ in 22% yield [49]. This is in contrast to earlier work at Hg cathodes in divided cells, where Si-O containing polymers and cyclotetrasilanes were obtained by other workers [50, 51]. Simultaneous ultrasonic irradiation at 47 kHz in the Japanese procedure increased the yield of the quoted polysilane to 33%, although without obvious or consistent effect on $\overline{M}_n$ or polydispersivity in the given systems.

The Japanese work is the subject of a number of patent applications assigned to the Osaka Gas Co Ltd, for disilanes obtained from $R^1R^2R^3SiX$ for a variety of R groups, also digermanes similarly These processes have been grouped into a European patent application [52], where the benefit is stated of switching the cell polarity at regular intervals as well as using ultrasound. In all the uses of this technique to prepare these materials it is said to be 'safe' and to cause no pollution. Variations on this theme concern the electrosynthesis of germane polymers, or silane-germane co-polymers, from dihalo derivatives. Alternatively, the use of trihalo derivatives ($RSiX_3$) allows the formation of network polysiloxanes [52]. In all these cases the mechanism of ultrasonic influence is not elucidated, and generally English language digests of this work give limited specific details of the ultrasonic system, its geometry and other relevant factors.

5.4.3 Synthetic Electrooxidations

Previously ultrasound had been employed in electroorganic systems to facilitate an expected reaction, but the possibility that ultrasound could be deliberately employed to alter the course of a reaction had been rarely studied prior to work from Coventry in the late 1980s. The chosen reaction was the electrooxidation of carboxylate anions (the Kolbé reaction), long-known as one of the earliest discovered organic reactions and the subject of a wealth of empirical data, regularly reviewed [2, 53] but where there still remains mechanistic controversy [54]. Different reaction pathways exist under different kinetic regimes and adsorption and other electrode phenomena are known to be important, all of which might be influenced to differing extents by ultrasound. Figure 5.3 gives the usual literature scheme to account for the range of products obtained from electrooxidation of a typical carboxylate anion ($RCOO^-$). The general mechanism breaks down into a pathway involving one electron per molecule of starting material, giving products from the radical intermediate, for example the dimer [R-R] (which is the actual Kolbé reaction); and a two-electron pathway per starting molecule, giving products from an intermediate cation, (the Hofer-Moest reaction) [8].

Empirical rules have been elaborated to account for competition between these pathways, depending on electrolysis conditions; and the Coventry group chose to examine a system almost at balance, where both pathways operate [55] in order to best identify any sonoelectrochemical effect on mechanism [56]. Table 5.1 shows product ratios (by GLC) from the electrooxidation of partially neutralised cyclohexanecarboxylate in methanol at platinum, at a current density of 200 mA cm^{-2}. The first column shows a substantial amount (49%) of the dimer bicyclohexyl from the one-electron pathway, together with cyclohexylmethylether, cyclohexanol and other products from the two-electron pathway (totalling $\sim$ 30%). The ester methyl cyclohexanoate (17%) is considered to arise from acid-catalysed chemical esterification of the starting material with methanol solvent, due to the quantity of protons produced around the anode – at the high current densities needed

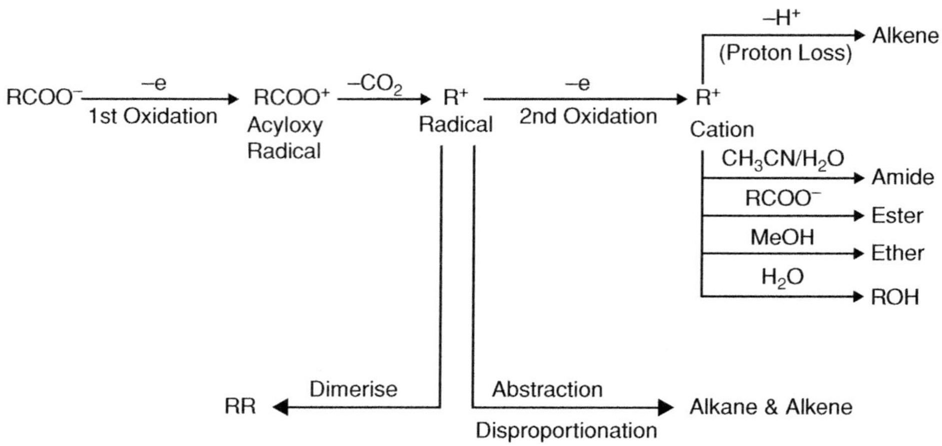

Figure 5.3 *Representative reaction pathways for the electrooxidation of carboxylates, leading to one-electron products (the Kolbe reaction – main product R-R dimer) and a range of two-electron products (the Hofer-Moest reaction).*

Table 5.1 *(a) Product ratios (% by GC) from cyclohexanoate electrooxidation after passage of 2.0 F mol^{-1}, at platinum in methanol: silent/stirred versus 38 and 850 kHz insonation. (b) Total ether-extracted quantities (g) from cyclohexanoate electrooxidation (from 4 g of starting material, after passage of 2.0 F mol^{-1}), effect of current density and percentage neutralisation of carboxylic acid.*

(a)

Product	*Silent*	38 kHz	850 kHz
Cyclohexane	7	3	3
Cyclohexene	12	22	29
Cyclohexyl methyl ether	3	31	31
Bicyclohexyl	43	11	6
Methyl cyclohexanoate	16	14	9
Cyclohexanol	18	12	21
Other products	1	3	1

(b)

Current density/mA cm^{-2}	% Partially neutralised	*Silent*	38 kHz	850 kHz
150	25	0.8	1.2	1.6
200	50	0.7	1.0	2.1

parasitic discharge of the solvent methanol, which produces protons, remains a competitive reaction [50]. Column 2 shows the effect on product ratio of ultrasonic irradiation from a Kerry Pulsatron 35 kHz (50 watts maximum power) cleaning bath during electrolysis. Now there is only ~8% of the bicyclohexyl dimeric one-electron product, with ~41% of two-electron products from nucleophilic capture of the intermediate carbocation, and a striking enhancement to 32% of cyclohexene, with only <3% of cyclohexane. The preponderance of cyclohexene over cyclohexane shows its formation by proton loss from the carbocation intermediate, since free-radical routes to cyclohexene (i.e. hydrogen atom abstraction) also produce cyclohexane in equal if not greater amounts [53, 55]. It is also noted in Column 2 that the parasitic formation of methyl cyclohexanoate ester is lessened under ultrasound, suggesting suppression of solvent discharge. There may also be an effect upon adsorption, since ultrasound is thought to enhance adsorption phenomena in some dissolving-metal chemical reactions [57].

Other procedural benefits in the electrooxidation of cyclohexane carboxylate under ultrasound include a drop in overall cell voltage from 8.3 to 7.3 V needed to maintain the same current density for the galvanostatic system, and it was also noticed that the reaction approached completion in a shorter time span despite the apparent switch to the two-electron process, confirming diminution of parasitic processes. Overall, ultrasound appears to favour the two-electron mechanism for the reaction, but the greatest effect of sonication upon product distribution was the substantial enhancement of alkene formation.

The Coventry group examined the use of higher ultrasonic frequency (850 kHz) on the electrooxidation of cyclohexanecarboxylate [58]. In Table 5.1, Column 3, the trend in product distribution from carboxylate electrooxidation at platinum in methanol was found to be similar to that under sonication in the 20 to 40 kHz frequency range. When first performed this was quite a striking result given the different cavitational effects to be expected (ultrasonic powers were matched within the limitation of the calorimetric technique). At the higher frequency the yields and current efficiencies were even better and there remained the lessened cell voltage requirements and other benefits. There were also fewer of the numerous low-yield methoxylated species and other side-products. This is evidenced visually by the clarity of the clear yellow electrolyte during the reaction and the absence of the build-up of a dark-brown coloration during electrolysis. It would seem that higher frequencies are more suitable for the insonation of these reaction systems; although this is not straightforward to explain, since cavitation phenomena, effects of acoustic impedance, cell geometry implications and other factors all change with frequency.

It may be an effect of the shorter wavelength at the higher frequency. Thus in the 20–60 kHz frequency region the half-wavelength is of the order of centimetres so that only a limited number of nodes and antinodes span the dimensions of the preparative electrolysis cell and electrode used; but in the 500–800 kHz region the distance between nodes and antinodes is only of the order of millimetres. If, as seems possible from some electroanalytical studies, certain sonoelectrochemical benefits originate from nodes or positive regions of the wave, but antinodes or negative regions of the wave do not interfere, then simplistically an increase in the number of nodes within the active region of the cell will produce more uniform and effective sonoelectrochemical phenomena. Sound waves also are reflected from interfaces, including that between a liquid and a solid, such as an electrode surface or the walls of a cell, and between a liquid and a gas, such as the interface at the surface of the electrolyte. These effects also vary with frequency and have been discussed by Pétrier and

Table 5.2 *Average product distribution from electrochemical oxidation of phenylacetate at platinum in methanol in absence and presence of ultrasound.*

Relative product distribution[a,b]	13% pyridine *Silent*	13% pyridine ultrasound	No pyridine[c,d] ultrasound
Toluene	0.4	1.5	3.1
Benzyl methyl ether	21.1	27.8	32.3
Methyl phenylacetate	10.2	4.2	6.2
Bibenzyl	59.8	51.3	52.7
Others	8.3	12.4	5.7

[a] Average potential difference across the cell for current density 100 mA cm^{-2} is 7.9 V in the absence and 6.6 V in the presence of ultrasound.
[b] Relative peak area % after passage of 1.1 F mol^{-1}.
[c] Without pyridine there is rapid cessation of reaction under silent conditions.
[d] Powder precipitate, additional product in 14% weight yield.

others [e.g. 59]. Such phenomena will affect the coupling of energy from insonation into the system.

Another carboxylate electrooxidation system was then studied where there is no proton-loss pathway from the intermediate carbocation, by using phenylacetate as substrate. Table 5.2 shows relative product ratios for phenylacetate in similar conditions to those used for cyclohexane carboxylate, using 35 kHz insonation from a bath, but employing 100 mA cm^{-2} current density [60]. There are three columns in the table (an implicit fourth is not shown) because here there is another complication. Simple electrolysis of the partially neutralised salt in methanol causes a very rapid increase in applied cell voltage due to the coating of the anode with a pale-coloured material that causes the reaction to cease. This well known [61] and it is necessary to add pyridine (up to 50% v/v) to keep the electrode clean, presumably by solubilising the inhibiting layer. The absent column in Table 3 would show 0% effective yields of products, while the actual Column 1 shows product ratios from silent reaction in the presence of 13% (v/v) of pyridine. Here there is 60% of the Kolbe dimer bibenzyl, 21% of the two-electron ether product and some 10% of the parasitic methyl ester. A major component of the remaining (8%) material is benzaldehyde, a persistent by-product of arylacetate electrooxidations. Reactions were terminated after 1.1 Faraday $mole^{-1}$ and were not run to completion.

Column 2 shows the effect of ultrasound upon the product ratio from methanol/pyridine. There is now 53% bibenzyl, 32% of methyl ether and 6% of methyl ester (with total 5% of other products) suggesting only a slight shift towards the two-electron products, but with an overall diminution of solvent discharge and side-reactions. However, phenylacetate electrooxidation is known to favour the one-electron route to bibenzyl in a wide range of conditions [62], and to be much less sensitive to mechanistic switches by manipulation of parameter than is cyclohexane carboxylate electrooxidation [53], and it is notable that this trend remains even under ultrasound.

Column 3 in Table 5.2 shows the product ratios under ultrasound in the absence of pyridine. Overall there is the same trend with sonication, namely a slight shift from the one-electron pathway towards the two-electron pathway, though here there seems to be a higher yield of benzaldehyde-derived by-products.

The most significant factor is that there is no evidence of electrode fouling and the reaction maintains current density at a steady and lower voltage. In addition, a fine white powder was formed during the electrolysis, which was isolated by filtration (yield 14% by weight). This appears to be a polymer which contains aromatic rings and also two types of methylene group (from NMR) and an aliphatic ester carbonyl (1735 cm^{-1} in the IR). This suggests a structure incorporating -C_6H_4-CH_2- units and -$C_6H_4CH_2COO$- units. It may be supposed that enhanced mass transport under ultrasound and the abrasion effect near the electrode surface has swept the inhibiting species into solution, thus keeping the electrode clean. There is much less of this powder under ultrasound in the presence of pyridine, suggesting that it is indeed solubility factors that demand the use of the cosolvent in silent conditions. It is very unusual to obtain products of carboxylate electrooxidation that originate from the acyloxy radical. Previous work has involved the trapping of this intermediate by intramolecular cyclisation reactions before it has time to decarboxylate [63] but here a linear product involving units from the first-formed radical has been isolated after intermolecular reaction.

The obviation of pyridine under ultrasound represents a significant procedural enhancement since its presence considerably hampers work-up and also causes problems on environmental and economic grounds. These results also have implications for electrolyses that are affected by electrode fouling problems. A further feature again is a lowering of the applied cell voltage from 7.9 to 6.6 V under ultrasound, representing an energy saving.

Shortly after our publications on the sonoelectrochemical oxidation of phenylacetate [60, 64], a parallel study was reported by Japanese workers [65] who employed crossed Kolbé electrolyses of phenylacetates and succinates, variously deuterated, to produced deuterated derivatives of 4-pehnylbutyric acids. In control experiments to produce deuterated bibenzyls from phenylacetate without succinate present they obtained 11% of dimer with pyridine present, in normal conditions, rising to a 47% yield and 41% of dimer under ultrasound from a cleaning bath, though here the reaction time is stated to be reduced threefold, but it is ambiguous whether this shortened reaction time benefit applies also to the reaction with pyridine but without ultrasound. The authors state that ultrasound 'helps to keep the electrode surface clean' and it would seem that in their conditions, which employ aqueous solution instead of methanol, the electrode is not completely 'switched off' by the insulating film in normal conditions. The authors did not examine the system with both pyridine present and ultrasound, but the observed yield drop from 47 to 41% might suggest the same trend towards the two-electron pathway under ultrasound, though other products were not identified and quantified.

The Coventry group also examined the behaviour of *p*-chlorophenyl acetate electrooxidation under ultrasound [60]. This substrate is known to favour markedly the two-electron mechanism [6] showing that choice of reaction pathway is more dependent on substrate nature than upon manipulation of electrolysis parameters. A further feature of this system is the appreciable yield of *p*-chlorobenzaldehyde-derived products. This is shown in Table 5.3, where it can be seen that sonication produces little change in relative product ratio although there is a marked increase in total yield and current efficiency after ether extraction, following the same passage of electrolysis charge (1.1 F mol^{-1}, i.e. not to completion). Thus 46% by weight of mixed product is obtained (unreacted acid is not recovered by this

Table 5.3 *Average product distribution from electrochemical oxidation of* p*-chlorophenylacetate in absence and presence of ultrasound.*

Relative product distribution[a,b]	*Silent*[c]	Sonicated[d]
p-Chlorobenzyl methyl ether	71.5	69.3
Methyl *p*-chlorophenylacetate	3.4	2.5
p-Chlorobenzaldehyde	9.1	4.2
p-Chlorobenzyl alcohol	3.5	9.0
p-Chlorobenzaldehyde acetal	10.8	11.2
Others	1.7	3.8

[a]Average potential difference across the cell for current density 100 mA cm^{-2} is 5.63 V in the absence and 4.83 V in the presence of ultrasound.
[b]Relative peak area % after passage of 1.1 F mol^{-1}.
[c]Total weight yield after extraction – 23%.
[d]Total weight yield after extraction – 46%.

procedure) with ultrasound, but only 23% by weight from the silent reaction. Carboxylate electrooxidations, which often require high current densities, and which promote solvent breakdown and other parasitic reactions, are notoriously inefficient, and here ultrasound doubles the observed efficiency.

A further unexpected advantage of sonication was found here. There was no formation of polymeric coating on the anode, further confirming that the free *para*-position of unsubstituted phenylacetate is necessary for production of the inhibitory species on the anode; but instead the cathode became coated in a black deposit as the electrolysis proceeds. In order to maintain current densities as high as 100 mA cm^{-2} in methanol it is necessary to have the working and counter electrodes close together in an undivided cell, to minimise resistance losses; and it is possible to envisage a reductive cleavage process occurring on the protonated acid molecule (the salt is only partially neutralised) with loss of chloride ion to yield species that could polymerise on the cathode, instead of the normal cathodic reaction of hydrogen evolution. Fouling on either electrode will increase cell resistance and energy loss, hence ultrasound here offers a benefit to the system by suppressing coating of the counter electrode.

The origin of ultrasonic effect upon carboxylate electrooxidation is not straightforward to establish in view of the complex mechanism of the reaction with different kinetic regimes, the loss of carbon dioxide and also the role of adsorption and other electrode phenomena. It has also been suggested that the second electron transfer to give the carbocation may not occur in bulk solution, although this requires methoxy radicals from solvent discharge or other species to act as redox mediators in the solution phase [55].

Ultrasound can provide appreciable microscopic heating effects, for example via cavitation phenomena. So to minimise possible contribution from bulk heating effects in carboxylate electrooxidation, the cells were cooled during sonoelectrolysis; but to examine other possible temperature effects of ultrasound the Coventry group performed electrolyses at 60 °C and found that any shift in product distribution so caused was not of the order observed under sonication at a lower bulk temperature. We have also confirmed the stability of reactants and products under ultrasound without application of an electric potential. This does not preclude microscopic heating effects being the cause of some sonoelectrochemical phenomena, since the 'hot spot' theory predicts large temperature gradients associated with cavitation [46].

Another multi-pathway sonoelectrochemical system where the frequency of insonation is a discriminating factor concerns the electrooxidation of alkylaromatics. A number of possible products may ensue, depending upon the substrate and the reactions system, particularly the nature of nucleophiles that may be present.

For hexamethylbenzene, with no free ring-position, only side-chain substitution must occur; and in acetonitrile with some 1% of acetate ion a competition ensues leading to either the acetamide derivative R-$CH_2NHCOCH_3$ on work-up or the acetate ester R-CH_2OCOCH_3 [6]. This is known to be affected by the nature of the electrolyte salt in the silent system and we repeated this reaction with insonation from either a 38 kHz bath or an 800 kHz bath, using a platinum anode in MeCN/CH_2Cl_2 containing either Bu_4NBF_4 or Bu_4NClO_4. We noticed some small effect of electrolyte salt as previously reported, but little major effect of insonation at either frequency [60, 66]. However, for pentamethylbenzene, with a free ring position, a much larger effect was seen. Here there is opportunity for nuclear substitution by another ring, leading to the product mixture shown. The major products are now a di-phenylmethane derivative produced in an overall one-electron process in which a reactive intermediate centred on a methyl group attacks the free ring position of another molecule, this being statistically favoured over dimerisation via the free ring positions on two adjacent species. Alternatively, attack by a nucleophile leads to product from an overall two-electron per molecule process. After the passage of 0.3 F mol^{-1} (i.e. not to completion), under potentiostatic control at +1.6 V versus SCE, we found that insonation from a 38 kHz bath favoured the acetamide product from nucleophilic capture of MeCN prior to aqueous work-up; but this was also the major product in a solution mechanically stirred by a magnetic follower, and overall there was a small shift in total amount of two-electron versus one-electron products by using ultrasound at this frequency. However, there is a greater effect with insonation at 800 kHz. Here the diphenylmethane compounds (there are several isomers which are not distinguished) become the major product in a switch to the one-electron mechanism. Thus insonation at 38 and 800 kHz (at powers matched by the calorimetric method) provide different effects upon the reaction [63]. Here it may be that high frequency ultrasound offers a route to a different hydrodynamic regime.

5.4.4 Sonoelectrochemically Produced Electrode Coatings: *Desirable* and *Undesirable*

Some examples of sonoelectrochemically produced electrode coatings have been discussed above, and in a separate chapter of this book a wide range of these systems is covered regarding the desirable formation of electrochemically produced polymers. Two main themes concern chain polymerisations [10, 67] and electrooxidatively formed electrically-conducting polymers [10], where ultrasound has been shown to have effects. However, it is worth noting the effect of insonation upon the polymerisation of vinylcarbazole [67–69], since this also involves electroorganic mechanistic opportunities. A product cross-linked through the aromatic rings forms a semi-conductive coating on the electrode surface, and this still forms even under ultrasound, but insonation does have a marked effect upon the

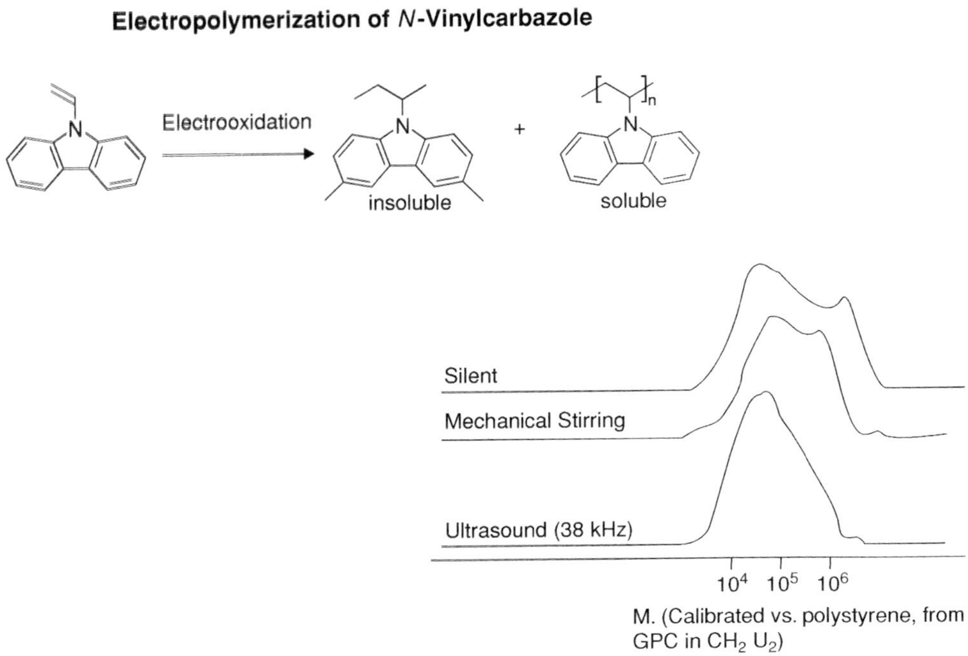

Figure 5.4 *Electropolymerisation of* N*-vinylcarbazole, either via exocyclic vinyl group or via direct aromatic coupling: effect of ultrasound on RMM control of soluble fraction.*

soluble fraction formed by polymerisation through the vinyl side-chains, which shows greatly improved relative molecular mass (RMM) control. Figure 5.4 shows these results.

Important from an electrosynthetic point of view are electroorganic systems that involves formation of an undesirable electrode coating. This section of the chapter concerns a number of these systems where ultrasound was employed to attempt mitigation of problematic formation of deleterious electrode coatings.

Thus the electroreduction of 6-phenyl-2,3-dihydro-1,4-diazepinium salts at a mercury pool in dimethylformamide gives high yields of interesting products, depending upon the substitution patterns in the starting materials. Thus the 6-phenyl derivative unsubstituted at least one of the annular nitrogen atoms gives pyrrolodiazepines, plausibly by dimerisation of the first-formed radicals followed by intramolecular attack with loss of 1,2-diaminoethane [69] (Figure 5.5).

If both the nitrogen atoms are substituted (e.g. the 1,4-dibenzyl derivative) then the first-formed dimer can be isolated as *dl* or meso stereoisomers, but in addition a further retrocycloaddition reaction can occur to give also bis-1,4-(tetrahydroimidazolyl)-butadiene side-products [68] as shown in Figure 5.6. In all this is a fruitful electrosynthetic system that gives a depth of products, but in the cause of obviating the toxic metal mercury electrode, we replaced this with solid platinum foil, only to find that an inhibitory coating formed over a period of some minutes so that the current died away and electrolysis ceased. This seemed a good opportunity to exploit ultrasound to avoid formation of these coatings

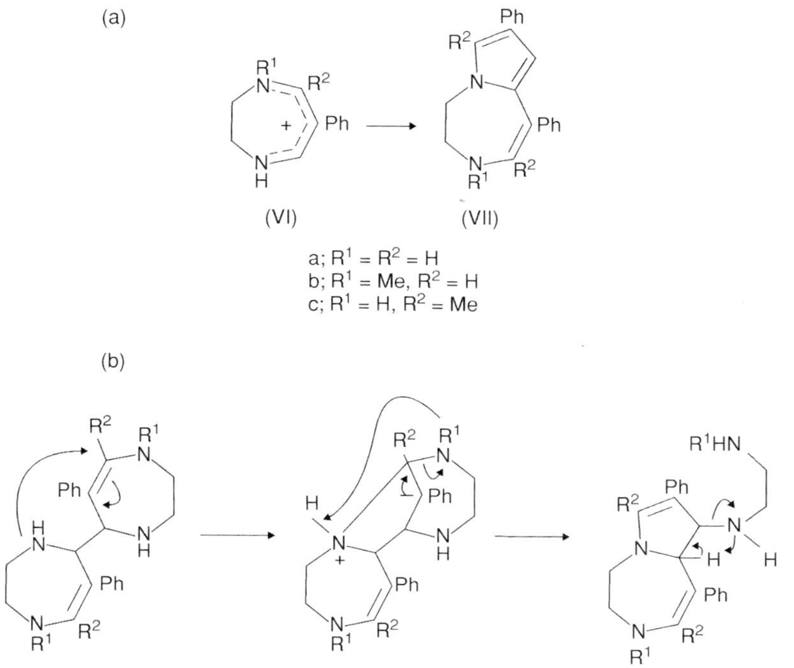

Figure 5.5 *Electroreductions of 6-phenyldihydrodiazapinium cations at mercury in dimethylformamide: (a) starting materials with one free ring -NH group, and pyrrolodiazepine products formed; (b) plausible mechanism for product formation.*

(and if the reaction continued then maybe also to modify the product mixtures too), but we found the opposite effect took place, and instead the current died away over a period of some tens of seconds, more rapidly than in the silent situation. Here ultrasound had enhanced the processes behind the formation of the insulating coating [68].

Another system of interest where there seems to be no major effect of insonation, at either 850 or 20 kHz frequencies concerns the electrooxidations of two similar types

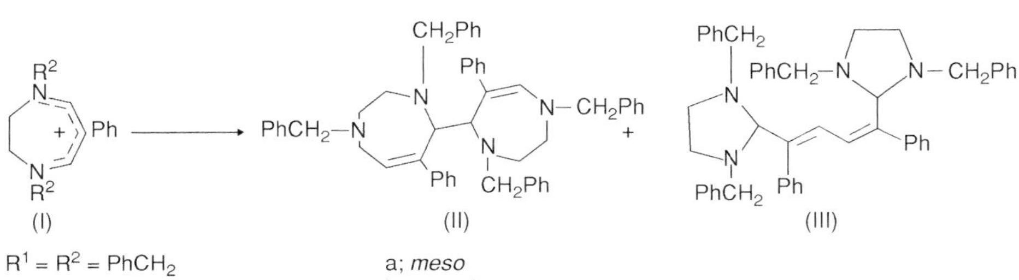

Figure 5.6 *Electroreductions of 6-phenyldihydrodiazapinium cations at mercury in dimethylformamide: products from starting materials substituted at both annular nitrogen atoms.*

of compound, tetraphenylthiophene-*S*-oxides and tetracyclones (tetraphenylcyclopentadienones). These molecules differ only in a single key atom, in that the tetracyclone has a central carbonyl (C=O) group, while the thiophene-monoxide derivative has a respective S=O central group [65].

The striking feature of silent electrolysis of tetracyclone in dry (but not super-dried) acetonitrile, at platinum with tetraalkylammonium salt electrolyte [70], is that the tetracyclone undergoes an electrooxidative Bayer-Villiger reaction, to eventually insert an oxygen atom (from adventitious water) giving the cyclic tetraphenylpyrone ester as shown in Figure 5.7, whereas the sulfur analogue gives a mixture of two unexpected products, a dibenzoylstilbene, and an unusual sulfinic acid derivative that has a new bond joining the aromatic rings present at positions 3 and 4, as shown in Figure 5.8, structures 3 and 4. These are in approximately equal ratios, and interestingly this is not affected by ultrasound (but see below). Also the voltammetry of both compound types (a range of substituted derivatives were tested) suffers from severe electrode fouling after the first sweep, and unexpectedly this is not ameliorated by insonation [70, 71]. Another method used by electrosynthetic chemists to keep an electrode surface active is from time to time to transiently pulse to a reverse (here reducing) potential before reapplying the desired (here oxidising) potential. This approach is used in entirely separate circumstances in reference [48], and here works well for tetracyclones and tetraphenylthiophenes too. The lack of effect of ultrasound is quite striking, since these are systems that have sufficient mechanistic complexity for one

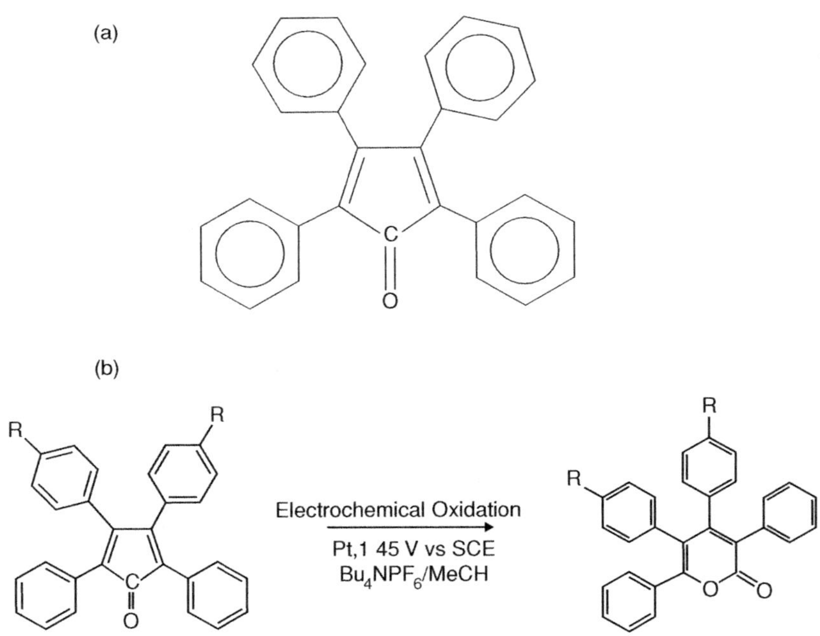

Figure 5.7 *Tetraphenylcyclopentadienone (tetracyclone) (a) structure, (b) pyrone reaction product from electrooxidation.*

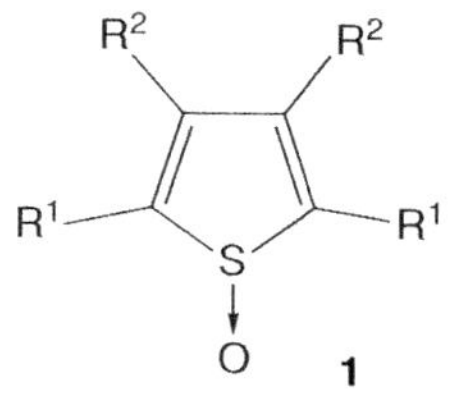

1

Thiophene-*S*-monoxide

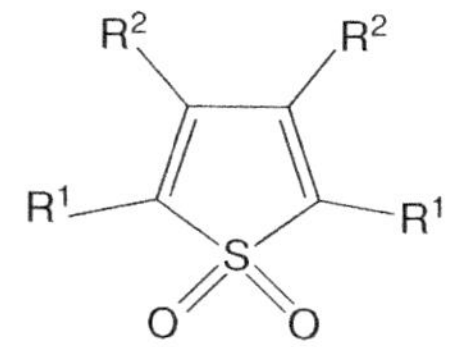

2

Thiophene-*S*,*S*-dioxide

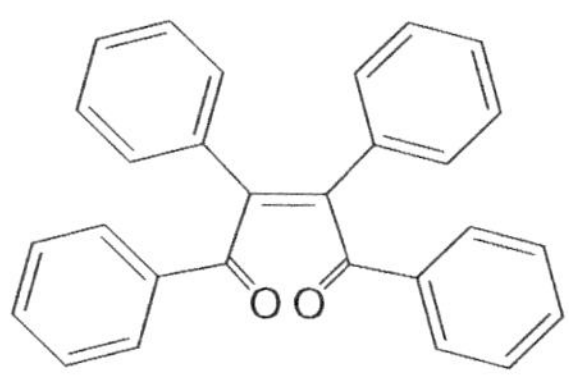

3

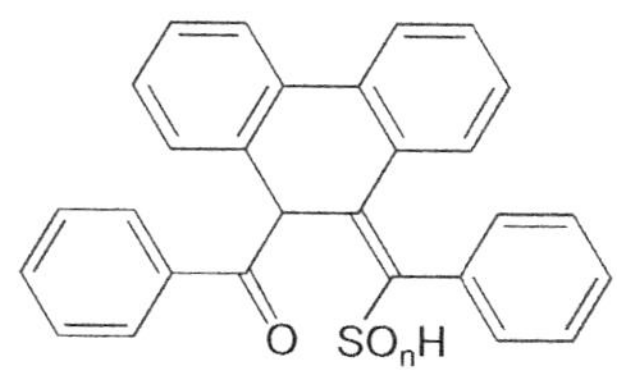

4

Figure 5.8 *Tetraphenyl thiophene-S-monoxide (structure 1, all R groups are phenyl), compared to commoner and more stable thiophene dioxide (structure 2). Structures 3 and 4, two main product species obtained from electrooxidations of tetraphenyl thiophene-S-monoxides.*

to expect an effect to occur. It can only be assumed that some mechanistic processes are intramolecular and/or have kinetic features such as not to be greatly perturbed by improved hydrodynamics in the cell. However, it is noted that in these sonoelectrolyses too, the cell energetics are improved by insonation, as has been found to be the general case.

There is a small but notable effect of ultrasound, in that insonation produces an increase in a minor product from the tetracyclone reaction, which is a derivative mono-acetamidated on a phenyl ring (the 'R' groups on phenyl rings at positions 3 or 4 in Figure 5.7). Normally such a product is formed by attack of acetonitrile solvent at a transient carbocation, followed by hydrolysis in the work-up of the so-formed nitrenium ion. Here an increase in this minor product seems the only outcome of insonation of this mechanistically complex system [71].

5.5 Other Systems

5.5.1 Hydrodynamics

In terms of generalities, while it is clear that insonation improves hydrodynamics in a cell, it seems that whether an electrosynthetic reaction will show a gross mechanistic switch depends upon the individual reaction system. There are other examples of mechanistic inference from insonation of electrolysis, such as the case of adamantyl systems [72], in which a shift from a one-electron mechanism to a two-electron one under ultrasound is inferred. A simple increase in hydrodynamics ought to favour the reverse switch, that is, from higher electron numbers per reactant molecule to lower ones, as is found in carbonyl reductions and other systems (described above). However, in the adamantyl reaction, insonated at 10 kHz, dimerisation is disfavoured by virtue of convective mixing promoted by ultrasound. In the case of carboxylate electrooxidation (described above), the mechanism also is switched in the reverse sense (from one -electron per mole to two electrons). It is notable that there is a gas-evolving step at the core of this mechanism, that is a process known to be influenced by ultrasound, which could alter the build-up of key intermediates, and it is therefore perhaps to be expected that insonation might affect the mechanistic route, although the apparent lack of an effect of ultrasonic frequency is notable. Apart from greatly improved current efficiencies and yields using 850 kHz insonation, the general product mix from carboxylate electrooxidation is similar to that obtained at 20 kHz. It was also notable in this work that the applied ultrasonic powers necessary to cause a mechanistic switch were not necessarily of great magnitude, and thus immersion in an ultrasonic bath of an electrolysis cell made of thick glass can still cause a substantial switch in electrolysis products.

The use of ultrasound for phase modification, such as emulsification, has been well studied, and obviation of, for example, undesirable solvents systems allows use of 'green' and environmentally more-benign systems [10, 73]. Recently the Tokyo group has addressed new approaches to anodic substitution reactions using acoustic emulsification [74]. Of course, greatly enhanced stirring of reactors need not necessarily be performed by ultrasound. Use of an Ultra-Turrax can achieve emulsification, such as for water/acetonitrile in the presence of salts, at concentrations that normally cause phase separation in the conditions employed [75]. But ultrasound is convenient to employ and need not necessarily be provided from a piezoelectric transducer. The use of a 'whistle reactor' to induce cavitation has been mentioned above [27].

5.5.2 Low-temperature Effects

As part of their extensive study into varied aspects of sonoelectrochemistry, the Compton group also addressed the effects of temperature. Various studies have considered the possibility of higher temperatures pertaining in local microenvironments, since localised higher temperatures and pressures are associated with cavitational phenomena, but here the authors addressed the employment of low temperatures in insonated systems. A feature of all electrolyses is that the energetics of electron transfer, controlled by applied potential, can be detached from the rates of chemical follow-up reactions (e.g. by heating or cooling a cell), thus allowing the intermediates and reactive species electrogenerated *in situ* to have different opportunities for chemical reaction, or for different physical processes to pertain, and the inclusion of ultrasound allows yet further independent control of reaction parameters. Thus in the sonocryoelectrochemical reduction of nitrobenzene derivatives in liquid ammonia (−35 to −70 °C) ultrasound was found to be beneficial by causing very fast mass transport, by enhancing mixing dissolution kinetics at low temperature, and by affecting the formation of solid products at the electrode surface [76]. In later work, this group studied the oxidation of anthracene at platinum at −40 °C, with emphasis upon the kinetics of follow-up processes. They noted that at room temperature under silent and insonated conditions 9,10-anthraquinone and bianthrone were major products, whereas at low temperature with acoustic streaming not only was anthracene 9,10-diol additionally produced, but there was improved current efficiency [77]. Sonocryoelectrochemistry remains relatively little studied but in principle it offers a promising means to probe electrochemical mechanisms.

5.6 Conclusions

The significance of insonation upon electroorganic systems is not just that new products may be produced, or that more efficient production occurs of expected products, or that cell energetics are enhanced, or that electrode fouling is minimised, or that when desired electrode coatings may be improved, or that various phase phenomena can be manipulated, it is that the additional perturbation method gives another set of parameters for mechanistic elucidation. Different reaction systems behave in different ways, and sonoelectrochemistry is an example of a 'multiperturbation' or 'hyphenated' methodology. Such systems have been the subject of targeted studies by (for example) European networks for some time. Thus COST D6 'Chemical Processes and Reactions Under Extreme or Non-Classic Conditions', COST D10 'Innovative Methods and Techniques for Chemical Transformations' and COST D32 'Chemistry in High-Energy Microenvironments' all contained components addressing advances in sonoelectrochemistry, and the reader is requested to examine the reports and outcomes from these initiatives for further information. The use of additional perturbations in electrosynthesis remains a fruitful area for study.

Acknowledgements

The author gratefully acknowledges the valued contributions from colleagues, collaborators, co-workers, assistants and all those who have inspired and supported my continuing interest in electrosynthesis. There are very many of these; I cannot list all, and I cannot thank them enough.

References

(1) Kolbe, W. (1849) Formation of symmetrical dimers by the electrolysis of carboxylates (decarboxylative dimerization). *Ann.*, **69**, 257.
(2) Baizer, M.M. and Lund, H. (eds) (1991) *Organic Electrochemistry*, 3rd edn, Marcel Dekker.
(3) Fry, A.J. (1972) *Synthetic Organic Electrochemistry*, Harper and Row.
(4) Kyriacou, D. (1994) *Modern Electroorganic Chemistry*, Springer-Verlag.
(5) Shono, T. (1991) *Electroorganic Synthesis, in Series of 'Best Synthetic Methods'*, Academic Press.
(6) Torii, S. (1985) *Electroorganic Syntheses - Part I Electrooxidations*, Kodansha; Torii, S. (1986) *Electroorganic Syntheses - Part II Electroreductions*, Kodansha.
(7) Weinberg, N.L. (ed.) *Techniques of Electroorganic Synthesis*, Wiley, Series Publisher.
(8) Walton, D.J. and Phull, S.S. (1996) *Sonoelectrochemistry, in Advances in Sonochemistry*, Vol **4** (ed. T.J. Mason), JAI Press.
(9) Walton, D.J. and Mason, T.J. (1998) *Chapter in 'Synthetic Organic Sonochemistry'* (ed. J.-L. Luche), Plenum Press, New York.

(10) Walton, D.J. (2002) Sonoelectrochemistry – The application of ultrasound to electrochemical systems. *Arkivoc*, (3), 198–218.

(11) Walton, D.J., Iniesta, J., Plattes, M., Mason, T.J., Lorimer, J.P., Ryley, S., Phull, S.S., Chyla, A., Heptinstall, J., Thiemann, T., Fuji, H., Mataka, S. and Tanaka, Y. (2003) Sonoelectrochemical effects in electro-organic systems, *Ultrason. Sonochem*, **10** (4), 209–216.

(12) Walton, D.J., Burke, L.D. and Murphy, M.M. (1996) Sonoelectrochemistry: Chlorine, hydrogen and oxygen evolution at platinized platinum. *Electrochim. Acta*, **41** (17), 2747–2751.

(13) Moriguchi, N.J. (1934) *Chem. Soc. Jpn.*, **55**, 751.

(14) Yeager, E. and Hovorka, F. (1953) Ultrasonic waves and electrochemistry. I. A Survey of the electrochemical applications of ultrasonic waves. *J. Acoust. Soc. Am.*, **25** (3), 443–455.

(15) Chanturiya, V.A., Zholshibekova, M.R. and Malakhov, Y.V. *Elektron Obrab Mater.* 276–77.

(16) Chanturiya, V.A., Dimitrieva, L.L., Zholshibekova, M.R. and Malakhov, Y.V. (1984) *Vestn Akad. Nauk Kaz SSR*, **9**, 47.

(17) Tsonkov, T., Paulapanski, M. and Djendova, S. (1994) *Oxid. Commun.* **17**, 321.

(18) Connors, T.F. and Rusling, J.F. (1984) Ultrasonically-assisted electrocatalytic dechlorination of polychlorinated biphenyls. *Chemosphere*, **13** (3), 415–420.

(19) Schweizer, S., Rusling, J.F. and Huang, G. (1994) Electrolytic dechlorination of DDT in a bicontinuous microemulsion. *Chemosphere*, **28** (5), 961–970.

(20) Zhang, S. and Rusling, J.F. (1993) Dechlorination of polychlorinated biphenyls by electrochemical catalysis in a bicontinuous microemulsion. *Environ. Sci. Technol.*, **27** (7), 1375–1380.

(21) Rusling, J.F. (1991) Controlling electrochemical catalysis with surfactant microstructures. *Acc. Chem. Res.*, **24** (3), 75–81.

(22) Rusling, J.F. (1994) *Electroanalytical Chemistry*, Vol **19** (ed. A.J. Bard), Marcel Dekker, New York, pp. 1–88.

(23) Rusling, J.F. (1994) *Modern Aspects of Electrochemistry*, No. **26** (eds B.E. Conway, J O'M Bockris), Plenum Press, New York, pp. 49–104.

(24) Lorimer, J.P., Mason, T.J., Plattes, M., Phull, S.S. and Walton, D.J. (2001) Degradation of dye effluent. *Pure Appl. Chem.*, **73** (12), 1957–1968.

(25) Gonzalez-Garcia, J., Saez, V., Esclapez, M.D., Bonete, P., Walton, D.J., Rehorek, A. and Louisnard, O. (2010) Sonochemical degradation of perchloroethylene. *Phys. Proc.*, **3** (1), 981–986.

(26) Oturan, M.A., Sires, I., Oturan, N., Perocheau, S., Laborde, J.-L. and Trevin, S. (2008) Sonoelectro-Fenton process: a novel hybrid technique for the destruction of organic pollutants in water. *J. Electroanal. Chem.*, **624**, 329–332.

(27) Chakinala, A.C., Bremner, D.H., Burgess, A.E. and Namkung, K.C. (2007) A modified advanced Fenton process for industrial wastewater treatment. *Water Sci. Technol.*, **55** (12), 59–65.

(28) Gonzalez-Garcia, J., Drouin, L., Banks, C.E., Sljukic, B. and Compton, R.G. (2007) At point of use sono-electrochemical generation of hydrogen peroxide for chemical synthesis: The green oxidation of benzonitrile to benzamide. *Ultrason. Sonochem.*, **14** (2), 113–116.

(29) Gigi, S., Paucescu, V. and Jurth, S. (1980) Romanian Patent RO 72382.

(30) Atobe, M., Naohiro, Y., Fuchigami, T. and Nonaka, T. (2003) Ultrasonic effects on electroorganic processes: Part 24. Rate and saturated amount of the self-assembled adsorption of thiolate species on a gold surface. *Electrochim. Acta*, **48**, 1759–1766.

(31) Kardos, N. and Luche, J.-L. (2001) Sonochemistry of carbohydrate compounds. *Carbohydr. Res.*, **332** (2), 115–131.

(32) Cheng, P.C., Nonaka, T. and Chou, T.C. (1991) Prediction of product-selectivity in an electroorganic process by mathematical modeling of reaction mechanism. A rotating cylinder electrode system. *Bull. Chem. Soc. Jpn.*, **64** (6), 1911–1915.

(33) Atobe, M., Matsuda, K. and Nonaka, T. (1996) Ultrasonic effects on electroorganic processes. Part 4: Theoretical and experimental studies on product-selectivity in electroreduction of benzaldehyde and benzoic acid. *Electroanalysis*, **8** (8–9), 784–788.

(34) Sunashima, K.T. and Nonaka, T. (1995) *Chem. Lett.*, 86.

(35) Atobe, M. and Nonaka, T. (1995) Ultrasonic effects on electroorganic processes. Electroreduction of benzaldehydes on ultrasound-vibrating electrodes. *Chem. Lett.*, **24** (8), 669.

(36) Durrant, A., Francois, W., Reisse, J. and Kirsch-De Mesmaeker, A. (1996) Sonoelectrochemistry: The effects of ultrasound on organic electrochemical reduction. *Electrochim. Acta*, **41** (2), 277–284.

(37) Atobe, M., Matsuda, K. and Nonaka, T. (1994) *Denki Kagaku Oyobi Kogyo Butsuri Kagaku*, **62**, 1298–9.

(38) Fry, A.J., Touster, J., Sirisoma, N.U. and Raimundo, B. (1991) *Electroorganic Synthesis (Manuel M Baizer Memorial Symposium)* (eds D.R. Little and N.L. Weinberg), Marcel Dekker.

(39) Fry, A.J. and Bujanauskas, P. (1978) Electrochemical and mercury-promoted reduction of α,α'-dibromophenylacetones in acetic acid. *J. Org. Chem.*, **43** (16), 3157–3163.

(40) Fry, A.J. and O'Dea, J. (1975) Electrochemical reduction of α,α'-dibromo ketones in acetic acid. Convenient synthetic route to highly branched α-acetoxy ketones. *J. Org. Chem.*, **40** (25), 3625–3631.

(41) Fry, A.J. and Herr, D. (1978) Reduction of α,α'-dibromo ketones by ultrasonically dispersed mercury in protic solvents. *Tetrahedron Lett.*, **19** (20), 1721–1724.

(42) Fry, A.J. and Lefor, A.T. (1979) Regioselectivity in the reductive acyloxylation of α,α'-dibromo ketones. *J. Org. Chem.*, **44** (8), 1270–1273.

(43) Lee, C.W., Compton, R.G., Eklund, J.C. and Waller, D.N. (1995) Mercury-electroplated platinum electrodes and microelectrodes for sonoelectrochemistry, *Ultason. Sonochem*, **2** (1), S59–S62.

(44) Gautheron, B., Tainturier, G. and Degrand, C. (1985) Ultrasound-induced electrochemical synthesis of the anions selenide (Se_2^{2-}, Se^{2-}), and telluride (Te_2^{2-}, Te^{2-}) . *J. Am. Chem. Soc.*, **107** (19), 5579–5581.

(45) Gautheron, B., Tainturier, G. and Degrand, C. (1986) Electrochemical synthesis and properties of $[\eta^5\text{-}C_5H_4R)_2TiSe_5$ (R = H, Me, Me_2CH). *Organometallics*, **5** (5) **14**, 942–946.

(46) Degrand, C. (1986) Electrochemical synthesis by the SRN 1 mechanism of 4-phenylselenobenzonitrile and 4-phenyltellurobenzonitrile. *J. Chem. Soc.*, 1113–1115.

(47) Degrand, C., Prest, R. and Compagnon, P.L. (1987) Electrochemical synthesis of (phenylseleno)benzophenones and (phenyltelluro)benzophenones by the SRN1 mechanism, using a redox catalyst. *J. Org. Chem.*, **52** (23), 5229–5233.

(48) Degrand, C., Prest, R. and Nour, M. (1987) Electrochemical synthesis of seleno and telluro derivatives in MeCN. *Phosphorus, Sulfur Silicon Relat. Elem.*, **38** (1–2), 201–209.

(49) Shono, T., Kashimura, S., Ishifune, M. and Nishida, R. (1990) Electroreductive formation of polysilanes. *J. Chem. Soc., Chem. Commun.*, **17**, 1160–1161.

(50) Hengge, E. and Firgo, H. (1981) An electrochemical method for the synthesis of silicon—silicon bonds. *J. Organomet. Chem.*, **212** (2), 151–166.

(51) Hengge, E. and Litscher, G. (1976) A new electrochemical method of forming Si-Si bonds. *Chem. Angew., Int. Ed. Engl.*, **15** (6), 370–372.

(52) Shono, T., Kashimura, S., Nishida, R. and Kawasaki, S. (1991) Eur. Pat. App. 17, EP 446578. Assigned to Osaka Gas Co Ltd.

(53) Eberson, L. and Utley, J.H.P. (1982) *Organic Electrochemistry*, 2nd edn (eds M.M. Baizer and H. Lund), Marcel Dekker.

(54) Vassiliev, Y.B. and Grinberg, V.A. (1991) Adsorption kinetics of electrode processes and the mechanism of Kolbe electrosynthesis. Part III. Mechanism of the process. *J. Electroanal. Chem.*, **308** (1–2), 1–16.

(55) Hawkes, G.E., Utley, J.H.P. and Yates, G.B. (1976) Electro-organic reactions. Part 1X.t mechanism of the Kolbe reaction stereochemistry of reaction of anodically generated cyclohexyl radicals and cations. *J. Chem. Soc., Perkin Trans.*, **2**, 1709–1716.

(56) Chyla, A., Lorimer, J.P., Mason, T.J., Smith, G. and Walton, D.J. (1989) Modifying effect of ultrasound upon the electrochemical oxidation of cyclohexanecarboxylate. *J. Chem. Soc., Chem. Commun.*, **9**, 603–604.

(57) Luche, J.L., Einhorn, C., Einhorn, J. and Sinisterra-Gago, J.V. (1990) Organic sonochenistry: A new interpretation and its consequences. *Tetrahedron Lett.*, **31** (29), 4125–4128.

(58) Walton, D.J., Phull, S.S., Geissler, U., Chyla, A., Durham, A., Ryley, S., Mason, T.J. and Lorimer, J.P. (2000) Sonoelectrochemistry—cyclohexanoate electrooxidation at 38 kHz and 850 kHz insonation frequencies compared. *Electrochem. Commun.*, **2** (6), 431–435.

(59) Benahcene, A., Petrier, C., Reverdy, G. and Labbe, P. (1995) Effects of ultrasonically induced cavitation on electrochemical processes, cell geometry and frequency effects. *New J. Chem.*, **19** (8–9), 989–995.

(60) Walton, D.J., Chyla, A., Lorimer, J.P. and Mason, T.J. (1990) Sonochemical enhancement of phenylacetate electrooxidation. *Synthetic Commun.*, **20** (12), 1843–1852.

(61) Coleman, J.P., Lines, R., Utley, J.H.P. and Weedon, B.C.L. (1974) Electro-organic reactions. Part II. Mechanism of the Kolbe electrolysis of substituted phenylacetate ions. *J. Chem. Soc., Perkin Trans.*, **2** (9), 1064–1069.

(62) Koehl, W.J. Jr (1967) Anodic oxidation of 3,3-diphenylacrylic acid and diphenylacetylene. *J. Org. Chem.*, **32** (3), 614–618.

(63) Bonner, W.A. and Mango, F.D. (1964) Kolbe electrolyses of 3-phenyl- and 3,3-diphenylpropanoic acids. *J. Org. Chem.*, **29** (2), 430–435.

(64) Walton, D.J., Geissler, U. and Durham, A. (1995) Proceedings of World Congress on Ultrasonics Berlin, September 1995, p. 659.

(65) Osawa, S., Ito, M. and Tanaka, K. (1992) The application of ultrasonic waves to the electrochemical polymerization of thiophene. *J. Polym. Sci. Part B: Polym. Phys.*, **30** (1), 19–24.

(66) Walton, D.J., Geissler, U., Ryley, S., Campbell, C. and Evans, J. (1996) Presentation to ISE 47th Meeting, Balatonfured, Hungary, September 1996.

(67) Aybar, S.P., Hacioglu, B. and Akbulut, U. (1991) Electroinitiated polymerization of butadiene sulfone. *J. Polym. Sci. Part A: Polym. Chem.*, **29** (13), 1971–1976.

(68) Walton, D.J., Iniesta, J., Plattes, M., Mason, T.J., Lorimer, J.P., Ryley, S., Phull, S.S., Chyla, A., Heptinstall, J., Thiemann, T., Fuji, H., Mataka, S. and Tanaka, Y. (2003) Sonoelectrochemical effects in electro-organic systems. *Ultrason. Chem.*, **10** (4–5), 209–216.

(69) Lloyd, D.M.G., Vincent, C.A. and Walton, D.J. (1981) The electrochemical reduction of *N,N′*-disubstituted 6-phenyl-2,3-dihydro-1,4-diazepinium salts: formation of bis (tetrahydrodiazepinyls) and a di-imidazolidinylbutadiene. *J. Chem. Soc., Perkin Trans. 2*, **5**, 801–806.

(70) Iniesta, J., Alcock, H., Walton, D.J., Watanabe, M., Mataka, S. and Thiemann, T. (2006) Electrochemical oxidation of tetracyclones and tetraphenylthiophene-*S*-oxide. *Electrochim. Acta*, **51** (26), 5682–5690.

(71) Valcarel, J.I., Walton, D.J., Fuji, H., Thiemann, T., Tanaka, Y., Mataka, S., Mason, T.J. and Lorimer, J.P. (2004) The sonoelectrooxidation of thiophene S-oxides. *Ultrason. Sonochem.*, **11**, 227–230.

(72) Paddon, C.A., Bhatti, F.L., Donohoe, T.J. and Compton, R.G. (2007) Electrosynthetic reduction of 1-iodoadamantane forming 1,1′-biadamantane and adamantane in aprotic solvents: Insonation switches the mechanism from dimerisation to exclusive monomer formation. *Ultrason. Sonochem.*, **14** (5), 502–508.

(73) Wadhawan, J., Marken, F. and Compton, R.G. (2001) Biphasic sonoelectrosynthesis. A review. *Pure Appl. Chem.*, **73** (12), 1947–1955.

(74) Asami, R., Fuchigami, T. and Atobe, M. (2008) Development of an anodic substitution reaction system using acoustic emulsification. *Chem. Commun.*, **2**, 244–246.
(75) Watkins, J.D., Amemiya, F., Atobe, M., Bulman-Page, P.C. and Marken, F. (2010) Liquid | liquid biphasic electrochemistry in ultra-turrax dispersed acetonitrile | aqueous electrolyte systems. *Electrochim. Acta*, **55** (28), 8808–8814.
(76) DelCampo, F.J., Neudeck, A., Compton, R.G. and Marken, F. (1999) Low-temperature sono-electrochemical processes Part 1: mass transport and cavitation effects of 20 kHz ultrasound in liquid ammonia. *J. Electroanal. Chem.*, **477** (1), 71–78.
(77) Paddon, C.A., Banks, C.E., Davies, I.G. and Compton, R.G. (2006) Oxidation of anthracene on platinum macro- and micro-electrodes: Sonoelectrochemical, cryoelectrochemical and sonocry-oelectrochemical studies. *Ultrason. Sonochem.*, **13**, 126.

6

Sonoelectrodeposition: The Use of Ultrasound in Metallic Coatings Deposition

Jean-Yves Hihn, Francis Touyeras, Marie-Laure Doche, Cédric Costa and Bruno G. Pollet

6.1 Introduction to Metal Plating

6.1.1 Why the Need to Cover Surfaces with Metals?

Since the mists of time, humans have strived to protect their homes, tools, industrial equipment and so on against the harmful effects of weathering over time. More recently, the development of new means of conveyance has created other requirements, since we now have to protect our vehicles and associated transport infrastructures! A natural inclination is to use a material that is intrinsically resistant, and frequently associated with specific mechanical or/and electrical properties. However, this is not particularly easy for a number of reasons. First, the required properties are not necessarily all present in the same metal. For example, gold is an inert metal, insensitive to the ravaging effects of the environment over time, but only has restricted mechanical characteristics. Furthermore, its resources are limited, generating problems of cost and availability. So why not use a more common metal like steel, which is very cost effective, easy to work with, exhibits high resistance to deformation, and when covered with a thin metallic deposit, provides the required surface properties? The story of surface coating is an exciting one, ranging from an ornamental notion to a protective function. In former times, humans discovered the possibilities of combining natural dyes stuck together using a cocktail of different substances whose combined effect gave the expected result! It is certainly possible that the very first objectives

Power Ultrasound in Electrochemistry: From Versatile Laboratory Tool to Engineering Solution, First Edition. Edited by Bruno G. Pollet.

were aesthetic only, for use in rituals or funerals. In 1500 BC, the golden calf idol cited in the *Hebrew Bible* was certainly not made of massive metal but of wood plated with a thin strip of gold. In the twentieth century, and particularly in the last decades, a wide range of coating techniques have been developed, offering considerable possibilities. These improvements are the result of new discoveries in organic and inorganic chemistry, advances in vacuum techniques, modern metallurgy, computerisation and process implementation, satisfying the requirements of increasingly stringent specifications. This natural evolution, which was slow for centuries because it was essentially based on empirical development, was nevertheless associated with better performances and a large field of applications. With the development of industrial societies, progress has accelerated to reach a scientific and technological breaking point. Surface engineering has been increasingly linked to the science of surfaces. In an industrial competitive context, it is a considerable asset to be able to manufacture products which do not wear out, and are corrosion resistant, while maintaining their electrical, optical and thermal properties and so on over time. Surface treatment, including metal plating, is absolutely essential today.

6.1.2 Process and Technology of Plating

Surface treatment is the name of the general operation consisting in influencing the surface of a material, generally down to a few microns and providing the surface with different properties to the bulk material. It is generally recognised that surface treatment can be carried out in a number of ways, as summarised in Figure 6.1. Possibilities are elaboration (wet or dry process), conversion, thermal diffusion and structural modifications.

Even though a large number of surface treatment methods should be concerned by ultrasonic enhancement, most studies are related to the wet process and particularly to electrolytic and electroless plating. This consists in electrochemically reducing a metal ion in solution to form a continuous layer at the substrate surface, using an electrical power supply in the case of electrolytic plating, or an electron provider in the electroless solution. Nevertheless, both the electrolytic and the electroless plating processes cannot be considered alone but as part of a global processing chain, including common steps of surface preparation or rinsing, as shown in Figure 6.2.

6.2 The Use of Ultrasound in Surface Treatment

6.2.1 Ultrasound in the Cleaning Step for Surface Treatment Processes

The most common use of ultrasound is in surface preparation during cleaning or degreasing steps. It is frequent to encounter such systems in surface treatment industries as ultrasound is able to replace electrocathodic degreasing advantageously, which generates problems mainly caused by hydrogen (H_2) production. Indeed, the degree of adhesion depends upon the deposit structure to the perfectly clean surface [1], and ultrasound is therefore helpful as it removes contamination from all forms of most hard, non-absorbent surfaces such as metals, plastics and ceramics. Ultrasound is often used to clean many industrial components such as jewellery, lenses or other optical parts, watches, dental and surgical instruments, and electronic equipment. Industrial ultrasonic cleaners are used in the automotive, sporting,

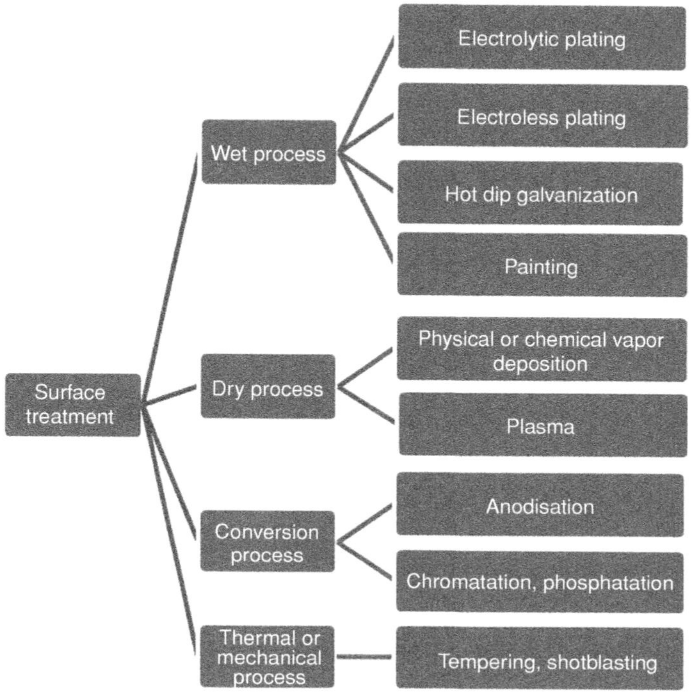

Figure 6.1 *Example of surface treatment methods.*

printing, marine, medical, pharmaceutical, electroplating, disk drive components, engineering and weapons industries. More recently, ultrasonic systems have been designed for large-scale systems such as boat cleaning [2].

The principle of the ultrasonic cleaning process could not be simpler: transducers vibrating in the range of 20 to 400 kHz are placed at the bottom of the cleaning tank, and wave propagation agitates the cleaning solution. In addition to the convective flow generated by

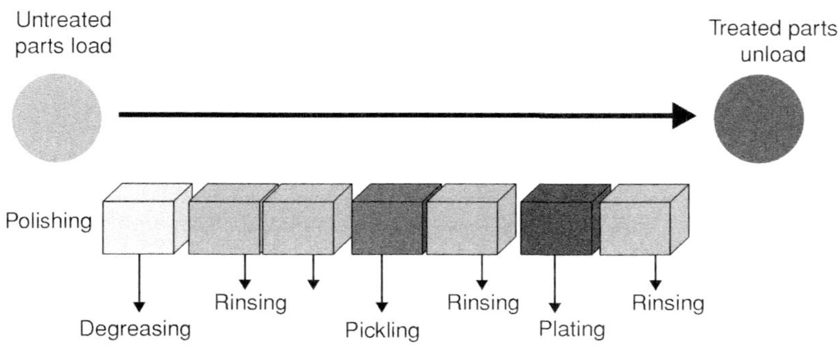

Figure 6.2 *Surface treatment steps.*

wave absorption in the liquid, acoustic cavitation occurs and acts on contaminants adhering to substrates such as metals, plastics, glass, rubber and ceramics. Millions of tiny bubbles, perfectly distributed in the cleaning tank, collapse and penetrate every orifice of the part being cleaned, removing dirt and grime within seconds. This action also penetrates blind holes, cracks, and recesses, removing all traces of contamination tightly adhering to or embedded onto solid surfaces. Its unique properties can be harnessed to clean items of all shapes, sizes and technical complexity, penetrating holes and cavities that are impossible to reach using ordinary cleaning methods.

It has already been experimentally shown that, while microscopic contaminants are best removed by lower frequency ultrasound, submicronic contaminants are often best separated at higher frequencies [3]. Moreover, the higher the frequency is, the smaller the nodes between the cavitation points are, allowing cleaning of more intricate geometries. Water or other solvents can be used, depending on the type of contamination and the work piece. Contaminants can include dust, dirt, oil, pigments, grease, polishing compounds, flux agents, fingerprints, soot, wax, mould release agents, biological, soil, blood, and so on. Ultrasonic cleaning can be used for a wide range of work piece shapes, sizes and materials, and may not require the part to be disassembled prior to cleaning [4].

Ultrasound can be used with water alone, but in most cases solvent is used to avoid the substrate from being chemically attacked by the cleaning fluid in presence of ultrasound. The cleaning solution contains additives designed to make cleaning more effective. As the cavitation activity is enhanced by the reduction of the surface tension, wetting agents (surfactants) are often incorporated in the solution. Aqueous cleaning solutions contain detergents, wetting agents and other components, which considerably affect the cleaning process. Correct composition of the solution greatly depends upon the item to be cleaned. Solutions are mostly used warm, at about 50–65 °C (122–149 °F). However, in medical applications, it is generally accepted that cleaning should be performed at temperatures below 38 °C (100 °F) to prevent protein coagulation. Water-based solutions are more limited in their ability to remove contaminants by chemical action alone than solvent solutions; for example, for delicate parts covered with thick grease. On another hand, studies on surface pretreatments, for example on anodizing film growth, are extremely rare and appear promising, but will not be discussed here [5]. Finally, this ability to clean continuously a surface has been successfully tested to prevent surface deposition in heat exchangers, and appears to be effective in avoiding crystallisation of minerals [6].

6.3 Ultrasound and Plating: Why Study Plating under Sonication?

There are a few industrial applications of ultrasonic enhancement in the plating step itself. However, it has been demonstrated at the laboratory scale that definite advantages may be gained by the use of ultrasound. Several authors such as Kochergin and Vyaselva [7], Shenoi *et al.* [1] and Walker and Walker [8] first reported the beneficial effect of ultrasound in metal deposition and plating. They also reported that plating in an ultrasonic field may produce electrodeposits with an increased hardness and brightness, better adhesion to substrate, a finer grain and a reduced porosity and internal stress – see Prasad *et al.* [9,10]. Moreover, and this is important from an industrial point of view, good quality deposits can be obtained at higher current density than normal, thus leading to a decrease in the deposition time required to produce a certain deposit thickness. Apart from the electrolytic deposition and plating of metals under sonication, many authors have explored the use of ultrasound in electroless metal plating on non-conductive substrates. For example, Touyeras *et al.* [11,12] studied the electroless copper coating of epoxy plates and Zhao *et al.* studied ceramics [13] in the presence of high-frequency ultrasound. They found that ultrasound can increase the plating rates and practical adhesion with an obvious decrease in internal stress. They also observed that there was a change of plating mechanism induced by sonication.

6.4 Electrodeposition Assisted by Ultrasound

6.4.1 The Electrodeposition Process

Electrolysis is a chemical process by which chemical reactions are produced electrically either in solutions or in molten salts. In the example of reduction at an electrode surface, five broad categories of reactions occur successively, from oxidant species in bulk solution to reduction products, as shown in Figure 6.3.

During plating processes, metallic species are present in cationic form in the solvent, and are reduced to zero oxidation state to form a metallic layer at the electrode surface. In this case, desorption and mass transfer steps towards the solution bulk are almost negligible.

6.4.1.1 Important Electrolysis Parameters

A number of parameters are important for performing electrolysis in optimum conditions. One of these is Faraday's law (see Appendix 1), consisting of a simple relationship between the amount of electricity flowing through an electrolytic cell and the amount of substance produced at the electrode. The quantity of electrical charge associated with 1 mole of electrons can be expressed in terms of coulombs (C). In electrolysis, 96 484.6 C mol^{-1} is often referred as the Faraday constant. The electrode potential then determines which electron-transfer reactions can occur, which is a major factor controlling current efficiency, space-time yield and product quality.

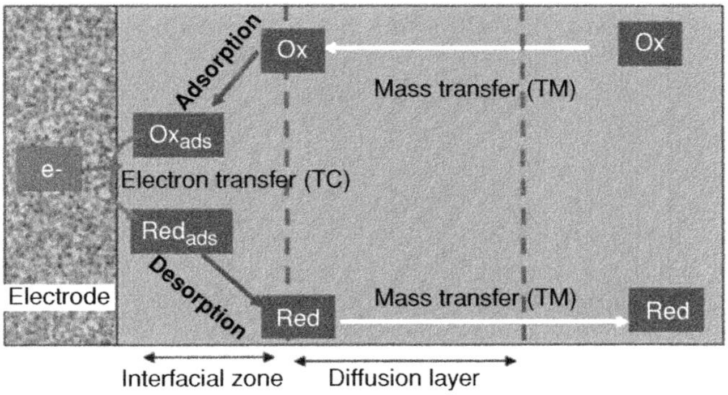

Figure 6.3 *Successive steps in electrochemical reactions.*

Special care should be taken with the electrode material. The 'ideal' electrode material for most processes should be totally inert and stable in the electrolysis medium, permitting the desired reaction(s) with a high current efficiency at low overpotentials. Often the best materials are expensive, and it is more common for the active material to be either a coating or an inert substrate (titanium or carbon for anodes, steels for cathodes). Electrode shape is also important for some specific processes. For example, electrodes are often constructed from meshes to maximize the surface area, reduce cost and weight, and enhance easily the release of gaseous products.

The properties of the electrolysis medium will be determined by the choice of solvent, electrolytes and complexing agents, pH and the concentration of each electroanalyte (its temperature is a parameter frequently used). Temperatures above ambient are employed due to their beneficial effects on the electrode kinetics and mass transport. For example, the diffusion coefficient (D) and exchange-current density (I_o) are increased.

Of particular importance when using ultrasound, the effectiveness of an electrolytic cell depends upon the nature of the mass transport employed, that is, the type of forced agitation used. Thus, the flow of electroanalyte from the bulk solution to the surface of the electrode has an important effect on the quality and the amount of electrodeposit obtained. It is known that when an electrolyte flows over an electrode, a boundary layer of solution develops. The formation of such a boundary layer has particular importance since the electrode reaction takes place within it. It has been shown that the development and scale of the boundary layer depends on the relative importance of the inertial and viscous forces. The ratio of inertial to viscous forces is known as the Reynolds number, Re, and is calculated using Equation (6.1):

$$Re = \frac{\rho l U}{\mu} = \frac{l U}{\upsilon} \tag{6.1}$$

where ρ is the density of the solution, μ its dynamic viscosity, υ its kinematic viscosity, U its limiting flow velocity and l the length of the electrode.

The boundary layer develops only below a critical value of the Reynolds number (Re_c), where viscous damping is sufficient to suppress any perturbations which arise – the flow is known to be laminar. However, above this critical Reynolds number, viscous damping is no longer predominant, and turbulence commences.

Experimentally, the Reynolds number may be determined provided that the rate of mass transfer is known, that is, that the limiting current density (i_{lim}) is known. It was shown that it is possible to derive an expression for the magnitude of the rate of mass transfer known as the Sherwood number, Sh, given by Equation (6.2):

$$Sh = H\,\text{Re}^{\text{a}}\,Sc^{\text{b}} = \frac{i_{\text{lim}} l}{nFDC^{*}} \tag{6.2}$$

where H, a and b are constants which may be obtained from experimental measurements of the limiting current density, and Sc, the Schmidt number, is defined as:

$$Sc = \frac{\mu}{\rho D} = \frac{\upsilon}{D} \tag{6.3}$$

D is the diffusion coefficient of the electroactive species. The following points should be noted:

1. For a disc electrode of radius r_e, the Reynolds number is $Re = Ur_e/\upsilon$ and the Sherwood number is $Sh = i_{lim}r_e/(nFDC^*)$
2. For a rotating disc electrode, $H = 0.620$, $a = 0.5$ and $b = 0.33$.
3. For a rotating cylinder electrode (RCE), $H = 0.079$, $a = 0.7$ and $b = 0.36$.

Therefore, using the simple, clear film model, it is possible to link the dimensionless Sherwood number, which describes mass transfer in many situations, thus allowing a comparison between different experiments in different laboratories or experimental setups. The raw recording is a current (measured in A). The result can be written in terms of current density (removing the electrode area influence), in terms of mass transfer coefficient k_d (removing the influence of the number of electrons exchanged and of electroactive species concentration), and, finally, in the Sherwood criterion (removing the influence of electrode size and electrolyte nature through the diffusion coefficient).

The immediate application and use is to be able to compare different conditions involving the use of ultrasound faced with other types of electrolyte stirring. As acoustic energy is not uniform in a sonoreactor, it is possible to map out the dissipated energy by systematically moving an electrode and by recording the mass transfer coefficient [14]. If the results are expressed as a Sherwood number, comparison between other experimental devices is possible.

6.4.1.2 *Important Characterisation Techniques*

When dealing with coating properties, it is important to be able to ensure good characterisation. Various methods are available, mostly complying with international standards, in views of obtaining coating characteristics. They are classified into four different sections [15].

(i) **Analytical methods:** all techniques able to determine the nature of the chemical elements, their concentrations and their distributions in the deposit, their valencies and, in some cases, their chemical bonds.
(ii) **Geometrical or topological methods:** determination of the roughness profile or/and three-dimensional studies to calculate physical and statistical criteria.
(iii) **Mechanical methods:** measurements of mechanical deformation, and, by means of calculation, determination of properties such as hardness and internal stress, as well as stress distribution in the deposit.
(iv) **Structural and textural methods:** observation of size and form of grains, their orientation and weight.

Additional characterisations can be performed, particularly for physical surface properties (electrical conductibility, magnetic susceptibility, etc.) as well as for sensory aspects (visual or feeling).

6.4.2 Ultrasonic Effects on Electrodeposited Coating Properties

As already shown in other chapters of this book, although there may be different origins for the various ultrasonic effects, one well-characterised effect of ultrasound irradiation is the generation and subsequent collapse of cavitation bubbles in an electrochemical cell, both within the electrolyte medium and near to the electrode surface. The electrode surface causes asymmetrical collapse of a bubble, which in turn leads to the formation of a high velocity jet of liquid being directed toward the surface. This jet, together with microstreaming, is thought to lead to the thinning of the mass transfer boundary layer at the electrode. This seems to be able to act on every step in the electrolytic process as described in Figure 6.3, and, moreover, on the overall mass transfer of the system and, consequently, on electrode reaction rates. This is experimentally observed by an increase in current and a modification in potential, and is at the origin of all measured effects. Application of ultrasound to electrodeposition has been the object of many claims since the 1950s, as acknowledged by the work of Walker and Walker [8]. Since then it has been demonstrated that there are many definite advantages to be gained by its use. Indeed, many researchers observed that plating in an ultrasonic field leads to an increase in brightness, hardness, reduced internal stress, porosity and adhesion to the substrate. Microscopic properties are affected too, with for example, a finer grain size texture. Furthermore, the possibility of obtaining good quality coatings while operating at higher current densities than normal opens new industrial applications and makes it possible to produce a defined deposit thickness in shorter time. Nevertheless, while testing various operating conditions, Walker and Walker [8] found that the effect of ultrasonic agitation on properties depends upon the metal deposited, the electrochemical parameters and the characteristics of the acoustic field used. However, it is possible to discern valuable trends, especially by grouping the results obtained for the same metal (see Tables 6.1, 6.2 and 6.3).

Table 6.1 also confirms the advantages of applying ultrasound to electrodeposition with improvement of the macroscopic properties of copper sonoelectrodeposits, the decrease in concentration overpotentials and, finally, the cathodic efficiency. Indeed, ultrasound reduces the thickness of the diffusion layer (δ) as it produces surface cavitation, microjetting and acoustic streaming. These three effects are known to assist the mass transfer of electroactive species from the bulk solution to the electrode surface [25]. Thus, under these conditions, more electroactive species are depleted at the electrode surface, which, in electrodeposition terms, means an increase in coating thickness. Cathodic efficiency is an important parameter in the electroplating industry since it indicates the percentage of current used for the electrodeposition of the metal. In many cases a cathodic efficiency of less than 100% is obtained. This is due to either the electrolysis of the background electrolyte (e.g. production of hydrogen) or the formation of a film at the electrode surface (e.g. metal oxide). It has been shown by Walker and Walker [8] that ultrasound is capable of removing hydrogen bubbles and oxide films present at the electrode surface, which may have shielded local sites required for electrodeposition, thereby providing increased efficiency. Apart from an increase in cathodic efficiency, porosity of the electrodeposit is improved in the presence of ultrasound. This is also due to the removal of hydrogen bubbles within the electrodeposit by the collapse of cavitation bubbles, and to the intense acoustic streaming at the electrode surface. An example of this capacity to reduce hydrogen evolution had already been highlighted for Zn-$NiCl_2$ systems [26]. Hydrogen evolution is the result of pit corrosion experienced by zinc, and its rate is the net result of surface kinetics, ohmic resistance or/and mass transfer contributions both inside and outside the pit. Ultrasound has been shown to remove the mass transfer contribution, eliminate the concentration gradient in the pit and ensure the process is ohmically controlled.

Table 6.1 *Macroscopic properties of copper sono-electrodeposits.*

	Advantages over *silent* conditions	Authors
Electroplated copper on iron from an acid sulfate bath	Cathodic efficiencies, hardness, brightness, and adherence of electrodeposition increase in comparison to still or magnetically stirred baths. Morphology is more compact and structured.	L. Martins, J.I. Martins; A.S. Romeira, M.E. Costa and M. Bazzaoui [16]
Electroplated copper on copper from a cupric-EDTA bath	Limiting current densities and cathodic current efficiencies both increase while grain size was reduced.	A. Chiba, W.C. Wu [17]
Electrodeposited copper at low temperature	Deposits are no longer powdery and are less scattered. The coating is more uniform and cleaner (absence of oxidised parts or adsorbed sulfurs).	A. Mallik, B.C. Ray [18]
Copper multilayer coating on stainless steel	Hardness decreases, whereas wear resistance and adhesion on the substrate are clearly enhanced. Corrosion resistance (NaCl 3%) is greatly improved	R.H. Cui, Z.M. Yu, Y.T. He and W.J. Shu [19]
Electroplated copper on copper from a cupric-EDTA bath	Deposition rate slightly increases (10%), hardness increases from 120 to 140 H, and grain size is reduced. Appearance of corrosion product is delayed in case of sonication.	W.C. Wu, A. Chiba, K. Nakanishi [20]
Electroforming micro components in copper under ultrasound	The layer thickness increases up to 40% more than regular electroformed copper, while the fracture strain remains constant.	Q.A. Liao, W.P. Li, H.C. Liu and L.Q. Zhu [21], Q.A. Liao, L.Q. Zhu, H.C. Liu and W.P. Li [22],
Wire bonding by copper electroplating	Cu layer with a homogeneous and compact granular structure has better wettability and higher adhesive strength between Cobalt based amorphous wire and the Cu layer.	J.S. Liu, J.F. Sun, D.W. Xing, X.A. Xue, S.L. Zhang, H.A. Wang [23]
Copper sonoelectrochemistry with sulfate based electrolytes (Quartz micro-balance)	Cathodic efficiency is increased from 84% to 100% (copper deposition), but not anodic efficiency (dissolution). Morphology is affected.	O. Schneider, S. Matic, C. Argirusis [24]

Table 6.2 *Macroscopic properties of nickel sono-electrodeposits.*

	Advantages over *silent* conditions	Authors
Electroplated nickel from an acid sulfate bath	Cathodic efficiency, hardness, brightness, and adherence of electrodeposition increase in comparison to still or stirred baths. Morphology is nanostructured.	N. Sulitanu, C. Pirghie, I. Sandu [27]
Electroplated nickel on brass from a Watt bath	Hardness increases, and internal stress is reduced, but depends on the sample location	F. Touyeras, J.-Y. Hihn, X. Bourgoin, B. Jacques, L. Hallez, V. Branger [28]
Electroplated nickel on copper from a Watt bath	Deposition rate increases, that is, exchange current densities increase and charge transfer reaction is promoted.	Q.A. Liao, W.P. Li, H.C. Liu, and L.Q.K. Kobayashi, A. Chiba, N. Minami [29]
Nickel alloy (Ni-Co and Ni-Fe) electrodeposition	Hardness of both alloys is increased, and their composition is still changed (cobalt content of Ni-Co is increased whereas iron content of Ni-Fe decreases or remains constant). Ni alloys are more ductile without loss of tensile strength	T.R. Mahmood, J.K. Dennis and P.L. Barrett [30]
Nickel-Co/Al_2O_3 composite coating by pulse reversed current	Composite coating becomes uniform, compact and the grain become fine, with a lower content of Al_2O_3. μ-hardness is improved first before decreasing when residual macrostress is increased	L.M. Chang, H.F. Guo, M.Z. An [31]
Electroplated and electroformed nickel from a sulfamate bath	Better and uniform filling of microelectroforms with high height to width microstructure features.	P.M. Ping, D. Zhu, Y.Y. Hu, Y.B. Zeng [32]
Electroplated nickel on brass from a Watt bath on brass millimeter-sized groove features	Topography of the Ni-deposit is markedly affected, and filling characteristics on 3D-patterned geometry are considerably improved	J.A.D. Jensen, P. Pocwiardowski, P.O.A. Persson. L. Hultman, P. Moller [33]

It can be seen that global macroscopic parameters are enhanced by ultrasonic exposure. Concentration overpotential is an important factor for electrochemical reactions. This is due to concentration changes in the near vicinity of the electrode surface. Ultrasound reduces concentration overpotentials, certainly by an increase in mass transfer of metal ions from the bulk solution to the electrode surface.

Table 6.3 Macroscopic properties of noble metal sonoelectrodeposits.

	Advantages over *silent* conditions	Authors
Electroplated silver on nickel	Silver covered the surface extensively, and the grains are finer	O. Yasushi, Y. Morihiro [34]
Recovery of silver from photographic processing solutions	Cathodic efficiency and mass transfer strongly increase. By continuously removing silver sulfur from the electrode; charge transfer is improved.	B. Pollet, J.P. Lorimer, S.S. Phull, T.J. Mason, D.J. Walton, J.-Y. Hihn, V. Ligier, M. Wéry [35], B. Pollet, J.P. Lorimer, S.S. Phull, J.-Y. Hihn [36], B. Pollet, J.P. Lorimer, J.-Y. Hihn, S.S. Phull, D. Feakins, D.J. Walton, T.J. Mason [37], B.G. Pollet, J.P. Lorimer, J.-Y. Hihn, F. Touyeras, T.J Mason, D.J. Walton [38]
Electroplating of iridium	Plating rate increases without increasing the applied potential in the negative direction (less hydrogen generation at high current densities; current efficiency is improved). No changes to hardness, but a reduction in internal stress and thus cracks, avoiding use of chemical additives.	T. Ohsaka, M. Isaka, K. Hirano, T. Ohishi [39]
Electroplating of various metals – a review of properties	Metals concerned are silver, copper, chromium, nickel, zinc, cadmium, lead, tin, on grain size, hardness, internal stress, porosity and brightness	C.T. Walker, R. Walker [8]

From Table 6.3, it is evident that, when examining the extensive electrochemical study on silver recovery [35–38], ultrasound will lead to great advantages in environmental applications [25,40,41] and in electrochemical removal of metals in dilute aqueous solutions.

6.4.3 Microscopic Effects of Ultrasound on Electrodeposited Metal Coatings

Many macroscopic properties of the electrodeposits are frequently explained by the formation of small-sized grains. Liao *et al.* [21] found that bulk crystals of electroformed copper layers are changed to nanocrystals of circa 10 nm in size. This is induced by the numerous collapses of bubbles on the surface, generating a large number of nuclei to start metallic crystal growth. This observation, although reported in many works, is not shared by all authors. Chiba *et al.* [17] for example, measured crystal grain sizes of circa 0.025 μm in silent as well as 40 kHz ultrasonic conditions. Present or not, the reduction of crystal sizes cannot account alone for all the enhanced properties under sonication, for example brightness according to Walker and Walker [42]. Indeed, although sonication usually increases surface smoothness and brightness, a careful comparison between SEM pictures and their roughness led Walker and Walker [42] to consider the action of shock waves and cavitation erosion on the growing surface of the electrodeposit. Growth perpendicular to the surface is therefore inhibited, resulting in a smoother finish of the deposit. Liao *et al.* [22] observed that the whole crystalline structure of thick copper electrodeposits may be affected by ultrasonic agitation, forming regular columnar crystals. This type of preferential layer orientation can be characterised by X-ray diffraction (XRD) and is found to grow preferentially along the (2,0,0) plane, instead of a (2,2,0) plane in silent conditions [22]. Similarly Cui *et al.* found that after sonication, the diffraction peaks of Cu (2,0,0) and Cu (1,1,1) were obviously larger than Cu (2,2,0) in silent conditions. Even in electroless conditions, Liu *et al.* [43] observed an increase in Cu (2,0,0) and Cu (1,1,1) orientations after ultrasound was introduced. The situation becomes more complicated as the crystal growth orientations may be influenced by various operating parameters. Chiba and Wu [17] presented an extensive study of the effect of ultrasonic irradiation on texture coefficients of copper electrodeposits from a cupric-EDTA bath, reminding us that current density, EDTA and copper concentration have to be taken into account. While increasing current density in silent conditions, they reported a decrease in Cu (1,1,1) orientations, while Cu (2,0,0) orientations increased. While the coefficients for these planes were equal at 15 mA cm^{-2}, Cu (2,0,0) was preponderant. In the presence of 40 kHz ultrasound, this value was completely reversed and the texture coefficients of Cu (1,1,1) increased while those of Cu (2,0,0) decreased. They then increased the EDTA/copper(II) ion ratio [Cu^{II} = 0.1 mol L^{-1} in the 0 to 5 range]. In silent conditions, the coefficients of the Cu (1,1,1) plane decreased, while those of the Cu (2,0,0) decreased with respect to the EDTA/Cu^{II} ratio. The values of Cu (1,1,1), very high for low ratios, dropped below Cu (2,0,0) when the EDTA/Cu^{II} ratio was 3.5 in silent conditions. This coefficient crossing occurred only for an EDTA/Cu^{II} of about 1.5 in presence of 40 kHz irradiation. Finally, while they reported that plane coefficient distribution is virtually constant in one sample, they observed a strong dependency of the respective position of the sample to the ultrasonic transducer. Another study by Kobayashi *et al.* [29] investigated electrolytic deposition of nickel on copper from Watts' bath. The found that substrate texture coefficients were 75% of the Cu (1,1,1) face and 10% of the Cu (2,0,0) face. In the absence of ultrasound, the texture coefficients of deposited film did not markedly change in the 10–200 mA cm^{-2} range of cathodic current density, although morphological variations were observed. Similarly, Chiba and Wu [17] also found that ultrasound greatly affects nickel film texture coefficients; they observed that when the cathodic current increased, Ni (2,0,0) decreased whereas Ni (1,1,1) increased. They also observed: (i) a dependency of the ultrasound frequency in the following order: silent conditions $<$ 28 kHz $<$ 100 kHz $<$ 45 kHz, and (ii) the crystal growth of the (1,1,1) face was faster than that of the (2,0,0) face, so that the (1,1,1) face was surrounded by the (2,0,0) face which disappeared during crystal growth. In their conditions, they concluded that ultrasonic irradiation increases the difference in the activation energies of the surface diffusion of Ni adions to the (1,1,1) face – 44.3 kJ mol^{-1} – and the (2,0,0) face – 119 kJ mol^{-1}.

Surprisingly, even if the literature contains a large number of publications devoted to the effects of ultrasound on electrochemical systems, only a few of them relate to the electrochemical studies for electrodeposition of metals under insonation. Kobayashi *et al.*

[29] observed for nickel electrodeposition, that the exchange current density (i_o) obtained by Tafel plots analysis increased in the presence of ultrasound, but the kinetic parameters, such as the transfer coefficient (α), did not really change. Using AC impedance, it has been shown that application of an ultrasonic irradiation resulted in increased charge transfer rate and solution resistance, while the capacitance of the double layer at the metal electrolyte interface was not really affected [29]. Once again, they found an ultrasonic frequency effect as charge transfer rates at 28 and 45 kHz were faster than those obtained under silent conditions. This obvious decrease in charge transfer resistance in the presence of ultrasound reveals an influence on charge transfer processes [29].

The Compton group [44] conducted an extensive potentiodynamic study (mainly by cyclic voltammetry) on mechanisms of zinc, lead, cobalt and mercury deposition on glassy carbon (GC) electrodes under 20 kHz ultrasound. Systems of relatively high metal concentration (1 M) and low concentration (10 mM) were compared. The objective of these studies was to examine the electrochemical behaviour of well-characterised metal deposition systems at various ultrasonic intensities, between silent to 238 W cm^{-2}, both above and below the threshold at which cavitation occurs. They noted that metal systems could be divided into two categories as a function of their concentration ranges, as the responses to ultrasound were found to vary considerably according to low and high metal contents [44]. High concentration systems (where deposition is under charge or interfacial control) showed little modifications to the deposition parts of their cyclic voltammograms, whereas their stripping peaks decreased as ultrasonic intensity increased [44]. They assigned their observations to the result of ablation of metal by ultrasound, removing material from the surface before it was stripped. Indeed, if the metal adheres satisfactorily to the electrode (as for example in the case of cobalt), the effect of acoustic streaming will outweigh the effect of ablation (caused by cavitation), and the ratio of stripping charge over deposition charge will be greater. Low concentration systems (in which deposition is 'purely' under diffusion control) showed a decrease in overpotentials required for deposition to start, and a marked increase in the total deposition current [44]. They concluded that, in electrodeposition/plating under ultrasound, the critical effect depends upon a mass transfer increase, which may be enough to change a diffusion-controlled system into a charge-controlled system. In this case, ultrasound also removes some material from the electrode surface, but may not have any effect on growth via charge transfer from the electrode to the metal ion.

This continuous modification of the surface by ultrasonic erosion had already been observed in other studies related, for example, to the recovery of silver from photographic processing solutions, which are the 'fix' (dilute aqueous $Na_2S_2O_3/NaHSO_3$) and the 'bleach-fix' solutions (dilute aqueous $Na_2S_2O_3/NaHSO_3/NaFeEDTA$) [38]. Thiosulfate ions, in the form of sodium or ammonium thiosulfate, are commonly used to remove any excess silver present in photographic film solutions during the development process. Electrochemical reduction of silver thiosulfate was studied potentiodynamically on platinum electrodes in the absence and presence of ultrasound (20 kHz). Pollet *et al.* [38] showed that this system was irreversible, and the reaction was both diffusion and kinetically controlled. In their conditions, ultrasound greatly improved mass transfer, which was explained to be due a change from diffusion to mainly convection mode. The slowest step was the kinetic reaction caused by the chemisorption of ions at the electrode surface attributed to the formation of an insoluble 'black' silver sulfide (Ag_2S) which prevented further electrodeposition of metallic silver. They also found that under ultrasound, the heterogeneous rate constant

(k_f) was increased by approximately 10-fold compared to silent conditions, mainly due to continuous abrasion of this sulfide film on the electrode surface [38].

The Compton group [45] studied and generated intrinsic details on mechanisms for cobalt electrodeposition on glassy carbon from aqueous sulfate solutions by using chronoamperometry and atomic force microscopy (AFM). They found that by using the Scharifker and Hills model for potentiodynamic growth (3D) at suitably high overpotentials, cobalt electrocrystallisation under silent conditions occurred via a diffusion-controlled mechanism. It is well known that the electrocrystallisation of cobalt proceeds via a 3D growth mechanism and three overpotential regions are usually defined as:

(i) Low overpotentials (−1.0 V versus SCE) where growth occurred under charge transfer control.
(ii) At −1.0 V versus SCE where nucleation is found to be instantaneous.
(iii) High overpotentials (−1.25 V versus SCE), negative enough to induce progressive nucleation.

Using Scharifker and Hills limiting cases indicate that, at low overpotentials, a 'progressive' nucleation behaviour is favoured, because an increase in overpotential promotes 'instantaneous' nucleation [45]. The rate of deposition increases versus overpotential, although hydrogen evolution is found to compete at potentials around −1.25 V versus SCE.

In the case of sonoelectrochemical experiments, Compton *et al.* [45] found that ultrasound affects the rate and mechanism of cobalt electrodeposition, as well as the surface morphology of the electrodeposit. Analysis of chronoamperometric data indicated that ultrasound strongly influences the nucleation behaviour of cobalt on glassy carbon electrodes compared to silent conditions. Irrespective of overpotential values, ultrasound systematically favoured instantaneous nucleation behaviour. The authors attributed this fact to the ablative effect of ultrasound that strips newly formed nuclei adhering poorly at the surface, combined with the enhancement growth of existing nuclei [45].

This ability to control diffusion layer characteristics without resorting to adding organic compounds to metal plating bath solutions is an important and promising aspect of ultrasound utilisation.

6.4.4 The Influence of Acoustic Energy Distribution on Coatings

The fact that researchers in the field have not used identical experimental conditions may explain why the results available in the literature can sometimes appear to be contradictory. Indeed, distribution of acoustic energy in a sonoreactor is a key point for understanding variation in electrochemical parameters. This distribution has been studied in great depth by the Besançon group, extensively mapping sonoreactors operating at low ultrasonic frequencies using either electrochemical sensors [46–48] or laser tomography [49] and particle image velocimetry (PIV) [50]. The problem is completely different at higher ultrasonic frequencies [51], and additional investigations have been carried out in the particular case of HIFU (high intensity focused transducers) [52, 53]. Most sonoelectrochemical experiments were performed using 'face-one' geometries (see previous chapters). In this case, the exact location of the sample in the acoustic field can generate important changes in the flow pattern/distribution at its surface, for example by studying nickel electrodeposition on brass substrates exposed perpendicularly to the ultrasonic field. Here, experiments

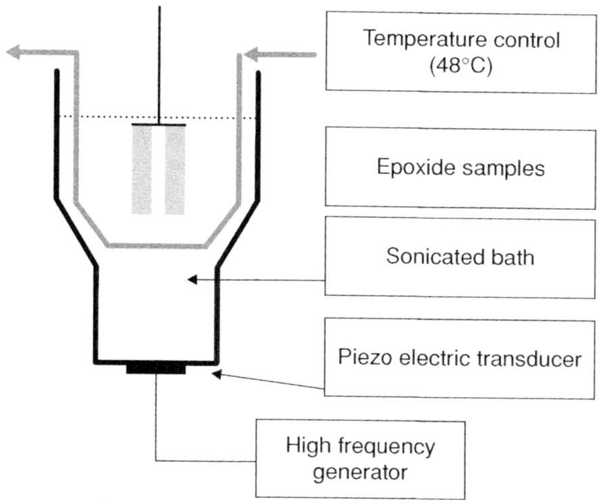

Figure 6.4 *Sonoreactor for metal coating applications.*

were performed at various ultrasonic frequencies (300 to 800 kHz) in a 500 mL cell. The transducer was insulated by a protective stuck-on glass plate. Ultrasound irradiation was generated parallel to the liquid surface, and the brass specimens were suspended (two at a time) vertically (Figure 6.4).

Parameters such as the acoustic power and stirring conditions were carefully determined in views of characterising exposure conditions. The acoustic power transmitted to the liquid electrolyte was determined by calorimetry [54], and mass transfer rate measurements were performed under sonication using the electrodiffusional method [55]. It was found [55] that: (i) the grain size varied as a function of the ultrasound power, for each ultrasonic frequency, and (ii) the grain distribution throughout the sample was altered especially when a brightener was used. At 500 kHz an alternation of bright and matt areas separated from each other by 1.5 mm were observed. This distance corresponded to a half wavelength (Figure 6.5). SEM pictures showed grain arrangement in the matt areas, with small, rounded and unorganised grains. In the bright areas, grains were organised in lines perpendicular to the wave propagation direction. Grain size depended upon the ultrasonic power, which led the authors to attribute the matt areas to pressure nodes, and the bright areas to pressure antinodes. Without the aid of a brightener, crystals exhibited a pyramidal organisation, less sensitive to sonication. These morphologies were comparable to the SEM pictures showing grain arrangement in silent conditions.

In general, it was found that other macroscopic properties are not so dependent on the effect of ultrasound. For example, deposit thickness is not clearly modified, whether or not a brightener is added. In their studies, internal stress was measured in two ways: (i) the foil method and (ii) the X-ray diffraction (XRD) method. The first method consists in calculating a global value from bending a sample of reduced thickness after plating only one face in the same conditions as the other samples [56]. Measurements are taken while removing the mask. However, this has the drawback of assuming that internal stress distribution is linear throughout the layer. The XRD method is based upon the determination

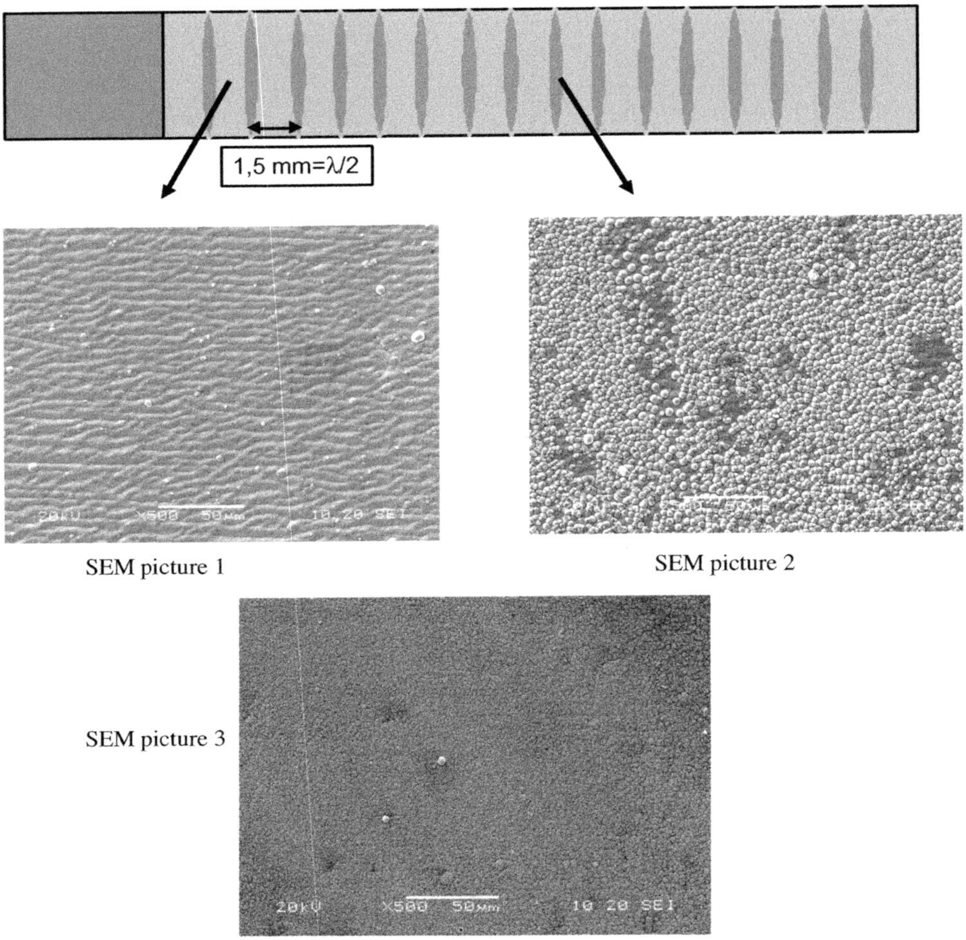

Figure 6.5 *Surface aspects of coatings produced under perpendicular ultrasound irradiation at 500 kHz: Picture 1: SEM in matt areas – Picture 2: SEM in bright areas – Picture 3: SEM in silent conditions (Reprinted with permission from [28] Copyright (2005) Elsevier Ltd).*

of crystalline network elastic deformation. Thus, for a crystalline material, the distances between two planes are a characteristic of the material in a given metallurgic state. If this peak is slightly displaced, this leads to a modification of this distance, which is linked to internal stress modification. Values from both methods are shown in Figure 6.6.

The influence of ultrasound is noticeable, and the global curve shape remains the same irrespective of the method employed. However, values obtained by the XRD method always remain higher than those obtained by the foil method. This can be accounted for by the fact that the XRD method relates to the crystalline phase, that is, in this case the crystal grains, whereas the whole layer using the foil method includes amorphous zones. Here, a conclusion can be drawn that, if internal stress relaxation takes place in amorphous zones, grains are more sensitive to ultrasonic irradiation. Then, in silent conditions, where

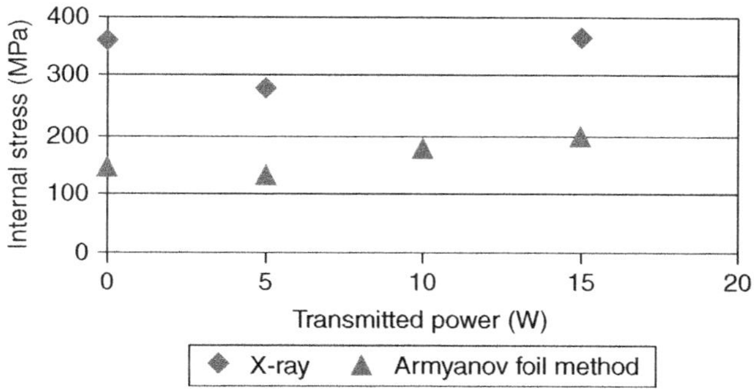

Figure 6.6 *Internal stress of copper deposits obtained at 500 kHz measured by both X-ray and Arnyanaov methods (Reprinted with permission from [28] Copyright (2005) Elsevier Ltd).*

amorphous zones are less present, relaxation will not occur to any great extent, leading to a large difference between global internal stress and internal stress in the grain compared to ultrasonic conditions.

In this investigation, hardness was measured with a micro-hardness HMW-M3 Shimadzu apparatus, and revealed dependence of ultrasonic frequency, power and the presence of brighteners. For example, in presence of a brightener and at 500 kHz, hardness measurements presented values ranging from 200 to 350 HV as shown in Figure 6.7. The figure shows that an alternation of zones of various hardnesses can be observed, in accordance with visual and SEM observations where zones of different grain sizes still exist. The step between two alternations is approximately 1.5 mm, directly connected to a half of the wavelength.

Thus it seems difficult to predict coating properties after exposure to ultrasound during their formations, in particular in industrial conditions, due to the main issues of monitoring and controlling ultrasonic power distribution and process scale-up. Nevertheless, technological progress in ultrasonic wave control and transducer design has shown, thanks to an original study, that these issues may be overcome [33].

Electroforming, which means creating electrodeposits of thick layers, is a limiting electrochemical process. Indeed, the depletion of electroanalyte ions near the cathode surface due to insufficient mass transfer may lead to unwanted side-reactions leading to the formation of metal hydroxides – the so-called 'burning' of the coating. This is particularly critical in the case of microstructure and 3D patterned geometries, for example for machined

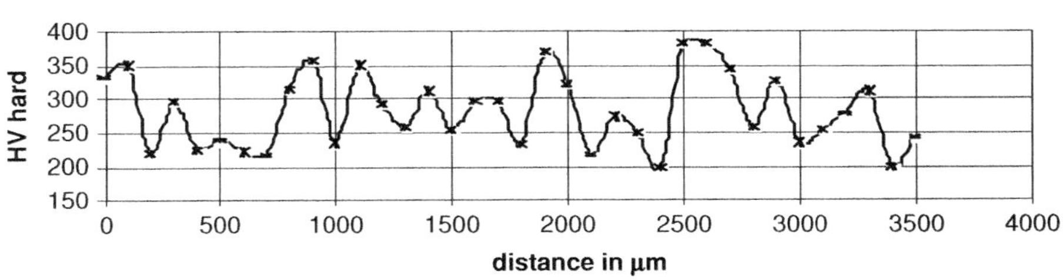

Figure 6.7 *Hardness profile in the wave propagation direction at 500 kHz and 15 W (Reprinted with permission from [28] Copyright (2005) Elsevier Ltd).*

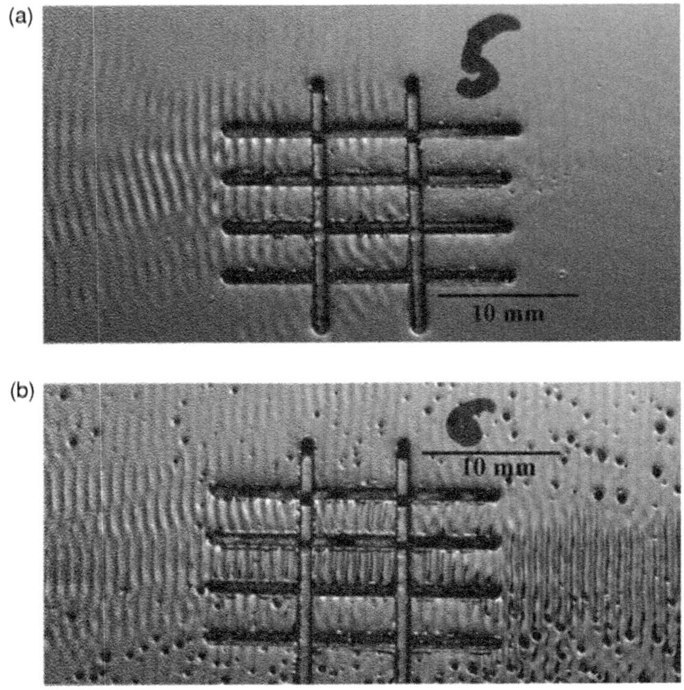

Figure 6.8 (a) Ni deposit morphology in presence of low power travelling waves (1 MHz). (b) Ni deposit morphology in low power standing waves (1 MHz) (Reprinted with permission from [33] Copyright (2003) Elsevier Ltd).

microstructures of millimetre sizes (three orders of magnitude larger than the thickness of the acoustic boundary layer at 1 MHz). Here, the originality of the investigation on nickel electroforming lies in the fact that the authors modulated the ultrasonic field to obtain travelling ultrasonic waves as well as ultrasonic standing waves. In their experiments, the set-up consisted of a low-power transducer (5–10 W, 1 MHz) sonoreactor where samples were inserted vertically into the electrolyte and inclined approximately by 10° (of the ultrasonic wave propagation). The resulting Ni deposit morphology is shown in Figure 6.8, where it can be seen that the ultrasonic wave, travelling from the right-hand side to the left-hand side of the sample, induces a fringe pattern separated by roughly half a wavelength (0.5 mm) [33]. In the second case, the Ni deposit formation resulted from the standing waves established between the two low-power transducers facing each other at a distance of 25 cm. Well-defined parallel fringes were observed at the surface in the region concerned by the acoustic field, and pits appeared, resulting certainly from a marked increase in local acoustic activity and, therefore, active cavitation events [33].

Typical results of cross-section are presented in Figure 6.9. The Ni layer elaborated in silent conditions shows a thick film at the bottom of the groove, and thin walls followed by thin films at the edges of the substrate surface. Conversely, the Ni-layer deposited in presence of sonication by travelling ultrasonic waves presents a uniform thickness and the

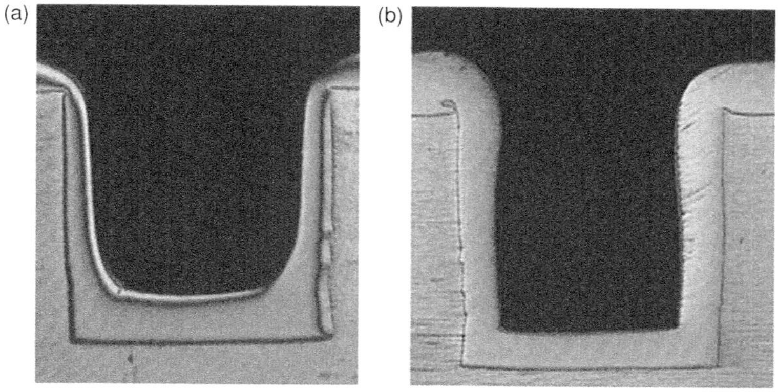

Figure 6.9 (a) Ni deposit morphology in silent conditions; (b) Ni deposit with low power (5 W) from one transducer (1 MHz traveling waves) (Reprinted with permission from [33] Copyright (2003) Elsevier Ltd).

sides of the grooves are determined by virtually 90° angles. No voids or inclusions were detected by optical microscopy and transmission electron microscopy (TEM).

Insonation generated near boundary ultrasonic streaming, thus enhancing electrolyte exchanges in the vicinity of the cathode surface, induces better functionality of the organic additive used (bytyne diole) and a reduction in its inhibiting effect at high current density areas like corners and edges [33]. This method successfully improved the uniformity of nickel film thickness in millimetre-sized tranches, a finding that could be of great importance in future industrial electroforming applications.

6.4.5 Influence of Ultrasound on Copper Electrodeposition in Unconventional Solvents

Many electrochemical processes, such as the electrodeposition of metals in aqueous solvents, are limited by reactions that control the potential window, which involve reduction and oxidation of the solvent. Several non-aqueous solvents, including acetonitrile, DMF, DMSO and so on, can be suitable as they exhibit a potential window that is 1.5–2.5 times wider than that of aqueous electrolytes. A wide range of salts can be used for aqueous electrolyte solutions, while salts for non-aqueous electrolyte solutions are more restricted and consist generally of large cations (e.g. tetraalkylammonium cations) and anions (e.g. hexafluorophosphate, tetrafluoroborate and perchlorate) to ensure full dissociation [57, 58]

In the last decade, room temperature ionic liquids (RTILs) that is, compounds made up of ions rather than molecular species, by combinations of organic and/or inorganic cations and anions, have attracted considerable interest for use as non-aqueous solvents [59–63] due to the following advantages:

(i) wide electrochemical windows of up to 4 V which allow electrodeposition of a number of metals that cannot be achieved in aqueous solutions
(ii) relatively high conductivities
(iii) relatively low viscosities.

For example, the electrodeposition of copper in RTILs has been widely studied [59,61]. It was found that the reduction of Cu^{II} to metallic Cu occurs in two one-electron steps: in the first step Cu^{+} is formed and in the second step the metal is deposited. However, the use of this type of ionic liquid has several disadvantages such as toxicity and cost, making them somewhat impractical for larger industrial applications such as metal electroplating and electrodeposition. An alternative to the above non-aqueous solvents are deep eutectic solvents (DES), which are a type of ionic solvent with special properties composed of a mixture forming an eutectic with a melting point far lower than either of the individual components [57]. Compared to ionic liquids, they share many characteristics, but are only ionic mixtures and not ionic compounds, and are known to be less toxic, air and moisture stable, biodegradable and economically viable to large-scale processes [64]. DESs are either urea or ethylene glycol and choline chloride-based ionic compounds [65]. A variety of applications are available, such as electropolishing [66] or selective metal extraction from oxides [67]. Recently, it has also been shown that DES based on choline chloride (as the quaternary ammonium salt) and either urea or ethylene glycol (as hydrogen bond donors) can be employed in the electrodeposition of zinc, tin and zinc-tin alloys. However, electrochemical reactions are rather slow due to the high viscosity of the electrolytes. This, in turn, reduces the mass transfer of the electroanalyte due to low diffusion coefficients [68].

Pollet *et al.* [57] showed that deposition of copper in both water and DES is greatly affected by ultrasound at the two frequencies of 20 and 850 kHz employed. Limiting current densities were obtained in both solvents under sonication at 20 and 850 kHz and a 10-fold and fivefold increase in currents in aqueous potassium chloride and glyceline 200 compared to silent conditions was observed, respectively. The difference in viscosity of water (KCl) and glyceline 200 was found to be a crucial parameter in the evaluation of limiting current densities. Nevertheless, it is important to note that at 20 kHz, current spikes were observed indicating that cavitation phenomena occurred. Also, current spikes are not observed below 41 W cm^{-2} but are more pronounced at 77 and 161 W cm^{-2} (20 kHz). This is an interesting finding, suggesting that a cavitation threshold exists, which is more pronounced for viscous electrolytes. It was shown by several workers in the field using high speed amperometry and voltammetry techniques in the presence of ultrasound and in aqueous solutions to study cavitational activities, that the observed current spikes are due to the collapse of asymmetric cavitation bubbles on the electrode surface [69–71]. Furthermore, it was also demonstrated that the experimental current spikes were not related to ultrasonic frequency induced by cavitation events and that if the electrode area was increased, current spikes (in the plateau region) were significantly reduced [72]. To verify that cavitation occurs in glyceline 200 (containing $CuCl_2$), a simple experiment was performed by inserting a piece of aluminium foil in a sonicated (20 kHz, maximum 'output' power) environment. The 'foil test' showed that cavitation occurred in the glyceline 200 as the piece of aluminium foil was violently destroyed into fine particulates after 3 minutes of sonication [57, 58].

During this work, it was shown that a large number of limitations stem from the cell design used in laboratories. For the first time, Costa *et al.* [73,74] showed it is possible to (i) sonoelectrodeposit metals in RTILs and (ii) use RTILs to study transport-limited currents in an especially designed microsonoreactor (Besançon cell, Figure 6.10) which was characterised at three low ultrasonic frequencies in presence of water, acetonitrile and imidazolium-based ionic liquids ($[BuMIm][(CF_3SO_2)_2N]$). This new sonoelectrochemical cell was developed for use in aqueous and the above cited 'exotic' solvents at various

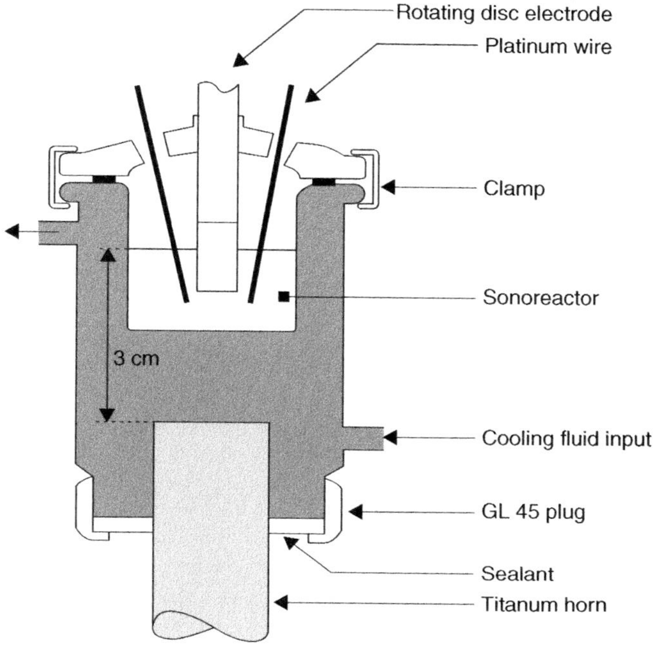

Figure 6.10 *The Besançon-cell – the double structured cell used in sonoelectrochemistry of metals at low frequencies.*

coolant pressures. This jacketed cooling 'microsonoreactor' is based on a particular design consisting of off-setting the ultrasonic probe out of the reaction volume (inner cell, V = 7 cm^3) in order to avoid contamination and to ensure perfect electrical insulation from the ultrasonic probe usually made of Ti alloy (Ti–6Al–4V). It was found that to ensure maximum ultrasonic effect, that is, ultrasonic intensity within the inner reaction volume, coolant pressure had to be increased up to 1.5 bars. Extensive characterisation and calibration were carried out, and a linear relationship was obtained between the ultrasonic (20 kHz) powers and the coolant pressures [74]. They conclude that it is possible to obtain 'acceptable' ultrasonic powers to undertake sonoelectrochemical experiments using this new sonoreactor. In particular, high mass transfer enhancements were observed, characterised by an average Sherwood number of 6500 while the value obtained with an electrode (RDE) rotating at 4500 rpm under silent conditions was only 1200 [74].

6.4.5.1 Investigations of Copper Reduction in Aqueous Solutions

The study of copper salt behaviour was conducted to enable data comparisons with previous work available in the literature using various solvents [73]. As a reference, the sonoelectrochemical reduction of copper sulfate (15 mM) in the presence of sodium sulfate (0.3 M, background electrolyte) in aqueous solutions (pH = 1) was performed. Linear sweep voltammograms (scan rate = 2 mV s^{-1}) were recorded both in silent (RDE 5000 rpm) and ultrasonic (40 kHz, transmitted power = 0.9 W cm^{-2}) conditions. Results shown in

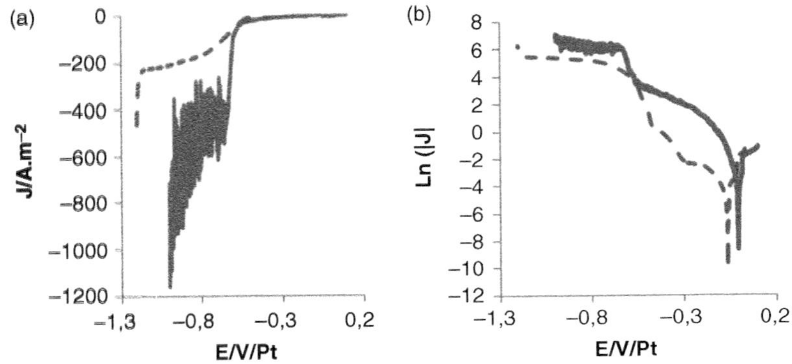

Figure 6.11 *(a) Polarisation curves at 2 mV s^{-1} $CuSO_4$ (15 mM) in aqueous media pH = 1 on a platinum disc in silent (– –) and ultrasonic (—) conditions. (b) Same curves plotted according to the Tafel method in silent (– –) and ultrasonic (—) conditions.*

Figure 6.11(a) indicate that the cathodic current density increases with increasing electrode potentials until a plateau is reached at high cathodic overpotentials. In the presence of ultrasound current densities measured at this plateau are twice as large as in silent conditions for an electrode rotation speed of 5000 rpm. It appears once again that ultrasound improves mass transfer significantly. Nevertheless, it is important to note that the plateau is not well established in the presence of ultrasound, as hydrogen release occurs at lower potentials than in silent conditions. Voltammetric curves were plotted according to the Tafel method to highlight kinetic mechanisms at low cathodic overpotentials [Figure 6.11(b)] [73]. In silent conditions, the Tafel curve presents three successive shoulders, suggesting a reduction mechanism more complicated than expected and referred to in the literature as the under potential deposition (UPD) and overpotential deposition (OPD) [73, 75].

The Tafel representation recorded under sonication [Figure 6.11(b)] indicates that the copper reduction mechanism still remains a two-step process. The general pattern suggests that the first step is governed by electron transfer (reduction of Cu^{2+} to Cu^{+}) whereas the second step is limited by mass transfer phenomena (reduction of Cu^{+} to $Cu^{0\circ}$). Moreover, the plot shows only two shoulders and the open circuit potential (OCP) of the working electrode increases from −0.065 V versus Pt (platinum used as a reference electrode) to −0.006 V versus Pt due to the influence of ultrasound. Through its mechanical action, ultrasound cleans off the electrode surface continuously and prevents sulfate adsorption (no adsorption peak of species was observed), accounting for a potential close to 0 V versus Pt. This observation is in good agreement with the sulfate adsorption mechanisms described by Wilms, and thus provides a good explanation for the effect of ultrasound in this redox system [76].

The study then focused on the formation of Cu coatings by chronoamperometry at −0.75 V versus Pt. The coulombic charge was fixed at $Q = 3$ C in order to obtain copper layer of circa 5 μm in thickness. Figures 6.12(a) and 6.12(b) show that Cu coating thicknesses are homogeneous at around 4.1 μm and the Cu electrodeposit is made of crystallites of more or less spherical shapes, with an average diameter of 2 μm. Here, the Cu film structure can be considered as orderly and present a bright aspect at a macroscopic scale.

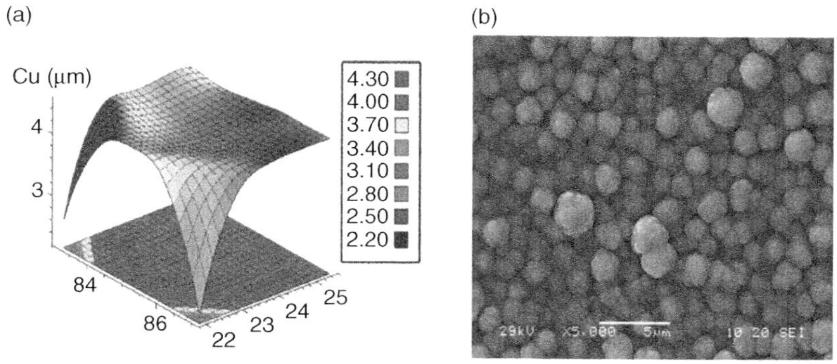

Figure 6.12 *(a) Mapping copper deposit thickness in silent conditions on a plate disc rotating at 2000 rpm in a solution [$CuSO_4$] = 15 mM [Na_2SO_4] = 0.1 M. (b) Image in scanning electron microscopy (SEM) of the deposit (×5000).*

In the presence of ultrasound, deposits were found to be thicker and exhibited lower thicknesses in the centre, probably due to the action of cavitation bubbles [Figure 6.13(a)]. Nevertheless, the crystallite structure was finer than that observed in silent conditions, showing a hexagonal pattern of smaller dimensions (μm range of magnitude). Furthermore, porosity was reduced, and appeared more compact [Figure 6.13(b)] and, at the macroscopic scale, the coating appeared less bright.

6.4.5.2 Investigations of Copper Reduction in Room Temperature Ionic Liquids (RTILs)

The application of ultrasound with RTILs for the electrodeposition of metals is a promising area. However, for such applications, problems still remain mainly due to the high viscosities of these solvents that slow down the molecules' motion within the solution. Another characteristic of RTILs is a pronounced hygroscopic nature which induces strong

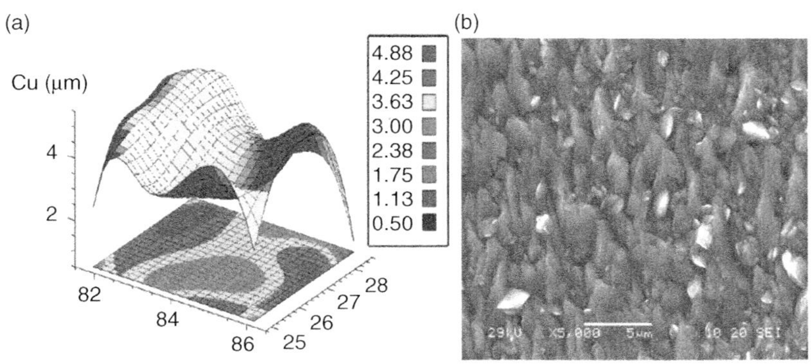

Figure 6.13 *(a) Mapping copper deposit thickness carried out under ultrasound on a plate disc in a solution [$CuSO_4$] = 15 mM [Na_2SO_4] = 0.1 M. (b) Image in scanning electron microscopy (SEM) of the deposit (×5000).*

dependence of physicochemical properties on their water and impurity contents. However, it was recently shown that ultrasound reduces water content [73]. The use of ultrasound in RTILs ensures beneficial effects, particularly through counteraction of their high viscosities, thanks to strong local agitation and limitations of electrode passivation (electrode fouling). During the micro-sonoreactor characterisation, it was observed that ultrasound induces a considerable increase in mass transfer as well as singular hydrodynamic behaviour of [Bmim] [Tf_2N] [74]. Due to the specific effects of ultrasound, hydrodynamic behaviours of various ionic liquids have been studied under insonation. Experimental mass transfer determinations were conducted using four ionic liquids ([Bmim] [Tf_2N], [Omim] [Tf_2N], [Bmpyr] [Tf_2N], [N1113] [Tf_2N]) as well as two types of synthetic electrolytes, namely acetonitrile/PEG and acetonitrile/[Omim] [Tf_2N] mixtures, to ensure variations in viscosity [77]. First, it was confirmed that cavitation occurs in all media, and ultrasound induces a large agitation effect, quoted as mass transfer coefficients (k_d) [77]. It was observed that at constant ultrasonic intensity of 1.1 W cm^{-2} (20 kHz), mass transfer coefficient values were always equal to those measured with an electrode rotating at 5000 rpm and up to 6 times higher than in the case of the [Omim] [Tf_2N]. As already observed in previous studies, a marked effect of viscosity was noticed. A comparison between ultrasonic and mechanical agitation (RDE) was conducted by calculating the mass transfer ratio for the four ionic liquids and synthetic fluids of various viscosities used. Results are shown in Figure 6.14 and are directly related to the type of solution employed.

Figure 6.14 clearly shows that the acetonitrile/PEG mixtures as well as [Bmpyr] [Tf_2N] and [N1113] [Tf_2N] ionic liquids are little affected by ultrasound (ratio about 1), contrarily to acetonitrile/[Omim] [Tf_2N] mixtures as well as [Bmim] [Tf_2N] and [Omim] [Tf_2N] ionic liquids [77]. Taking into account the two components induced by ultrasonic agitation, it appears that the differences in mass transfer cannot be ascribed to solely to convective flow, since the variations of parameters such as density and viscosity from one media to another does not follow a similar trend. Similarly, the rheological behaviour of the fluid does not

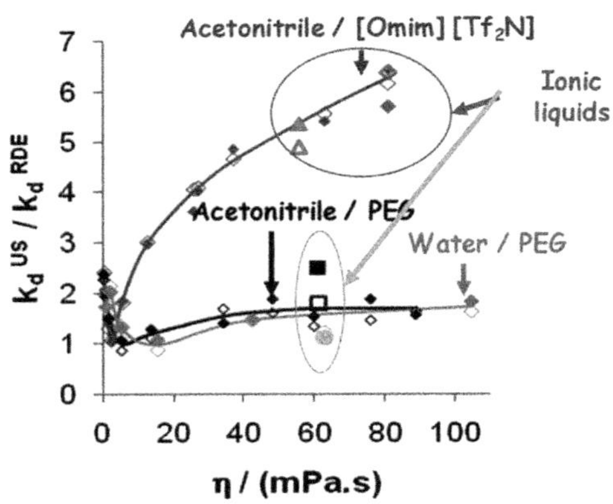

Figure 6.14 *Evolution of the ratio k_{dus}/k_{drde} of ultrasonic agitation (20 and 40 kHz) upon an electrode rotating at 5000 rpm versus dynamic viscosity (η). 20 kHz (solid line, no marker) – 40 kHz (no symbols). [N1113] [Tf_2N] (□); [Bmpyr] [Tf_2N] (○); [Bmim] [Tf_2N] (Δ); [Omim] [Tf_2N]; (◇) Acetonitrile [Omim] [Tf_2N] mixtures (+); acetonitrile/PEG mixtures (x) (Reprinted with permission from [77] Copyright (2010) Elsevier Ltd).*

seem to be of great importance (acetonitrile and all synthetic mixtures are known to be non-Newtonian fluids whereas [Bmim] [Tf_2N], for example, is Newtonian). Cavitation collapse alone cannot be the only explanation for this observation, as high viscosity and low vapour pressure systems do not allow the formation of large bubbles. The most probable hypothesis is the change in coordination sphere around the electroactive species, which greatly depends upon the type of ionic liquid and is reduced by ultrasonic irradiation, thereby improving its global mobility [77].

It was thus decided to work with [Bmim] [Tf_2N], which can be considered as a reference material. Indeed, it is the least viscous ionic liquid and exhibits high sensitivity to ultrasonic agitation (mass transfer to the electrode is up to 5 times higher than for a RDE rotating at 5000 rpm under silent conditions) [74]. Choice of metallic salt is more complex as dissolving copper salts in ionic liquids proves to be a challenge [73]. In this study, the same (Tf_2N) anion was kept and electrolytes were produced by dissolving 15 mM of copper triflimide ($Cu(Tf_2N)_2$) into the [Bmim] [Tf_2N] ionic liquid [73]. Results shown in Figure 6.15(a) indicate that the cathodic current density increases as the electrode potential increases until a plateau is reached at high cathodic overpotentials, although not well established in the presence of ultrasound. Furthermore under insonation, current densities measured at this plateau are considerably greater than those in silent conditions for an electrode rotating at 4000 rpm, confirming the significant effect of ultrasound on mass transfer in [Bmim] [Tf_2N]. Here, Tafel curves present two successive shoulders, according to the two-step reduction mechanism (Figure 6.15(b)). The first step is governed by one-electron transfer (reduction of Cu^{II} to Cu^{I}) whereas the second step is limited by the mass transfer phenomenon (reduction of Cu^{I} to Cu^{0}). The figure also shows that two plateaus are clearly visible, indicating that the first step may be diffusion controlled.

Coatings were then elaborated by chronoamperometry in the [Bmim] [Tf_2N] ionic liquid in the same conditions as in aqueous media under silent conditions and at a rotation speed of 2000 rpm. Figure 6.16(a) shows that coating thicknesses are really small indicating that other possible mechanisms may occur during electrodeposition. Crystal growth is limited by the low mass transfer and exhibits small dimensions (roughly 0.1 μm, Figure 6.16(b)).

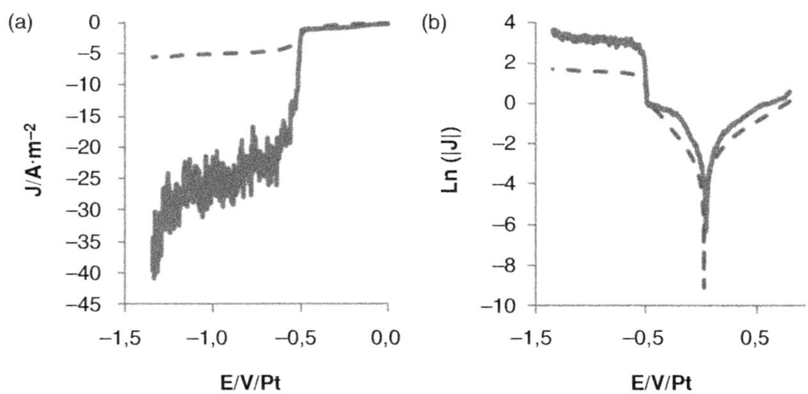

Figure 6.15 *(a) Polarisation curves at 2 mV s^{-1} ($Cu(Tf_2N)_2$) (15 mM) in [Bmim] [Tf_2N] on a platinum disc in silent (– –) and ultrasonic (—) conditions. (b) Same curves plotted according to the Tafel method in silent (– –) and ultrasonic (—) conditions.*

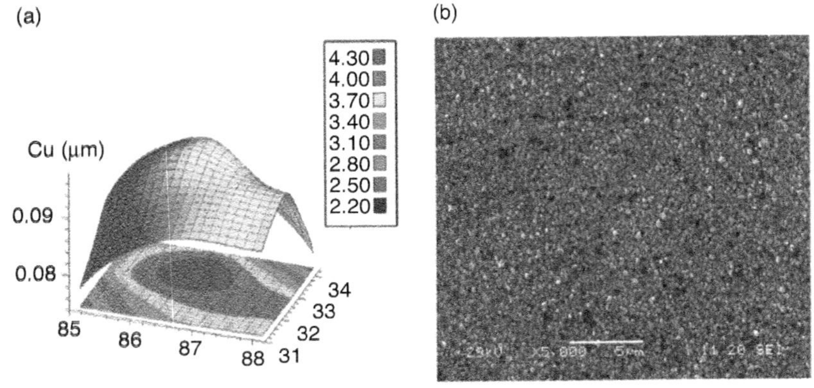

Figure 6.16 *(a) Mapping copper deposit thickness in silent conditions on a plate disc rotating at 2000 rpm in a solution ($Cu(Tf_2N)_2$) = 15 mM in [Bmim] [Tf_2N] (b) Image in scanning electron microscopy (SEM) of the deposit (×5000).*

Crystallites are unable to thrive easily but germination is favoured with growth and the morphology remains close to the one observed in aqueous electrolytes, while presenting a black macroscopic appearance.

However, in the presence of ultrasound (Figure 6.17), the deposit thickness is significantly improved and, as expected by mass transfer measurements, ultrasonic agitation is sufficient to allow the deposit growth.

Crystallites exhibited larger dimensions than those in silent conditions (about 2 μm), but their morphologies were quite different as they seem to consist of agglomerates of thinner crystallites (flower-like of 0.5 μm diameter), highlighting another organisation and germination mechanism (Figure 6.17(b)). Finally, film thickness exhibited a decrease at the centre, as shown in the figure. The high viscosity of the ionic liquid induces a higher

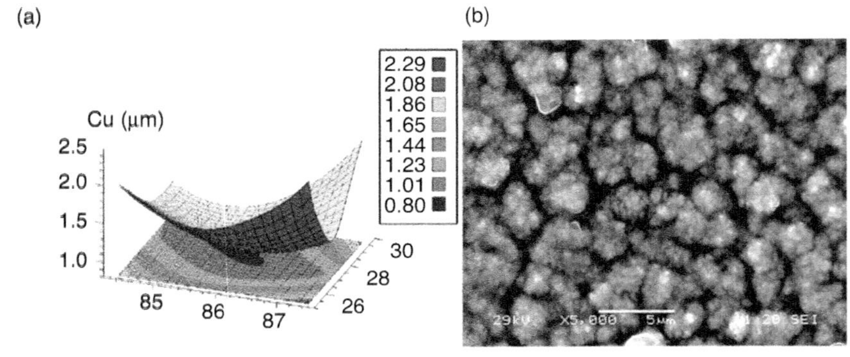

Figure 6.17 *(a) Mapping copper deposit thickness carried out under ultrasound on a plate disc in a solution ($Cu(Tf_2N)_2$) = 15 mM in [Bmim] [Tf_2N] (b) Image in scanning electron microscopy (SEM) of the deposit (×5000).*

energy transmission, with numerous bubble impacts at the centre, generating more erosion. Porosity is greater giving an intense black appearance at macroscopic level.

6.4.6 Incorporation of Particles Assisted by Ultrasound

In the past decades and in many studies, composites coatings obtained by incorporation of various types of particles have shown considerable improvement in the mechanical and physicochemical properties of the material. Incorporation of particles based on polytetrafluoroethylene (PTFE) is of particular interest because of mechanical (low friction coefficients), chemical (inert) and self-lubrication properties. Nevertheless, many issues exist in this method, mainly due to poor incorporation rates and poor heterogeneous distribution of the particles in the coating layer, which contributes to disappointing properties. In order to improve the above method and coating properties, ultrasound was applied to the system.

In this study, coatings were prepared in a 500 mL cylinder double-vessel sonoreactor, filled with a gold-nickel electrolyte (10 g L^{-1} of Au, 1.5 g L^{-1} of Ni, 60 g L^{-1} of 0.2 μm PTFE particles) maintained at 55 °C and at pH = 4.5 [78]. The dispersion was stabilised using a non-ionic wetting agent and, during the electroplating process, the use of a turbine rotating at 70 rpm was required to avoid sedimentation. The working electrode consisted of a 2 cm^2 copper plate polarised cathodically, covered with a thin nickel layer to avoid diffusion of gold into copper. The amount of incorporated particles varied with the operating conditions, in particular with the current density and the stirring conditions [8]. A range of current densities between 2 and 10 A dm^{-2} was explored and the stirring conditions were varied by changing the ultrasonic power from 2 to 10 W at a constant wave frequency of 500 kHz. The samples were positioned parallel to the irradiation direction (4 cm from the transducer) and each sample was electrodeposited at a constant polarisation time of 3 min in the absence and presence of ultrasound. The presence of PTFE in the fabricated coatings was determined by IR analysis conducted by specular reflection spectroscopy (infra red reflection absorption spectroscopy, IRRAS). This technique allowed qualitative and quantitative analysis of the coatings [79] as the fluoropolymer presents strong absorption bands.

Tribological measurements consisted in imposing a constant load on the metallic ball (the wiper: 5 mm diameter steel ball) in contact with the plane surface of the sample, which was positioned on a rotating stand. The tangential stress generated by friction during rotation (0.2 to 1 rps) was used to determine the friction coefficient which corresponded to the ratio of the normal load over tangential stress. It was then possible to monitor the evolution of the friction coefficient as a function of the number of cycles (i.e. sample revolution) [78].

As previously observed for the production of metallic coatings under ultrasound, the deposition rates can be either increased or decreased. In silent conditions, that is, under mechanical stirring only, the deposition rate increases slowly with current density, to reach a maximum value at 6 A dm^{-2}. For upper current values, the competitive hydrogen evolution reaction induces a decrease in deposit yield. When ultrasound is applied, only the low transmitted ultrasonic powers (2–4 W) cause significant improvement of the deposition rate in the current ranges employed, with an optimum value around 7 A dm^{-2}. At this stage, a thickness of 2.5 μm was recorded, that is, 60% higher than that under silent conditions

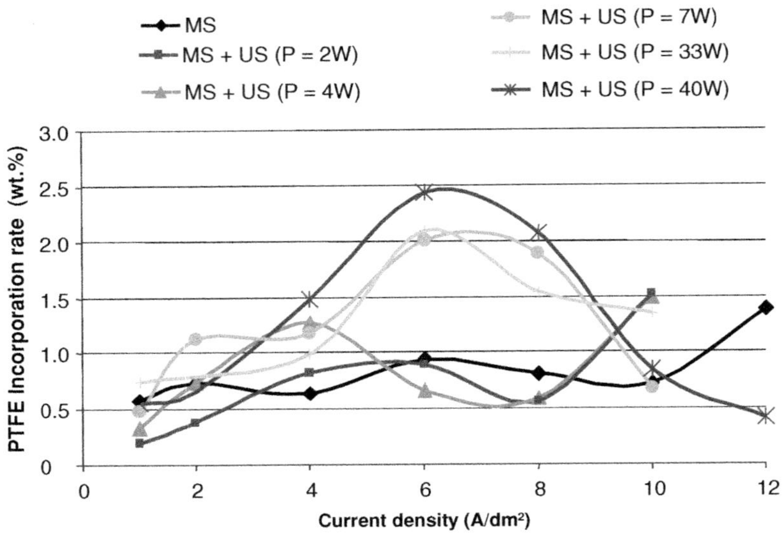

Figure 6.18 *PTFE incorporation rates versus cathodic current densities at various ultrasonic powers (Reprinted with permission from [78] Copyright (2005) Elsevier Ltd).*

[78]. From the polarisation curves, the diffusion plateaux previously observed virtually disappears [78]. This effect of ultrasound on the mass transfer controlled processes has already been widely reported for metallic coating electrodeposition [8] and, more generally, for quasi-reversible electrochemical systems [78].

Figure 6.18 shows plots of PTFE incorporation rates versus current densities at various ultrasonic powers. The electrodeposits obtained under mechanical stirring exhibit an inclusion rate that is virtually constant (0.75 wt %) whatever the current density used. Under sonication, the PTFE incorporation rates increase as the transmitted ultrasonic power increases, as expected. The best values were obtained for ultrasonic powers ranging from 7 to 40 W in the current density range of 6–7 A dm^{-2} [78].

It was shown that particle incorporation rates were directly related to the agitation level in the range of magnitude of ultrasonic powers used in the study. However, no direct relationship was observed between the particle incorporation rates and the operating conditions. Nevertheless, even if the influence of hydrodynamic regimes was difficult to evaluate, the results show that an optimum particle incorporation rate value is obtained for a given current density at the maximum transmitted ultrasonic power. Finally, tribological properties of the coating were evaluated and Figure 6.19 shows the evolution of friction coefficients of the coatings in silent conditions. Starting with a very low friction coefficient value of ~0.2, it is shown that friction coefficients increase steadily with the number of cycles (sample revolution) up to a value of 0.34 at a current density of 6 A dm^{-2} [78].

It was found that resistance to wear, that is, the continuous degradation of the electrodeposit properties, was lower for the coating obtained at 1 and 4 A dm^{-2} due to the increase in friction coefficients at those current densities. It is noteworthy to mention the fact that these

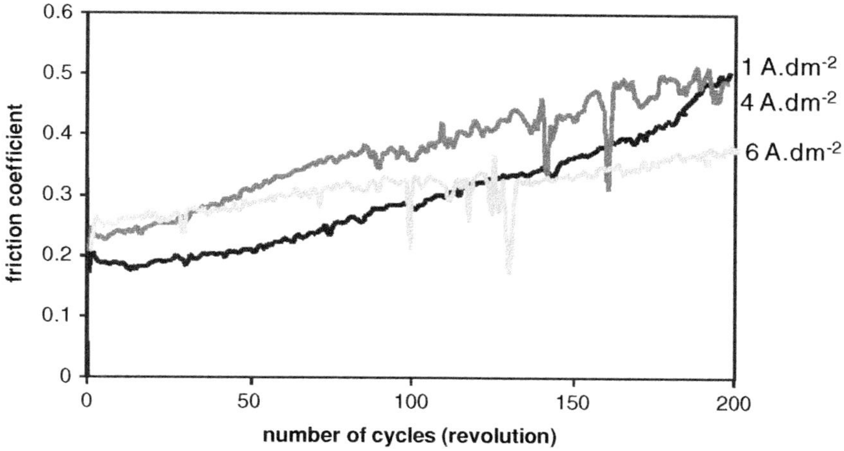

Figure 6.19 *Friction coefficient versus number of cycles in silent conditions (Reprinted with permission from [78] Copyright (2005) Elsevier Ltd).*

results relate to the incorporation rates, and also that a maximum of PTFE particles incorporation leads to the best properties. In ultrasonic conditions, the situation is different; while friction coefficients of fabricated electrodeposits under 40 kHz ultrasound present equivalent initial values, they vary little in the course of the cycling test (Figure 6.20), for example, at a current density of 4 A dm^{-2}, the friction coefficient remains almost constant [78].

Friction coefficients are not only improved by ultrasound, but also present an optimum value which does not correspond to the maximum of the PTFE particle incorporation rate (see Figure 6.18); although it can be noticed that in this case that the abraded thickness after 200 cycles is roughly 0.5 μm, confirming a good coating resistance that is certainly due to good particle distribution in the layer.

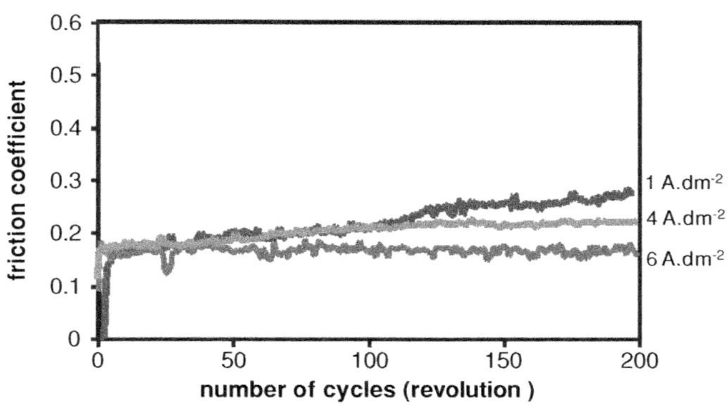

Figure 6.20 *Friction coefficient versus number of cycles under 40 kHz ultrasound (Reprinted with permission from [78] Copyright (2005) Elsevier Ltd).*

6.5 Electroless Coating Assisted by Ultrasound

6.5.1 The Electroless Process

Electroless coating on metals, polymers and ceramics is nowadays an extremely popular method. It allows non-conductive substrates to combine light weight and elasticity with tribological properties, conductivity (electrical, magnetic and thermal) while keeping and obtaining aesthetic aspects normally specific to metallic parts. In the case of specific products such as connectors, transducers and electronic shields, the largest electroless coating applications are found in up-market finishing in the luxury goods and automotive industries. This method involves numerous objects used in everyday life, such as perfume flasks, bathroom plumbing and car fittings. Ever since it was first discovered in the 1950s, the electroless metallisation process has continued to progress. Nevertheless, current techniques and methods have found their limitations, for example, low plating rates as well as reduced properties such as thickness, internal stress and adhesion. In other words, the electroless process is time consuming, and its mean lifetime short, particularly if coatings are subjected to severe mechanical or climatic constraints. The process itself is very simple, both from a chemical and a technological point of view, as it consists of careful surface preparations, followed by reduction of the metal by a catalyst with an electron provider present in the solution. However, the optimisation of the electroless process depends upon three main steps:

(i) good control of the surface roughness
(ii) good activity of the catalyst (i.e. palladium)
(iii) good degassing of hydrogen as a result of proper chemical reaction of metal reduction.

6.5.2 Ultrasound Effects upon Electroless Coating Properties

Electroplating rates are usually low in electroless processes, and coatings frequently exhibit high porosity and weak adhesion strengths, leading to poor reliability of the electrodeposits. The use of ultrasound can therefore be considered as a promising tool, and many interesting studies have been carried out for various applications, in terms of metallic layers as well as non-conductive substrates. Some investigations focus on the preparation of nickel nanoparticles on various substrates [80–82], but nickel coating electrodeposits on metallic substrates have mainly been tested because they are known to increase corrosion protection, and give good wear characteristics as well as micro hardness on parts possessing complex geometries. Even if pioneer studies by Rich [83] and Mallory [84] have shown that ultrasonic irradiation improves the Ni electroless process, most of the papers deal with very different operating conditions, with a particular emphasis on the properties obtained (Table 6.4).

Furthermore, only a few investigations consider the mechanisms involved, and only a few trends have been highlighted, such as an increase in plating rate with little alterations in crystal structure. However, reduced resistance to corrosion is frequently reported, which is often attributed to cracks present on the coating surface; this in turn leads to serious issues, because this property is an important parameter on the final behaviour of the parts considered. Thus, this critical parameter must be investigated further as it strongly depends

Table 6.4 *Electroless nickel deposition.*

	Advantages over *silent* conditions	Examples
Nickel-Phosphorus coating on steel – 20 kHz	Increase in plating rates (30%) and micro-hardness (10%). Change of crystal structure from amorphous to micro-crystalline. Decrease in phosphorus content and corrosion resistance (in H_2SO_4)	L.X. Yang, W.T. Hou and Y.S. Wu [85]
Nickel-hypophosphite-citrate bath – 28 kHz to 100 kHz	Electrochemical studies show that the increase in plating rates stems from an enhancement in diffusion of Ni^{2+}-citrate complex.	K. Kobayashi, A. Chiba and N. Minami [29]
Nickel hypophosphite- and dimethylamine-borane bath – 45 kHz to 1.6 MHz	Plating rate is increased, but resistance to corrosion (in NaCl 3%) is reduced by the presence of cracks.	A Chiba, H. Haijima and K. Kobayashi [86]
Diamond or cubic boron nitride particles co-deposited with Nickel-P on abrasive steel wheels 15.5 kHz	Plating rate is increased up to 1.5 times, and cohesion between nickel and particles is much better, thus increasing lifetime	M. Okumiya, Y. Tsunekawa, T. Saida and R. Ichino [87]

upon the operating conditions. Very few papers deal with other type of metallic systems, but one may cite the interesting work by Liu *et al.* [88] who investigated the effect of ultrasound on electroless deposition of Ag on PET. Electromagnetic measurements performed in addition to traditional coating characterisations revealed that shielding effectiveness is clearly enhanced by ultrasound. An original work by Wang *et al.* [89] focused on a simple but effective method for the synthesis of well-defined silver dendritic nanostructures, in the presence of isopropanol as a reducing agent and polyethylene glycol as a disperser. They showed that sonication alters crystallisation from a silver spheroidal nanostructure to a single crystal structure with cubic phase, aggregating jointly into silver dendrites.

Further information is available for copper coatings on non-conductive substrates, which is the major industrial application of electroless copper processes, where it frequently constitutes the first step in recovery, followed by electrolytically obtained layers. This requires an electron donor in the solution and a particular and specific surface preparation. It should be emphasised that it is possible to obtain spontaneous copper reduction by transfer if the metallic substrate is of lower potential. Table 6.5 shows some examples of the advantages of using ultrasound in electroless copper deposition.

It has been shown that major benefits are possible by the use of ultrasound, and although the absolute properties achieved remain low, all improvements are welcome in the field of electroless plating. The case of electroless plating on non-conductive substrates is particularly interesting, because the effect of ultrasound can be observed on the coating itself as well as during previous steps of surface preparation.

Table 6.5 *Electroless copper deposition.*

	Advantages over *silent* conditions	Examples
Copper coating on ceramics, tartrate-formaldehyde bath – 28.2 kHz and 1020 kHz	More uniform and well bonded coating with a faster coating rate. Resistance to thermal cycling, and a more uniform and finer microstructure. Whiskers are removed, and porosity decreases.	Y. Zhao, C. Bao, R. Feng and Z. Chen [13]
ABS etching prior to copper coating elaboration – 28 kHz	Adhesion strengths are higher due to greater roughness, which increases with irradiation time, while total etching mass is lowered. A suggested mechanism is the production of uniform cavities on the surface.	Y. Zhao, C. Bao, R. Feng and R. Liu [90]
Silver coating on Poly(ethylene terephtalate (PET) modified by silanes	Compact silver films coated on the modified PET surfaces with high adhesion strengths (> 40 N cm^{-1}) in the presence of ultrasound. Conductivity and mechanical properties are maintained after simulated body fluid (SBF) treatment.	Y. Lua, S. Jianga and Y. Huang [91]
Copper coating on epoxide formaldehyde bath – 300 kHz to 800 kHz	The use of ultrasound during the plating step increases plating rates by 25% and practical adhesion by 30% while reducing internal stress. Synergetic effects are observed if ultrasound is applied to that activation step. Mechanisms are discussed as a function of ultrasonic powers and frequencies.	F. Touyeras, J.Y. Hihn, M.L. Doche, X. Roizard [11], F. Touyeras, J.Y. Hihn, S. Delalande, R. Viennet, M.L. Doche [12], F. Touyeras, J.Y. Hihn, X. Bourgoin, B. Jacques, L. Hallez and V. Branger [28]
Copper coating on Ta/SiO_2/Si substrates by $PdCl_2$/BOE/HNO_3 bath	Copper coverage decreases gradually as activation time increases (from 30 to 150 s) whereas a high coverage is maintained in the presence of ultrasound. Crystalline organization is seriously affected with a preferential diffraction intensity Cu (1,1,1) 4.53 times greater than Cu (2,0,0).	D.L. Liu, Z.G. Yang, J. Wang, C. Zhang [43]
Copper coating on magnesium alloy	Copper is deposited on magnesium by transfer, and ultrasound enhancement of magnesium anodic dissolution maintains a continuous driving force for cathodic copper reduction.	L.X. Yang, B. Luan, W.J. Cheong, D. Shoesmith [92]
Composite Cu-Ti_3SiC_2 powder	Ultrasound allows coverage of the powder surface by copper nanoparticles, due to the increase in wetting property between the copper matrix and Ti_3SiC_2	Z.B. Zhang, S.F. Xu [93]

6.5.3 Copper Coating on Non-conductive Substrates under Insonation

6.5.3.1 Procedure for Surface Preparation and Coatings

Electroplating or metallisation of non-conductive substrates is usually carried out with formaldehyde in view to provide the required electrons for copper reduction. However, to limit this reduction to resin substrates, this reaction must be catalysed, which calls upon a precise surface preparation consisting of different steps [11, 13]. Epoxy resin plates are successively polished (600 grade), cleaned, and then sensitised and activated, as shown in Figure 6.21.

It was shown in the case of steel samples that temperature and immersion times have to be optimised throughout fractional factorial design experiments [94]. During sensitisation, the active sites are produced by dipping the substrate in a reductive solution based on $SnCl_2$, promoting nucleation and inlay of precious metallic catalyst particles. Usually the sensitisation bath contains $SnCl_2$ and hydrochloric acid. Tin is added to the solution to avoid direct transformation of the divalent form of tin and tin ions into tetravalent tin (the activation step then consists in producing the catalysts). The substrate is immersed in a bath containing precious metal ions (usually palladium ions), and the net reaction is:

$$Sn^{2+} + Pd^{2+} \rightarrow Sn^{4+} + Pd \tag{6.4}$$

The sensitisation and activation steps can be carried out in a single step. In this case, the sensitisation-activation step is performed using a Pd solution stabilised by a colloidal protective suspension composed of divalent Sn and Pd ions and HCl. Some tetravalent Sn is then added to facilitate the hydrophobic substrate wettability. The main reactions are shown in Equations (6.5) and (6.6):

$$Pd^{2+} + Sn^{2+} + 4Cl^{-} \rightarrow Sn^{2+}(PdCl_4)^{2-} \tag{6.5}$$

$$Sn^{2+} + Pd^{2+} \rightarrow Sn^{4+} + Pd \tag{6.6}$$

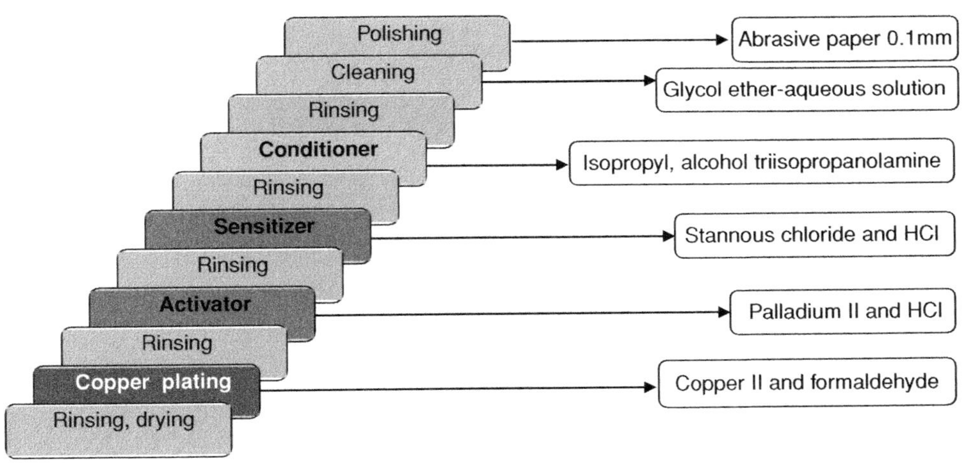

Figure 6.21 *Surface preparation steps of epoxy resin plates.*

In this case, palladium is required at the substrate surface to reduce Cu^{2+} ions in the presence of formaldehyde and the higher the palladium surface concentration, the higher the plating rate [Equation (6.7)].

$$Cu^{2+} + 2HCHO + 4OH^- \rightarrow Cu + 2HCO_2^- + H_2 + 2H_2O \tag{6.7}$$

It is important to note that, due to the low formaldehyde vapour pressure and reasonable ultrasonic powers [11], the formaldehyde concentration is not affected by ultrasound.

6.5.3.2 Sonoreactor and Ultrasonic Applications

Ultrasonic operating conditions, such as the ultrasonic frequency, power, irradiation time and sonoreactor geometries, as well as the use of ultrasound for pre-treating surfaces (prior to deposition) are important parameters. For example, previous studies [90] have shown that ultrasound is capable of etching ABS surfaces leading to better electrodeposit final properties. The mechanical role of ultrasound is important in the case of ABS, because it eliminates the acidic Cr^{VI} etching step, which has very harmful health and environmental effects.

Furthermore, it is of great importance to optimise the sonoreactor in view of obtaining better electrodeposit properties. For example, experiments were conducted at various ultrasonic frequencies (300 to 800 kHz) and in a filled 500 mL cell (Figure 6.4). Here, the transducer was insulated by a protective stuck-on glass plate. Ultrasound irradiation was produced parallel to the liquid surface, and the epoxy specimens were suspended (two at a time) vertically. Parameters such as the acoustic power and stirring conditions were determined to carefully characterise exposure conditions. For this purpose, sonovoltammograms of ferri/ferrocyanide (a quasi-reversible redox couple) on a stationary Pt electrode were generated [55]. Characterisation results were plotted in terms of mass transfer coefficient versus transmitted ultrasonic powers, as shown in Figure 6.22, showing an obvious dependency on ultrasonic frequency. For each ultrasonic frequency, it was found that the mass transfer coefficient increased with the transmitted ultrasonic power, but the highest value was obtained for 500 kHz. At this frequency, ultrasonic agitation is clearly not a linear

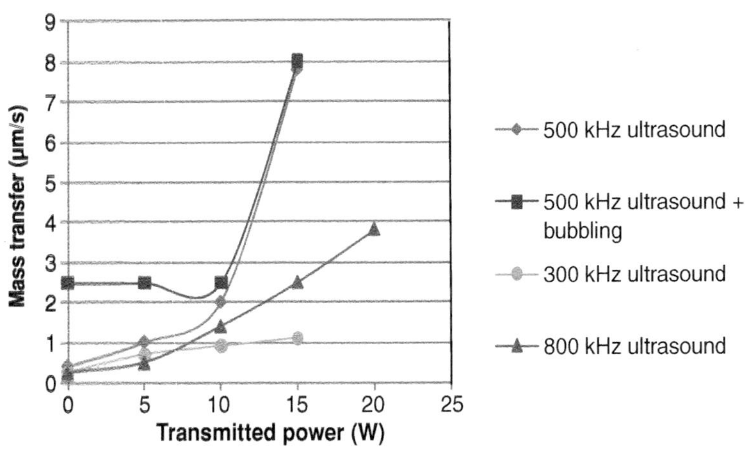

Figure 6.22 *Mass transfer coefficients on vertical samples versus transmitted ultrasonic power at various ultrasonic frequencies.*

function of the transmitted ultrasonic power, in the presence of gas bubbling (generating significant 'stirring').

6.5.3.3 *Influence of Ultrasonic Power and Ultrasonic Sequences on Plating Rates*

From the literature, it appears that the whole electroless deposition process exhibits two different behaviours due to strong agitation induced by ultrasound. During the ignition phase, that is, when the first nuclei is formed at the surface, agitation has a great influence on reaction rates through mass transfer enhancement and the cleaning effects on the catalyst surface. Following the growth of a continuous copper layer at the surface, the reaction is controlled by the 'autocatalytic' properties of copper, whereby the influence of agitation drops drastically.

In this case, plating rates are determined in two phases, either by applying ultrasound in the first few minutes of the plating step or during the last period after the ignition time under silent conditions. Argon bubbling is either added or removed to eliminate its influence (see Table 6.6). This is repeated for various input ultrasonic powers [11].

During the irradiation sequence, that is, the exact moment when ultrasound is applied, it shows a positive influence on plating rates. Although, if ultrasound does not occur until the end of the plating, no further increase in plating rates are observed. However, ultrasound has a significant effect when applied during the very first few minutes of the metal plating step. This confirms that the ignition period of plating is fundamental. During this period, the first copper layers, which affect the overall coating properties, are influenced by the catalyst activity, which in turn is affected by ultrasound. Furthermore, plating rates are dependent upon the transmitted ultrasonic power. The highest value of the coating thickness obtained after 1 hour corresponds to the lowest acoustic power applied at 5 W [11]. This value is significantly higher than that obtained under silent conditions. While ultrasonic agitation allows efficient electroactive species mixing close to the interface and a decrease in diffusion layer resistance, excessive agitation could lead to poor catalyst anchoring and recovery and thus to lower plating rates. To quantify this effect, plating rates were plotted as a function of mass transfer coefficients (Figure 6.23).

The Figure 6.23 shows that minimum agitation is required to ensure good mixing of the electroactive species at the interface, but strong turbulences have negative impact on the plating rates and mass transfer coefficients. Nevertheless, when observing results obtained using gas bubbling, it appears that all plating rates reach higher values, and that the

Table 6.6 *Plating rates versus various ultrasonic powers and sequences.*

	Transmitted ultrasonic power (W)			
Plating rates ($\mu m\ h^{-1}$)	0	5	10	15
Without bubbling (5 mins with ultrasound + 55 mins *silent*)	2.1	2.8	2.1	1.8
Without bubbling (5 mins *silent* + 55 mins ultrasound)	–	2.1	–	–
With bubbling (5 mins with ultrasound + 55 mins *silent*)	2.7	3.4	3.1	2.7

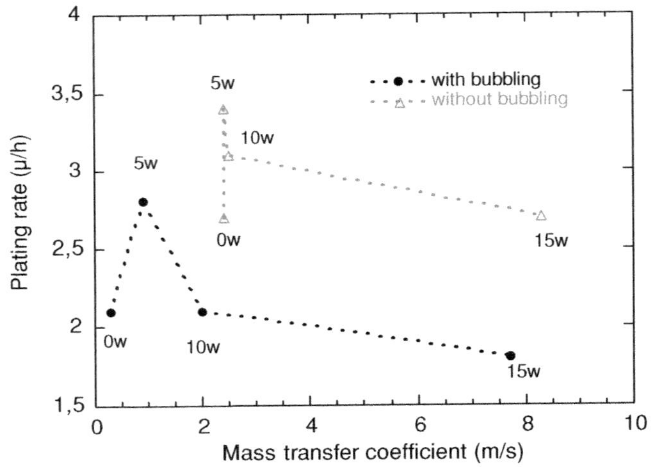

Figure 6.23 *Plating rates versus mass transfer coefficients at 500 kHz in the absence and presence of gas bubbling (Reprinted with permission from [11] Copyright (2001) Elsevier Ltd).*

optimum (at 5 W) has a value of $k_d = 2.7$ m s^{-1}. However, in respect of the above findings, the following remarks should be taken into account: for example, present stirring levels originate from global mass transfer coefficients, which are not necessarily representative of cavitation bubble implosion. Micro-jets impinging the surface allow high, but very located agitation. Although it seems difficult to compare the two agitation modes using this method, the average mass transfer coefficient measurement still remains a very useful tool for the characterization of operating conditions under sonication. These observations confirm the importance of the ignition phase and, thereby, of potential sensitivity of the activation step to ultrasound. In this study, a series of ultrasonic powers were then applied throughout immersion of the samples into the activation solution (2 min), followed by regular copper plating for 1 hour. It was observed that the plating rate increases with ultrasonic power to reach a maximum of 3.8 μm h^{-1} at 15 W compared to 2.5 μm h^{-1} under silent conditions. This ~50% increase may be explained by the removal of some contaminants, like tin(II), which form complexes on the palladium surface due to ultrasonic agitation (see Figure 6.24). Asymmetric collapse of cavitation bubbles induces micro-jets at the surface that seem to be well adapted to surface cleaning. Thus, when applied during the activation step, ultrasound leads to a larger specific catalyst area and thus an increase in efficiency.

To provide further information on the mechanisms involved during surface preparation, contact angle measurements and XPS studies were performed in a variety of preparation conditions [12, 28]. The contact angle of epoxy plates and distilled water was measured using a Kruss angle-meter, and the XPS analysis was conducted using a KRATOS Axis Ultra, with monochromatised Al K_α radiation at the PSA Peugeot-Citroën Belchamp Research Centre. The conditions were as follows:

(i) epoxy plates without surface preparation
(ii) epoxy plates activated and sensitised in one step under silent conditions

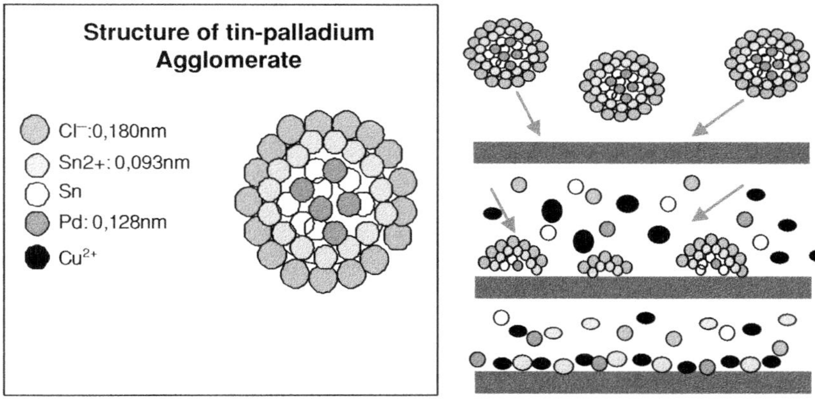

Figure 6.24 *Structure of tin-palladium agglomerates and dissolution mechanisms. Particles with negative charges are adsorbed on the polymer surface (Reprinted with permission from [28] Copyright (2005) Elsevier Ltd).*

(iii) epoxy plates activated and sensitised in one step under 15 W ultrasound
(iv) Epoxy plates activated and sensitised in two steps under silent conditions
(v) Epoxy plate activated and sensitised in two steps with 15 W ultrasound during the activation step.

The results are represented in Table 6.7.

The Table 6.7 shows that the contact angle increases in the presence of palladium and can be accounted for by the high palladium surface energy. During the activation step, palladium atoms are deposited onto the epoxy surface, thus increasing its surface and energy. When ultrasound is applied, palladium is more present or more visible in its metallic form. This is confirmed by XPS measurements, regardless of the preparation method chosen (in one or two steps), and a slight increase in palladium concentration corresponds systematically to higher plating rates. The mechanism involved is certainly due to strong agitation, linked to very high values of mass transfer coefficient k_d [12,28]. The role of Sn is more ambiguous because it does not take part in the reaction in its metallic form. The correlations have

Table 6.7 *Results of contact angles and atomic concentration on epoxy surface prior to plating.*

Preparation sequence	Contact angle	Pd atomic concentration measured by XPS	Sn atomic concentration measured by XPS	O atomic concentration measured by XPS
Sequence (i)	61	0	–	–
Sequence (ii)	44	2.05	5.69	33.21
Sequence (iii)	22	2.69	5.53	40.15
Sequence (iv)	–	0.95	4.66	39.35
Sequence (v)	–	2.03	5.97	31.41

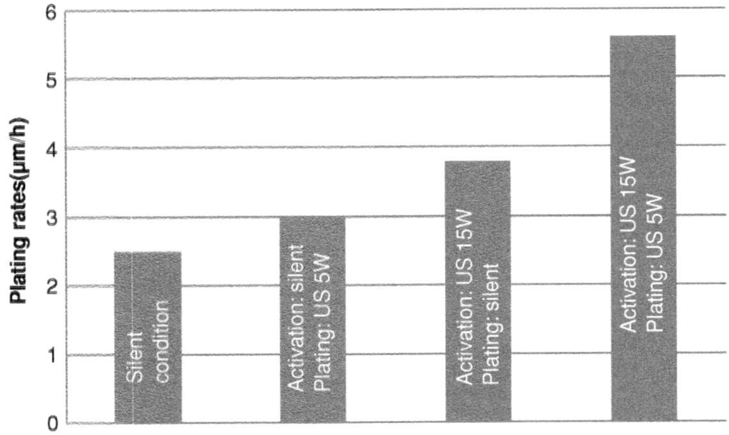

Figure 6.25 *Evolution of the plating rates versus ultrasonic powers used during activation and plating phases.*

been checked between oxygen and the Sn elements at the surface, in order to elucidate the distribution between Sn^{2+} and Sn^{4+} cation concentrations in the solution, which is directly related to the preparation mode (in one or two steps), as seen in the mechanisms described above. Finally, the combination of the optimum conditions under ultrasound during plating and during the activation phase comes naturally, and the results are grouped in Figure 6.25.

While ultrasound provides beneficial effects if applied right at the start of plating (+40%) or during the sensitisation/activation step (+52%), the combination of both actions induces synergetic effects yielding a large increase (+124%), allowing a plating rate of 5.6 $\mu m\ h^{-1}$, which is considerable for the electroless copper process. In this case, more effects than local stirring on palladium complexes have to be taken into account. The key factor is certainly the presence of hydrogen [95,96]. Indeed, water might be dissociated as follows during the plating step (Figure 6.26):

$$H_2O \rightarrow H_{[ads]} + HO_{[ads]} \quad (6.8)$$

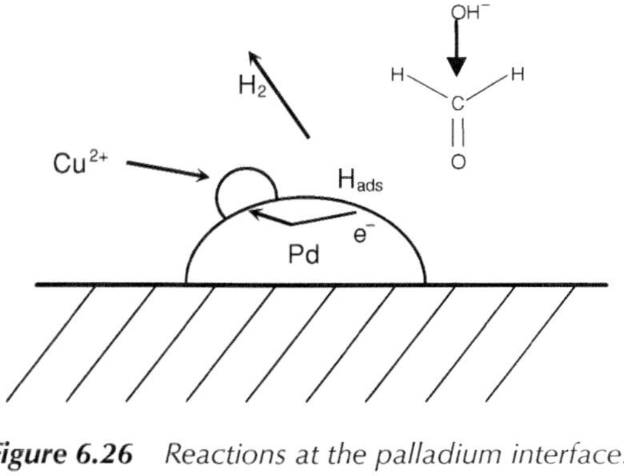

Figure 6.26 *Reactions at the palladium interface.*

and the following reactions may also occur:

$$H_{[ads]} + H_{[ads]} \rightarrow H_2 \tag{6.9}$$

$$Pd^{2+} + 2\,H_{[ads]} \rightarrow Pd + 2H^+ \tag{6.10}$$

Palladium plays a major role at the start of the coating process; hydrogen is adsorbed onto palladium in its ionic form, thus leading to the formation of active palladium in its metallic form. This adsorption decreases the presence of hydrogen in the solution, therefore limiting the competition between copper and hydrogen reductions. Adsorption would also lead to an increase in the plating rate at the start of the deposit. In the meantime, more copper nuclei are formed on the surface by ultrasound, and the mechanical anchor of the copper in the fine pits on the resin is favoured.

Immediately after the start of the coating process, when the epoxy surface is completely covered with copper, ultrasound cannot act on the catalyst efficiency. The effective catalyst area thus remains virtually unchanged, and the plating rate constant.

6.5.3.4 *Influence of Ultrasound on Coating Properties*

One of the major challenges of metallic coating on a non-conductive substrate is its sustainability. Thus, the presence of high level internal stress, and, consequently, of low adhesion strength, is an important indicator.

The internal stress average is derived from a balance of:

(i) The plating stress, resulting from atomic lattice and crystal distortion produced in the electrodeposit during the formation.
(ii) The thermal stress generated by major differences between the substrate and the coating thermal expansion coefficients. Whenever work is carried out at high temperatures, as in copper plating (48 °C), thermal stress is greater than the plating stress.
(iii) The stress due to the hydrogen content which appears to be the most important factor affecting the copper layer since it generates high tensile strength. At the interface, the presence of hydrogen results in a decrease in resin surface free energy and in its wettability. Contact area is reduced, generating loss of adhesion and stress enhancement. In addition, this amount of hydrogen occluded in the coating leads to high porosity.

Internal stress is usually measured using the foil method: a special epoxy sample (one side only covered with masking tape) of reduced thickness (150 μm) is immersed in the plating simultaneously with the other samples. The stress value is determined from material bending after curvature measurement, when removing the mask [11]. The adhesion strength is measured using a three-point flexure test (Instron 1121 apparatus), consisting in bending a test piece by applying a deformation at constant speed. The adhesion strength is defined before rupture [11] and, to obtain reliable results, it is necessary to acquire statistical values from at least approximately 30 measurements for each set of operating conditions. Here, as the magnitude of internal stress is very high with respect to coating adhesion evolution for electroless coatings, both parameters were measured in the case of sensitization and activation in the two steps under silent conditions, followed by 5 min sonication at the start of the plating step at various transmitted ultrasonic powers. Figure 6.27 shows that an inverse relationship between the stress value and the force of adhesion is obtained.

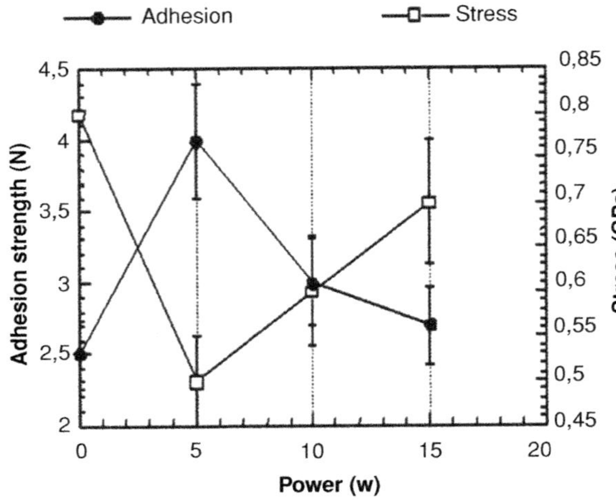

Figure 6.27 Internal stress and adhesion strength versus ultrasonic power during the plating step (Reprinted with permission from [11] Copyright (2001) Elsevier Ltd).

The Figure 6.27 shows that the coating with minimum residual stresses exhibits the strongest adhesion force. Once again, optimum conditions are obtained for 5 W transmitted ultrasonic power, but in all cases, ultrasound yields beneficial effects. Practical adhesion and internal stress were respectively increased and decreased by ~35% compared to silent conditions. To discriminate the influence of agitation by the global convection originating from cavitation effects, with or without gas bubbling, the adhesion strength versus the mass transfer coefficients are plotted (Figure 6.28).

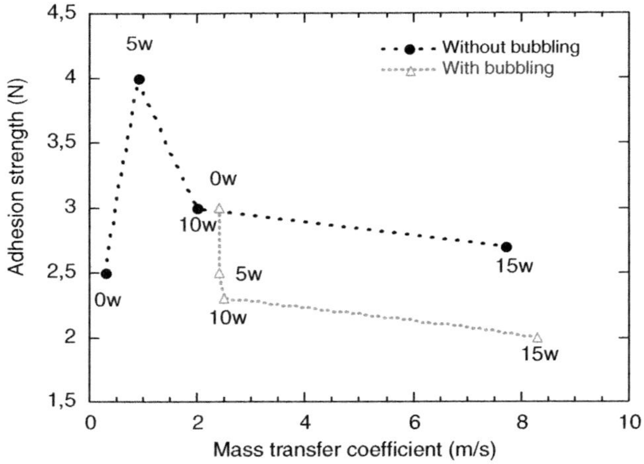

Figure 6.28 Adhesion strength versus mass transfer coefficient plots at various ultrasonic powers and in the absence/presence of gas bubbling (Reprinted with permission from [11] Copyright (2001) Elsevier Ltd).

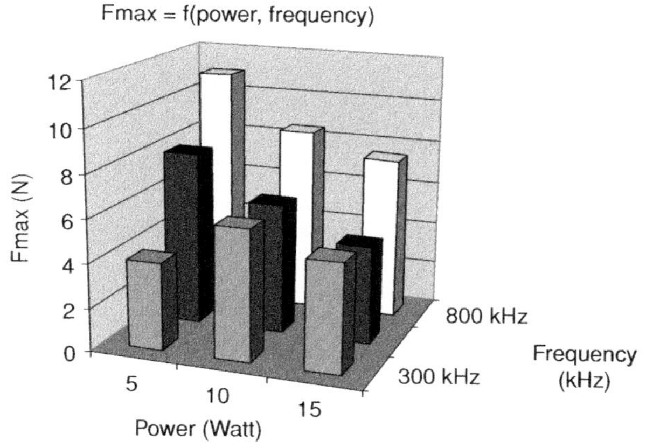

Figure 6.29 *Adhesion strength versus transmitted ultrasonic powers at various ultrasonic frequencies (Reprinted with permission from [28] Copyright (2005) Elsevier Ltd).*

Surprisingly, although gas bubbling alone has a positive effect, the combination of ultrasound and gas bubbling induces lower adhesion values. The positive contribution of ultrasound is not only due to the induced agitation, but also to a specific ultrasonic effect. This observation could be due to the following: numerous copper nuclei are formed on the surface, and the mechanical anchor of the copper in the fine pits on the resin surface is thus favoured and the copper-epoxy binding is more frequent and improved. In addition, due to ultrasonic agitation, hydrogen degassing is enhanced and removed by cavitation bubble collapse, which is not the case for gas bubbling agitation. Beyond 5 W, there is simultaneously an enhancement of internal stress and a loss of adhesion. Turbulent agitation is detrimental to the deposition process because the stress reduction induced by hydrogen is compensated by an increase in plating stress which occurs during the coating formation. However, minimum agitation is required to ensure better dissolution of palladium agglomerates in the alkaline media, but high convection leads to a negative impact on the metallic bonding formation. To complete this investigation, practical adhesion was measured as a function of the ultrasonic power and frequency, as shown in Figure 6.29.

The Figure 6.29 clearly shows that for the 500 and 800 kHz ultrasonic frequencies, the lower the ultrasonic power, the higher the adhesion strength, and 5 W is once again found as the optimal value. For 300 kHz, the maximum adhesion strength value is obtained at 10 W.

References

(1) Shenoi, B.A., Indira, K.S. and Subramanian, R. (1970) Ultrasonics in metal finishing, a comprehensive review. *Metal Finishing*, **July**, 40–50.

(2) Mazue, G., Viennet, R., Hihn, J.-Y., Carpentier, L., Devidal, P. and Albaïna, I. (2011) Large-scale ultrasonic cleaning system: Design of a multi-transducer device for boat cleaning (20 kHz). *Ultrasonics Sonochemistry*, **18** (4), 895–900.

(3) MacQueen, D.H. (1986) Frequency dependence of ultrasonic cleaning. *Ultrasonics*, **24**, 273–280.

(4) Todd, R.H., Allen, D.K. and Alting, L. (1994) *Manufacturing Processes Reference Guide*, Industrial Press Inc., New York.
(5) Ximei, W., Liqun, Z., Huicong, L. and Weiping, L. (2008) Influence of surface pretreatment on the anodizing film of Mg alloy and the mechanism of the ultrasound during the pretreatment. *Surface and Coating Technology*, **202**, 4210–4217.
(6) Ashley, M.J. (1974) Preventing deposition on heat exchange surfaces with ultrasound. *Ultrasonics*, 215–221.
(7) Kochergin, S.M. and Vyaselva, G.Y. (1966) *Electrodeposition of Metals in Ultrasonic Fields*, Consultants Bureau, New York, p. 19.
(8) Walker, C.T. and Walker, R. (1973) Effect of ultrasonic agitation on some properties of electrodeposits. *Electrodeposition and Surface Treatment*, **1**, 457–469.
(9) Prasad, R., Vasudevan, P. and Seshadri, S.K. (1993) Ultrasonic agitation during electrodeposition. *Transactions Indian Metallurgy*, **46** (4), 247.
(10) Prasad, R., Vasudevan, P. and Seshadri, S.K. (1994) *Indian Journal of Engineering and Material Sciences*, **1** (3), 178.
(11) Touyeras, F., Hihn, J.Y., Doche, M.-L. and Roizard, X. (2001) Electroless copper coating of epoxide plates in an ultrasonic field. *Ultrasonics Sonochemistry*, **8**, 285–290.
(12) Touyeras, F., Hihn, J.Y., Delalande, S., Viennet, R. and Doche, M.L. (2003) Ultrasound influence on the activation step before electroless coating. *Ultrasonics Sonochemistry*, **10**, 363–368.
(13) Zhao, Y., Bao, C., Feng, R. and Chen, Z. (1995) Electroless coating on ceramic in an ultrasonic field. *Ultrasonics Sonochemistry*, **2** (2) 99–103.
(14) Faïd, F., Romdhane, M., Gourdon, C., Wilhelm, A.M. and Delmas, H. (1998) A comparative study of local sensors of power ultrasound effects: electrochemical, thermoelectrical and chemical probes. *Ultrasonics Sonochemistry*, **5** (2), 63–68.
(15) Béranger, G. and Mazille, H. (2011) Approche scientifique des surfaces. Caractérisation et propriétés. *Techniques de l'Ingénieur*, **M 1 425**, 1–21.
(16) Martins, L., Martins, J.I., Romeira, A.S., Costa, M.E. and Bazzaoui, M. (2004) Morphology of copper coatings electroplated in an ultrasonic field. *Advanced Materials Forum II*, **455–456**, 844–848.
(17) Chiba, A. and Wu, W.C. (1992) Ultrasonic agitation effects on the electrodeposition of copper from a cupric-EDTA bath. *Plating and Surface Finishing*, **79** (12), 62–66.
(18) Mallik, A. and Ray, B.C. (2009) Morphological study of electrodeposited copper under the influence of ultrasound and low temperature. *Thin Solid Films*, **517** (24), 6612–6616.
(19) Cui, R.H., Yu, Z.M., He, Y.T. and Shu, W.J. (2010) Copper multilayer coating prepared by ultrasonic-electrodeposition. *Manufacturing Science and Engineering, Advanced Material Research*, **1–5** (97–101), 1348–1351.
(20) Wu, W.C. and Chiba, A. (1993) Effect of electrodeposition in an ultrasonic field on the corrosion of electrodeposited coatings. *Journal of Materials Science Letters*, **12**, 794–796.
(21) Liao, Q.A., Li, W.P., Liu, H.C. and Zhu, L.Q. (2010) Fabrication of nanostructured electroforming copper layer by means of an ultrasonic-assisted mechanical treatment. *Chinese Journal of Aeronautics*, **23** (5), 599–603.
(22) Liao, Q.A., Zhu, L.Q., Liu, H.C. and Li, W.P. (2010) Effects of microstructure on tensile strength of electroformed copper. *International Journal of Physics B*, **24** (15–16), 3118–3123.
(23) Liu, J.S., Sun, J.F., Xing, D.W., Xue, X.A., Zhang, S.L. and Wang, H.A. (2011) Experimental study on the effect of wire bonding by Cu electroplating on GMI stability of Co-based amorphous wire. *Physica Status Solid I – Applications and Material Science*, **208** (3), 530–534.
(24) Schneider, O., Matic, S. and Argirusis, C. (2008) Application of the electrochemical quartz crystal microbalance technique to copper sonoelectrochemistry, part 1 sulfate-based electrolytes. *Electrochimica Acta*, **53** (17), 5485–5495.

(25) Walker, R. (1997) Ultrasound improves electrolytic recovery of metals. *Ultrasonics Sonochemistry*, **4**, 39–43.

(26) Banerjee, S., Kumar, R. and Gandhi, K.S. (1995) Analysis of ultrasonically enhanced hydrogen evolution for Zn-$NiCl_2$ systems. *Chemical Engineering Science*, **50** (15), 2409–2418.

(27) Sulitanu, N., Pirghie, C. and Sandu, I. (2007) Effects of ultrasound irradiation on the electrolytic growth of nanocrystalline Ni films. *Revista de Chemie*, **58** (1), 20–24.

(28) Touyeras, F., Hihn, J.Y, Bourgoin, X., Jacques, B., Hallez, L. and Branger, V. (2005) Effects of ultrasonic irradiation on the properties of coatings obtained by electroless plating and electroplating. *Ultrasonics Sonochemistry*, **12**, 13–19.

(29) Kobayashi, K., Chiba, A. and Minami, N. (2000) Effects of ultrasound on both electrolytic and electroless nickel depositions. *Ultrasonics*, **38**, 676–681.

(30) Mahmood, T.R., Dennis, J.K. and Barrett, P.L. (1984) Effect of ultrasonic agitation on Ni-Co and Ni-Fe deposition. *Surface Technology*, **22**, 219–239.

(31) Chang, L.M., Guo, H.F. and An, M.Z. (2008) Electrodeposition of Ni-Co/Al_2O_3 composite coating by pulsed reverse method under ultrasonic conditions. *Material Letters*, **62** (19), 3313–3315.

(32) Ping, P.M., Zhu, D., Hu, Y.Y. and Zeng, Y.B. (2008) Experimental research on microelectroforming with ultrasonic agitation. *Advances in Machining and Manufacturing Technology IX*, **375–376**, 253–257.

(33) Jensen, J.A.D., Pocwiardowski, P., Persson, P.O.A. and Hultman, L. (2003) Acoustic streaming enhanced electrodeposition of nickel. *Chemical Physics Letters*, **368**, 732–737.

(34) Yasushi, O. and Morihiro, Y. (2002) Effect of ultrasonic irradiation for silver immersion plating. *Journal of the Surface Finishing Society of Japan*, **53** (10), 700–701.

(35) Pollet, B., Lorimer, J.-P., Phull, S.-S., Mason, T., Walton, D.J., Hihn, J.-Y., Ligier, V. and Wéry, M. (1999) The effects of ultrasonic frequency and intensity upon electrode kinetic parameters for the Ag(S_2O_3)3-2/Ag redox couple. *Journal of Applied Electrochemistry*, **29** (12), 1359–1366.

(36) Pollet, B., Lorimer, J.-P., Phull, S.-S. and Hihn, J.-Y. (2000) Sono-electrochemical removal of silver from photographic processing solutions. *Ultrasonics Sonochemistry*, **7** (2), 69–76.

(37) Pollet, B., Lorimer, J.P., Hihn, J.Y., Phull, S.S., Feakins, D., Walton, D.J. and Mason, T.J. (2002) The effect of ultrasound upon the oxidation of thiosulphate upon stainless steel and platinum electrodes. *Ultrasonics Sonochemistry*, **9** (5), 37–44.

(38) Pollet, B.G., Lorimer, J.-P., Hihn, J.Y., Touyeras, F., Mason, T.J. and Walton, D.J. (2005) Ultrasound effects on the kinetic reaction step in the silver thiosulphate reduction on platinum electrodes. *Ultrasonics Sonochemistry*, **12** (1–2), 7–11.

(39) Ohsaka, T., Isaka, M., Hirano, K. and Ohishi, T. (2008) Effect of ultrasound sonication on electroplating of iridium. *Ultrasonics Sonochemistry*, **15**, 283–288.

(40) Yaqub, A. and Ajab, H. (2009) Electrochemical removal of copper and lead from industrial waste-water: mass transport enhancement. *Water Quality Research Journal of Canada*, **44** (2), 183–188.

(41) Xie, F., Li, H., Ma, Y., Li, C., Cai, T., Huang, Z. and Yuan, G. (2009) The ultrasonically assisted recovery treatment of printed circuit board waste sludge by leaching separation. *Journal of Hazardous Materials*, **170** (1), 430–435.

(42) Walker, R. and Walker, C.T. (1975) New explanation for the brightness of electrodeposits produced by ultrasound. *Ultrasonics*, 79–82.

(43) Liu, D.L., Yang, Z.G., Wang, J. and Zhang, C. (2010) Effect of ultrasonic on electroless copper deposition on Ta/SiO_2/Si. *Rare Metal Materials and Engineering*, **39** (8), 1454–1459.

(44) Hyde, M.E. and Compton, R.G. (2002) How ultrasound influence the electrodeposition of metals. *Journal of Electroanalytical Chemistry*, **531**, 19–24.

(45) Floate, S., Hyde, M., Compton, R.G. (2002) Electrochemical and AFM studies of the electrodeposition of cobalt on glassy carbon: an analysis of the effect of ultrasound. *Journal of Electroanalytical Chemistry*, **523**, 49–63.
(46) Pollet, B.G., Hihn, J.Y., Doche, M.L., Lorimer, J.P., Mandroyan, A. and Mason, T.J. (2007) Transport limited currents close to an ultrasonic horn equivalent flow velocity determination. *Journal of Electrochemical Society*, **154** (10), E131–E138.
(47) Mandroyan, A., Hihn, J.-Y., Doche, M.-L. and Pothier, J.M. (2010) A predictive model obtained by identification for the ultrasonic "equivalent" flow velocity at surface vicinity. *Ultrasonics Sonochemistry*, **17** (6), 965–977.
(48) Hihn, J.-Y., Doche, M.-L., Mandroyan, A., Hallez, L. and Pollet, B.G. (2011) Respective contribution of cavitation and convective flow to local stirring in sonoreactors. *Ultrasonics Sonochemistry*, **18** (4), 881–887.
(49) Mandroyan, A., Viennet, R., Bailly, Y., Doche, M.-L. and Hihn, J.-Y. (2009) Modification of the ultrasound induced activity by the presence of an electrode in a sonoreactor working at two low frequencies (20 and 40 kHz). Part I: Active zone visualization by laser tomography. *Ultrasonics Sonochemistry*, **16** (1), 88–96.
(50) Mandroyan, A., Doche, M.L., Hihn, J.Y., Viennet, R., Bailly, Y. and Simonin, L. (2009) Modification of the ultrasound induced activity by the presence of an electrode in a sono-reactor working at two low frequencies (20 and 40 kHz). Part II: Mapping flow velocities by particle image velocimetry (PIV). *Ultrasonics Sonochemistry*, **16** (1), 97–104.
(51) Viennet, R., Ligier, V., Hihn, J.Y., Bereziat, D., Nika, P. and Doche, M.L. (2004) Visualisation and electrochemical determination of the actives zones in an ultrasonic reactor using 20 and 500 kHz frequencies. *Ultrasonics Sonochemistry*, **11** (3–4), 125–129.
(52) Hallez, L., Touyeras, F., Hihn, J.Y. and Klima, J. (2007) Energetic balance in an ultrasonic reactor using focused or flat high frequency transducers. *Ultrasonics Sonochemistry*, **14** (6), 739–749.
(53) Hallez, L., Touyeras, F., Hihn, J.-Y., Klima, J., Guey, J.-L., Spajer, M. and Bailly, Y. (2010) Characterization of HIFU transducers designed for sonochemistry application: Cavitation distribution and quantification. *Ultrasonics*, **50** (2), 310–317.
(54) Mason, T.J., Lorimer, J.P., Bates, D.M. and Zhao, Y. (1994) Dosimetry in sonochemistry: the use of aqueous terephthalate ion as a fluorescence monitor. *Ultrasonics Sonochemistry*, **1** (2), 91–95.
(55) Pollet, B.G., Hihn, J.-Y., Doche, M.-L., Lorimer, J.P., Mandroyan, A. and Mason, T.J. (2007) Transport limited currents close to an ultrasonic horn equivalent flow velocity determination. *Journal of the Electrochemical Society*, **154**, E131–E138.
(56) Arnyanaov, S.A. (1993) Internal stress in electrodeposited cobalt, nickel and their alloys: Part 3. *Metal Finishing*, **91** (4), 59.
(57) Pollet, B.G., Hihn, J.Y. and Mason, T.J. (2008) Sonoelectrodeposition (20 and 850 kHz) of copper in aqueous and deep eutectic solvents. *Electrochimica Acta*, **53**, 4248–4256.
(58) Pollet, B.G. and Hihn, J.Y. (2011) Sonoelectrochemistry: From theory to applications. *Handbook of Applications of Ultrasound and Sonochemistry*, **24**, 623–657.
(59) Endres, F. and Schweizer, A. (2000) The electrodeposition of copper on Au(111) and on HOPG from the 66/34 mol% aluminum chloride/1-butyl-3-methylimidazolium chloride room temperature molten salt: an EC-STM study. *Physical Chemistry Chemical Physics*, **2**, 5455–5462.
(60) Wassersheid, P. and Welton, T. (2003) *Ionic Liquid in Synthesis*, Wiley-VCH Verlag, Weinheim Germany.
(61) Endres, F. (2004) Ionic liquids: promising solvents for electrochemistry. *Zeitschriftfür Physikalische Chemie*, 218–255.
(62) Buzzeo, M.C., Evans, R.G. and Compton, R.G. (2004) Non-haloaluminate room-temperature ionic liquids in electrochemistry. *Chemical Physics and Physical Chemistry*, **5** (8), 1106–1120.

(63) Galinski, M., Lewandowski, A. and Stepniak, I. (2006) Ionic liquid as electrolytes. *Electrochimica Acta*, **51** (26), 5567–5580.

(64) Abbott, A.P., Capper, G., Davies, D.L., Rasheed, R.K. and Tambyrajah, V. (2003) Novel solvent properties of choline chloride/urea mixtures. *Chemical Communications*, **1**, 70–71.

(65) Abbott, A.P., Boothby, D., Capper, G., Davies, D.L. and Rasheed, R.K. (2004) Deep eutectic solvents formed between choline chloride and carboxylic acids. *Journal of the American Chemical Society*, **126**, 9142–9147.

(66) Abbott, A.P., Capper, G., Swain, B.G. and Wheeler, D.A. (2005) Electropolishing of stainless steel in an ionic liquid. *Transactions of the Institute of Metal Finishing*, **83**, 51–53.

(67) Abbott, A.P., Capper, G., Davies, D.L., Rasheed, R.K. and Shikotra, P. (2005) Selective extraction of metals from mixed oxide matrices using choline based ionic liquids. *Inorganic Chemistry*, **44**, 6497–6499.

(68) Abbott, A.P., Capper, G., McKenzie, K.J. and Ryder, K.S. (2007) Electrodeposition of zinc tin alloys from deep eutectic solvents based on choline chloride. *Journal of Electroanalytical Chemistry*, **599**, 288–294.

(69) Maisonhaute, E., White, P.C. and Compton, R.G. (2002) Surface acoustic cavitation understood via nanosecond electrochemistry. 2. The motion of acoustic bubbles. *Journal of Physical Chemistry B*, **106** (12), 3166–3172.

(70) Birkin, P.R., Offin, D.G. and Leighton, T.G. (2004) Electrochemical measurements of the effects of inertial acoustic cavitation by means of a novel dual microelectrode. *Electrochemistry Communications*, **6** (11), 1174–1179.

(71) Hihn, J.-Y., Doche, M.-L., Mandroyan, A., Pollet, B., Viennet, R., Maisonhaute, E. and Bailly, Y. (2007) *Electrochemical Measurements of Ultrasonic Wind and Asymmetric Cavitation*. 19th International Congress on Acoustics – ICA2007 Madrid.

(72) Ligier, V., Hihn, J.Y., Wéry, M. and Tachez, M. (2001) The effects of 20 kHz and 500 kHz ultrasound on the corrosion of zinc precoated steels in [Cl^-] [$SO_4{}^{2-}$] [$HCO_3{}^-$] [H_2O_2] electrolytes. *Journal of Applied Electrochemistry*, **31** (2), 213–222.

(73) C. Costa (2009) Influence des ultrasons de puissance sur les procédés électrochimiques en milieu liquide ionique. PhD Thesis, University of Franche-Comté.

(74) Costa, C., Hihn, J.Y., Rebetez, M., Bisel, I., Moisy, P. and Doche, M.L. (2008) Transport-limited current and microsonoreactor characterizations at 3 low-frequencies in the presence of water, acetonitrile and imidazolium-based ionic liquids. *Physical Chemistry Chemical Physics*, **10** (16), 2149–2158.

(75) Quinet, M., Lallemand, F., Ricq, L., Hihn, J.Y., Delobelle, P., Arnould, C. and Mekhalif, Z. (2009) Influence of organic additives on the initial stages of copper electrodeposition on polycrystalline platinum. *Electrochimica Acta*, **54** (5), 1529–1536.

(76) Wilms, M., Broekmann, P., Stuhlmann, C. and Wandelt, K. (1998) In-situ STM investigation of adsorbate structures on Cu(111) in sulfuric acid electrolyte. *Surface Science*, **4161** (1–2), 121–140.

(77) Costa, C., Doche, M.L., Hihn;, J.Y., Bisel, I., Moisy, P. and Leveque, J.M. (2010) Hydrodynamic sono-voltammetry of ferrocene in [Tf_2N]$^-$ based ionic liquid media. *Ultrasonics*, **50** (2), 323–328.

(78) Rezrazi, E.M., Doche, M.L., Berçot, P. and Hihn, J.Y. (2004) Au-PTFE composite coatings elaborated under ultrasonic stirring. *Surface and Coatings Technology*, **192** (1), 124–1130.

(79) Serhal, Z., Berçot, P., Morvan, J., Rezrazi, M. and Pagetti, J. (2002) Quantitative characterisation of Au–Co–PTFE composite coatings by differential enthalpic analysis. *Surface and Coatings Technology*, **150** (2–3), 290–296.

(80) Ge, S.H., Wu, Z.J., Zhang, M.H. and Li, W. (2006) Preparation of Ni-B/MgO catalyst by ultrasonic wave assisted electroless plating and its catalytic hydrogenation performance. *Chinese Journal of Catalysis*, **27** (8), 690–696.

(81) Wu, Z.J., Zhang, M.H., Ge, S.H., Zhang, Z.L., Li, W. and Tao, K.Y. (2005) Synthesis and characterization of a porous amorphous Ni-B catalysts on titania by silver-catalysed electroless plating. *Journal of Materials Chemistry*, **15** (46), 4928–4933.
(82) Wu, Z.J., Ge, S.H., Zhang, M.H., Li, W. and Tao, K.Y. (2009) Synthesis of nanoparticles supported on metal oxides using electroless plating: controlling the dispersion and size of nickel nanoparticles. *Journal of Colloid and Interface Science*, **330** (2), 359–366.
(83) Rich, S.R. (1955) Improvement in electroplating due to ultrasonics. *Plating*, **42**, 1407.
(84) Mallory, G.O. (1978) *Transaction IMF*, **56**, 82.
(85) Yang, L.X., Hou, W.T. and Wu, Y.S. (1997) Influence of ultrasound on the properties of electroless plated nickel. *Transaction IMF*, **75** (4), 131–133.
(86) Chiba, A. and Haijima, H. (2003) Effect of sonication and vibration on the electroless Ni-B deposited film from acid bath. *Surface and Coating Technology*, **169–170**, 104–107.
(87) Okumiya, M., Tsunekawa, Y., Saida, T. and Ichino, R. (2003) Creation of high strength bonded abrasive wheel with ultrasonic aided composite plating. *Surface and Coating Technology*, **169–170**, 112–115.
(88) Liu, Y., Jiang, S. and Huang, Y. (2010) Ultrasonic-assisted electroless deposition of Ag on PET fabric with low silver content for EMI shielding. *Surface and Coating Technology*, **204**, 2829–2833.
(89) Wang, X.-K., Shao, L., Guo, W.-L., Wang, J.-G., Zhu, Y.-P. and Wang, C. (2009) Synthesis of dendritic silver nanostructures by means of ultrasonic irradiation. *Ultrasonics Sonochemistry*, **16**, 747–751.
(90) Zhao, Y., Bao, C., Feng, R. and Liu, R. (1998) A new method of etching ABS plastic for plating by ultrasound. *Plating and Surface Finishing*, **9**, 98–100.
(91) Lua, Y., Jianga, S. and Huang, Y. (2010) Fabrication of high adhesive silver-coated PET electrode challenged by simulated body fluid (SBF) soaking. *Synthetic Metals*, **160** (5–6), 419–424.
(92) Yang, L.X., Luan, B., Cheong, W.J. and Shoesmith, D. (2005) Sono-immersion deposition on magnesium alloy. *Journal of the Electrochemical Society*, **152** (3), C131–C136.
(93) Zhang, Z.B. and Xu, S.F. (2007) Copper-Ti_3SiC_2 composite powder prepared by electroless plating under ultrasonic environment. *Rare Metals*, **26** (4), 359–364.
(94) Jeudy, A., Touyeras, F., Hihn, J.Y. and Roizard, X. (1997) Influence des ultrasons sur l'adhérence d'un dépôt de cuivre chimique sur un substrat polymère. *Le Vide, Science Technique et Applications*, **53** (284), 66–69.
(95) Cuevas, F. and Hirscher, M. (2000) The hydrogen desorption kinetics of Pd-coated $LaNi_5$-type films. *Journal of Alloys and Compounds*, **313** (1–2), 269–275.
(96) Yamanaka, S., Fujikane, M., Uno, M., Murakami, H. and Miura, O. (2004) Hydrogen content and desorption of carbon nano-structures. *Journal of Alloys and Compounds*, **366** (1–2), 264–268.

7

Influence of Ultrasound on Corrosion Kinetics and its Application to Corrosion Tests

Marie-Laure Doche and Jean-Yves Hihn

7.1 Introduction to Metal Corrosion

7.1.1 What Exactly is Corrosion?

According to the International Organization for Standardization, ISO 8044-198 [1], corrosion is defined as: '*A physicochemical interaction between a metal and its environment which results in changes in the properties of the metal and which may often lead to impairment of the function of the metal, the environment, or the technical system of which these form a part*'. The NACE definition [2] is: 'The deterioration of a material, usually a metal, that results from a reaction with its environment'. Although shorter, this definition suggests that this term, initially associated with metals, is now extended to all types of materials susceptible to degradation including polymers and ceramics. However, in this chapter we will limit our discussion to the electrochemical degradation of metallic materials.

7.1.2 Why Do Metals Corrode?

Metals corrode because they are chemically unstable in various environments. Only copper and precious metals (gold, silver, platinum, etc.) are found in nature in their metallic states. All other metals, including iron, the most commonly used metal, are processed from minerals or ores into metals which are inherently unstable in their environments. The driving force that causes metals to corrode is then a natural consequence of their temporary

Power Ultrasound in Electrochemistry: From Versatile Laboratory Tool to Engineering Solution, First Edition. Edited by Bruno G. Pollet.

existence in a metallic form. To reach this metallic state from their occurrence in nature in the form of various stable mineral forms (ores), it is necessary for them to absorb and store up Gibbs energy. As a result, the metal has a driving force that is sufficiently strong for it to return to its low energy equilibrium oxidation state. The return of this metal to its native oxidation state, via an electrochemical process, is what we name corrosion.

7.1.3 The Price to Pay: the Economical Impact of Corrosion

In thermodynamic terms, corrosion appears inevitable and leads to metal degradation. Apart from minor cases where the oxidative process of metal is required, corrosion damage is a real issue in modern industrial environments. Corrosion can lead to failures in plant infrastructure and machines, which are usually costly to repair, in terms of lost or contaminated products, environmental damage, and possibly of human safety.

Thus, it comes as no surprise that the costs associated with corrosion are extremely high. Several studies carried out over the past 30 years [3] have shown that the annual direct cost of corrosion for an industrial economy is approximately 3.1% of the country's Gross National Product (GNP). This figure is considered to be a conservative estimate since only well-documented costs have been used in the study. The indirect cost of corrosion (linked to corrosion prevention) was conservatively estimated to be equal to the direct cost, giving a total direct plus indirect cost of $552 billion or 6% of the GNP [4]. Other studies in Europe and Asia report corrosion costs ranging from approximately 1 to 5% of the GNP. Usually, the highest segments of corrosion cost are associated with utilities, transportation and infrastructure.

However, national statistics do not provide companies with information on their own corrosion costs and ways they can control them. For a company, calculation of the overall cost of corrosion consists in summing all the costs due to corrosion from its prevention to its consequences on residual injuries, throughout the company activity chain. The overall cost of corrosion usually integrates both direct and indirect costs. Direct costs are made up of two main components:

(i) The costs of design, manufacturing and construction. Designing, manufacturing and construction costs include preventive and protective actions such as: material selection (use of stainless steel instead of carbon steel), additional material and overdesign of the supporting structure (increasing wall thickness), material used to mitigate or prevent corrosion, such as coatings, sealants, corrosion inhibitors and cathodic protection.
(ii) The cost of management which includes repairs due to corrosion, replacement of corroded parts, rehabilitation, but also corrosion-related inspection, corrosion-related maintenance and inspection.

The indirect costs of corrosion, although they usually account for the major part of the total cost, are more difficult to calculate. These costs are related to halts in production, loss of production time, energy losses and so on.

Of course, environmental and safety consequences of corrosion take precedence over the purely economical calculations. Today, no company can tolerate major corrosion failures, especially those involving personal injuries, fatalities, unscheduled shutdowns and environmental contaminations. For this reason, considerable efforts are generally expended in corrosion control at the design stage and in the operational phase.

7.1.4 Corrosion Control Technology: the Need for Reliable Corrosion Tests

To make the 'best' of a 'bad' situation, Shaw and Kelly present their corrosion prevention humoristic philosophy: *'Like death and taxes, corrosion is something we hope to avoid; but ultimately it is something we must learn to deal with'* [5]. Going back to more fundamental aspects, substantial barriers (corrosion control methods) are available to prevent or minimize corrosion. This is the main purpose of Corrosion Engineering (or Corrosion Science). In that sense, the term 'Anti-Corrosion' Engineering should be preferred.

It is the rate of the corrosion process that is often the parameter of interest. This rate is controlled not only by the nature of the metal surface, but also by the nature of the environment as well as the evolution of both. The basis of corrosion control requires the ability to anticipate the useful life of a component (or part of it) in its service use.

7.1.4.1 Corrosion Testing [6]

Corrosion testing consists in developing test methods for determining the corrosion resistance of a material in specific environments and conditions. A large number of factors affect corrosion behaviour, and therefore there is no universal corrosion test. Typically, there are two major types of corrosion tests: laboratory testing and service testing (or field testing in the case of atmospheric corrosion).

Naturally, the most reliable indication of corrosion behaviour is given by service and field testing, but this is time consuming and sometimes impossible to carry out. On the contrary, laboratory testing allows predictive information to be obtained outside of service history (usually at laboratory scale) and in an acceptable time interval (typically a few weeks). Over time, accelerated corrosion testing has proven to be the most reliable and simple means for generating information, providing that the laboratory environment is similar to the service environment. Before implementing any corrosion tests, care must be taken in the selection of the corrosive media and in the preparation of the test specimen. Accelerating tests often represent the worst case scenario, requiring very aggressive corroding factors. However, caution must be exercised to gain meaningful results and to ensure that corrosion mechanisms are not altered. This is a difficult task and, consequently, accelerated tests cannot always be used to predict lifetime or corrosion rates, but rather to qualitatively select the most appropriate materials for use in specific applications.

7.1.4.2 Corrosion Acceleration Factors

Testing does not have to correlate exactly to the service environment as long as corrosion mechanisms remain the same [6]. However, the results of accelerated testing should correlate to results from more reliable sources, for example, service experience and field testing. The method for accelerating corrosion testing will depend upon the material being examined, the environment and the type of corrosion mechanism. The most common methods for acceleration include increasing:

(i) temperature (the most common and greatest accelerating factor),
(ii) pressure (autoclave corrosion test),
(iii) acidity, concentration and aeration in immersion tests,
(iv) the amount of NO_2 and SO_2 for atmospheric tests, and
(v) relative humidity (RH) for atmospheric tests.

7.1.4.3 *Laboratory Corrosion Tests*

There are many publications derived from national standard organizations (AFNOR in France, ASTM in the USA, BSI in the UK, DIN in Germany, JISC in Japan, SAC in China etc.) or international organizations (ISO), describing standard test methods for different types of corrosion testing in various atmospheres. The *Corrosion Tests and Standards Handbook* [7] subdivides laboratory corrosion tests into several categories.

(i) **Salt Spray testing:** This method refers to tests conducted in closed cabinets where the conditions of exposure are controlled and mostly designed to accelerate specific corrosion situations, while trying to simulate as closely as possible the corrosion mechanisms at play. Cabinet tests are generally used to determine the corrosion performance of materials intended for use in natural atmospheres. To correlate test results with field performance, it is necessary to establish acceleration factors and to verify that the corrosion mechanisms do indeed follow the same paths.

(ii) **Immersion tests:** Historically, immersion tests have been extensively used to generate uniform corrosion data for alloys used in process industries under immersion conditions. They can be divided into two categories:
- *Simple immersion tests* basically consist in exposure of small sections of the candidate material to the test medium. Loss of weight of the material is measured over a period of time. Although these tests form the quickest and most economical means for providing a preliminary selection of best suited materials there is no simple way to extrapolate the results obtained from these simple tests to the prediction of system lifetime.
- *Alternate immersion tests* (deep dry test) form a variation of the immersion test. In the cyclic procedure, a test specimen is immersed for a period of time in a test environment, then removed and dried before being re-immersed to continue the cycle. This test allows shorter response times due to acceleration of the corrosion mechanism in cyclic conditions.

(iii) **Electrochemical tests:** These tests are related to immersion tests coupled with electrochemical instrumentation. Indeed, the rate of corrosion can be measured by electrochemical methods because aqueous corrosion is an electrochemical phenomenon. All electrochemical tests use some fundamental model of electrode kinetics associated with corrosion processes to quantify corrosion rates. Amplification of the electrical signals generated during these tests has allowed very precise and sensitive measurements to be taken.

(iv) **Mechanically assisted corrosion tests:** There are also some tests specially designed to simulate mechanically assisted types of corrosion. The goal is to accelerate the test as far as possible, while maintaining the same test corrosion mechanisms as for service corrosion [8]. For example, some common standard tests are related to stress corrosion cracking or corrosion fatigue:
- *Stress corrosion cracking* (SCC) tests essentially describe a specimen and a loading method that stresses the specimen while exposed to the solution. Susceptibility to SCC is then assessed by the time taken for failure of the specimen, or the development of cracks in the surface of the specimen.
- *Corrosion fatigue testing* implies the same strategy as SCC testing but using the cyclic stress procedure.

7.1.5 Why Study Corrosion Under Sonication?

7.1.5.1 Cavitation Corrosion Definition

Amongst the mechanically influenced forms of corrosion (see later in this chapter), erosion-corrosion is caused by the relative movement between a corrosive fluid and a metal surface. It usually occurs at locations where turbulence develops in a fluid-circulating system. Turbulence can be caused by excessive velocity, sharp changes in direction and flow obstacles such as burrs and solder blobs. This phenomenon leads to typical corrosion attacks like grooves, valleys, wavy surfaces, holes, with a characteristic directional appearance (comet tails, horseshoe marks, etc.) [9].

Cavitation-corrosion is a particular form of erosion-corrosion caused by the 'implosion' of gas bubbles on a metal surface [10, 11]. As explained in Chapter 1, cavitation refers to the nucleation and growth of cavities, containing vapour and dissolved gas, in a vibrating or turbulent fast-flowing liquid. It occurs when local pressure in the liquid drops below a critical level, and is often associated with sudden variations in pressure related to the hydrodynamic parameters of the fluid. After they migrate toward high pressure regions, bubbles will implode. This collapse is associated with shock waves and microjet formation, both phenomena resulting in cavitation impingements at a solid surface. Repetition of cavitation bubble implosion (localized corrosion events) can eventually result in mass loss, fatigue failure and severe damage, which are often observed on hydraulic turbine blades, heat exchangers, ship propellers, stirrer blades, and so on.

7.1.5.2 Cavitation-Erosion Testing

A common way to generate cavitation in a fluid and to perform cavitation-erosion testing is to use an ultrasonic device immersed in distilled water [12]. The ASTM G32-92 standard recommends the use of a 20 kHz ultrasonic apparatus to perform cavitation-erosion testing, either in the direct or indirect configuration (see Section 7.2.4.2). This test method is recommended to estimate the relative resistance of materials to cavitation erosion as it may be encountered in pumps or hydraulic turbines, and valves. In the direct configuration the specimen tip is screwed onto the ultrasonic probe and then immersed in the test liquid at 25 °C (usually deionized water).

Research papers describing sonication of a specimen in distilled water usually refer to cavitation-erosion testing and mechanisms. These tests are usually performed by scientists and engineers mostly concerned by mechanical and wear properties, and rarely take into account the chemical composition of media. The role of chemistry and electrochemistry is often ignored and not fully understood [13]. In fact such investigations deal with cavitation-erosion tests rather than cavitation-corrosion tests but confusion still remains in the literature.

7.1.5.3 The Synergetic Effect of Corrosion and Cavitation

Tomlinson *et al.* [12] were amongst the first to point out the synergetic effect of mechanical (cavitation) and electrochemical effects. However, studies considering the synergetic effect between cavitation and corrosion on metal damage are not so frequent. Corrosion-cavitation mechanisms are far from being fully investigated, principally because each metal/media pairs behave differently to corrosion (general corrosion, localized corrosion due to passivity breakdown, etc.). At the beginning of the twenty-first century, an attempt to define an international cavitation-erosion test was made through the International Cavitation Erosion Test (ICET) project [13]. The experimental protocol consisted in testing six selected materials (from pure aluminium to cast iron and plastics) in various laboratory devices (cavitation tunnel, ultrasound vibratory rigs, rotating disks, etc.). This study was used to define a set of relevant parameters to characterize cavitation-erosion damage and mechanisms. It also led to standardization of experimental techniques and the development of a database. Part of this study is easily transposable to cavitation-corrosion testing. Kwok and Man [14] also made an important contribution by studying the synergetic effect between ultrasonically induced cavitation and corrosion on a wide variety of metals and alloys (copper, brass, bronze, iron, steels and stainless steels) in 3.5% NaCl solution.

The aim of this chapter is to provide substantial information on both corrosion-cavitation mechanisms and testing, performed using ultrasonic devices.

(i) In the literature, most research and technical papers deal with the influence of ultrasound on the corrosion mechanisms of various metal/media pairs. This approach will be referred to as sono-corrosion (or acoustic-corrosion mechanisms). A large part of the studies concerns ultrasonically enhanced corrosion of stainless steel and cavitation-induced passivity breakdown. Similar studies concern aluminium behaviour. A few papers focus on the generalized corrosion mechanisms of zinc, copper, bronze, iron and steel under ultrasound.

(ii) Papers of the second type are those using ultrasound to develop a laboratory accelerated corrosion test. In this case ultrasound can be used either to simulate the real corrosion conditions (cavitation-corrosion) or simply as a tool to accelerate the deterioration mechanism. Tests performed with this goal in mind are varied and concern atmospheric corrosion of zinc, corrosion of steel in concrete, of stainless steel for surgical applications, and of stainless steel in exhaust system condensate.

7.1.6 Corrosion and Corrosion-Cavitation Mechanisms

7.1.6.1 Electrochemical Principles of General Corrosion

Corrosion consists of an oxidation reaction coupled with a reduction reaction at the surface of the corroding material [10]. The oxidation reaction generates metal ions and electrons, which are consumed in the reduction reaction. A corroding surface can then be compared to a short-circuited battery, where the dissolution reaction at the anode supplies electrons for the reduction reaction at the cathode. For example iron in an acidic solution will corrode according to the following reactions:

$$\mathrm{Fe} \rightarrow \mathrm{Fe}^{2+} + 2\mathrm{e}^{-} \quad \text{Anodic half-reaction} \tag{7.1}$$

$$2\mathrm{H}^{+} + 2\mathrm{e}^{-} \rightarrow \mathrm{H}_2 \quad \text{Cathodic half-reaction} \tag{7.2}$$

$$\mathrm{Fe} + 2\mathrm{H}^{+} \rightarrow \mathrm{H}_2 + \mathrm{Fe}^{2+} \rightarrow \text{Net reaction} \tag{7.3}$$

For environments with water present, including moisture in the air, corrosion is mainly due to oxygen, and electrons are consumed by converting oxygen and water to hydroxide ions. In iron and many iron alloys, these hydroxide ions in turn combine with iron ions to

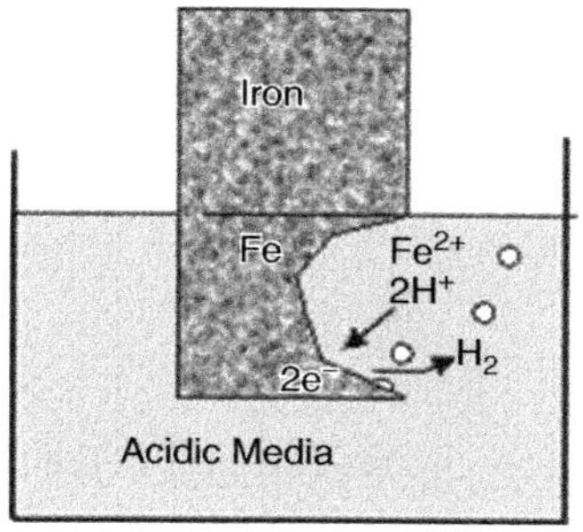

Figure 7.1 *Diagram of the corrosion mechanism of iron in an acid solution.*

form a hydrated oxide Fe(OOH). Subsequent reactions form a mix of magnetite (Fe_3O_4) and hematite (Fe_2O_3). This red-brown mixture of iron oxides is rust. The reaction pathway below illustrates corrosion by oxygen:

$$Fe + 3OH^- \rightarrow FeOOH + 3e^- + H_2O \quad \text{Anodic half-reaction} \tag{7.4}$$

$$O_2 + 2H_2O + 4e^- \rightarrow 4OH^- \quad \text{Cathodic half-reaction} \tag{7.5}$$

$$4Fe + 3O_2 + 2\,H_2O \rightarrow 4FeOOH \quad \text{Net reaction} \tag{7.6}$$

Figure 7.1 illustrates the basic oxidation/reduction reaction of corrosion.

7.1.7 Corrosion Rate

Since corrosion reactions involve transfer of electrons and ions between the metal and the solution, the corrosion rate can be expressed in terms of current density thanks to Faraday's law. Faraday's law for a given current flowing through a conductor I, is expressed by:

$$I = \mathrm{d}Q/\mathrm{d}t = nF\,\mathrm{d}n_i/\mathrm{d}t \tag{7.7}$$

where I is the current (A), Q is the charge (C = A s), F is the Faraday constant (96 500 C mol^{-1} of electrons), n is the number of moles of electrons and $\mathrm{d}n_i/\mathrm{d}t$ = the reaction rate (mol s^{-1}).

Using Faraday's law, corrosion rate can be expressed in:

(i) mass variation per unit of time and surface
(ii) number of moles per unit of time and surface
(iii) depth corroded per unit of time
(iv) current density.

The following expression is the most commonly used for this purpose:

$$i = I/A = nFm/MAt \tag{7.8}$$

where i is the current density (A m^{-2}), F is the Faraday constant (96 500 C mol^{-1} of electrons), n is the number of moles of electrons, m is the weight loss (g), M is the atomic mass (g mol^{-1}) and A is the electrode surface (m^2).

In practice, the composition and the physical form of the corrosion products greatly influence the corrosion rate of the exposed sample. There are three possible types of corrosion products, namely:

(i) dissolved ions
(ii) porous layer
(iii) compact film.

In acidic media, metal corrosion produces hydrated cations which have normally no influence on the corrosion rate. In neutral or basic media, corrosion products become less soluble, and precipitate on the metal surface in the form of hydroxide or salts. A superficial porous layer can then form, which does not protect the metal against corrosion but which helps slow down the process. It typically corresponds to rust formation on iron exposed to a wet atmosphere. On the contrary, the formation of a dense surface oxide film is critical in mitigating the rate of metal dissolution. Such thin (< 10 nm) native oxide films, called passive films, make the technological use of metallic materials possible by serving as barriers to dissolution. Some metals are able to exhibit a passive behaviour: stainless steel, aluminium exposed to neutral solution or wet air.

Apart from corrosion product influence, Faraday's law is useful in that it gives a corrosion rate only in the case of general corrosion (uniform or galvanic). It is meaningless to describe corrosion damage in the case of localized corrosion since weight loss is often negligible. However, electrochemical techniques can be used in all situations in order to give at least a qualitative evaluation of corrosion resistance.

7.1.8 Electrochemical Study of Corrosion Reactions

The study of corrosion processes involves the use of many of the same tools that are used in physical and analytical electrochemistry. Application of mixed potential theory to corrosion was originally presented by Wagner and Traud [15]. In 1957, Stern and Geary theoretically analysed the shape of polarization curves, providing the basis for the primary experimental technique (electrochemical polarization) used in electrochemical studies of corrosion [16].

Basically, electrochemical tests present three main advantages: electrochemical corrosion rate measurements can be made almost instantly (typically within a few minutes), long before a conventional weight change would be measurable. This is virtually a non-destructive testing, but the tests also allow corrosion damage (such as pitting corrosion) to be induced.

Electrochemical corrosion tests are usually based on the following techniques:

(i) Potentiodynamic polarization methods: this involves changing the potential of the working electrode and monitoring the induced current as a function of time or potential. These techniques can provide significant useful information on corrosion mechanisms and corrosion rate in specific environments.
(ii) Linear polarization resistance (LPR): the polarization resistance of a material is defined as the slope of the potential-current density curve ($\Delta E/\Delta i$) at the 'free' corrosion potential, yielding the polarization resistance R_p that can be related (for reactions under activation control) to the corrosion current by the Stern-Geary equation.

(iii) Electrochemical impedance Spectroscopy (EIS): EIS is used to access the charge transfer resistance or polarization resistance that is proportional to the corrosion rate at the monitored interface. EIS results have to be interpreted with the help of a circuit model of the interface. A major advantage of EIS over other laboratory techniques is the possibility of using very small amplitude signals (between 5 and 50 mV over a range of frequencies of 0.001 to 100 000 Hz) without significantly interfering with the properties being measured.

(iv) Electrochemical noise (EN): this consists in measuring the fluctuations in potential or current of a corroding metallic sample and in correlating this phenomenon to corrosion processes. Localized corrosion noise is believed to be generated by a combination of stochastic processes, such as passivation breakdown and repassivation events, and deterministic processes, which can be caused by film formation or pit propagation processes. For example, the study of corrosion potential fluctuations is applied to monitor the onset of events characterizing localized corrosion such as pitting, exfoliation, erosion-corrosion in either laboratory or diverse and complex industrial environments.

7.1.9 Forms of Corrosion

Traditionally, corrosion is classified into eight categories based on the morphology of the attack, as well as the type of environment to which the material is exposed [10]. There are other equally valid methods for classifying corrosion, and no universally accepted terminology is in use. Note that a given situation may lead to several forms of corrosion on the same piece of material. Table 7.1 gives an overview of the most common forms of corrosion.

The ASTM classification, in Table 7.2, allows a complementary classification of corrosion [7]. Various corrosion processes are classified according to their mechanism of attack. Table 7.2 shows that some forms of corrosion affect relatively large areas, while other mechanisms are local in nature. Other corrosion mechanisms are facilitated by metallurgical factors, while still others depend on the movement of two structures relative to each other or to components subjected to moving fluids or slurries. The final category identifies corrosion mechanisms that result from the inspection of a 'stressed' material in the presence of a corrosive agent.

7.1.10 Cavitation-Corrosion

7.1.10.1 Origin of the Damage

In most papers published on the subject [11], the mechanism responsible for cavitation-corrosion is the existence, on the metallic surfaces, of tiny asymmetrically imploding bubbles ejecting a very powerful jet of liquid towards the surface. As stated earlier, cavitation is the growth and collapse of bubbles inside a liquid submitted to high pressure variations. It is caused by the local and sudden drop of internal pressure below a critical level which allows liquid to vaporize. Cavitation damage is a serious technical problem for many devices; from ship propellers, to hydroelectric turbines, car combustion chambers and even high power ultrasonic vibrators, all these devices are subjected to cavitation-corrosion damage.

Table 7.1 *Traditional corrosion classification [10].*

Uniform general	Uniform or general corrosion is the most prevalent form of corrosion. It is characterized by the existence of several individual electrochemical processes that occur uniformly over the whole of the surface considered. The consequences of uniform corrosion are a decrease in metal thickness per unit time (or a weight loss per unit area per unit time) if the corrosion products are soluble, or a more or less uniform deposit of these products if they are insoluble.	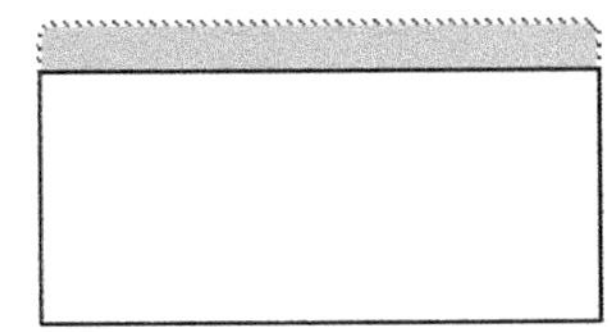
Galvanic corrosion	Galvanic corrosion can be defined simply as being the effect resulting from contact between two different metals or alloys in a conducting corrosive environment. Another term employed is galvanic coupling. This leads to fast dissolution of the less noble metal which operates as the anode in the (-) terminal of the battery. The more noble metal (the cathode) is hence protected.	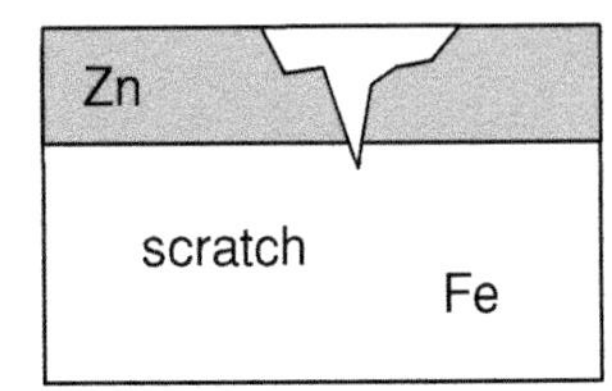
Crevice corrosion	This form of attack (rather generalized) is generally associated with the presence of small volumes of stagnant solution in occluded interstices, beneath deposits and seals, or in crevices. Deposits of sand, dust, scale and corrosion products can all create zones where the liquid can only be renewed with great difficulty. The mechanism is based on the fact that, in an interstice, convection in the liquid is strongly impeded, and the dissolved oxygen is locally rapidly exhausted. This creates a 'differential aeration cell' between the small deaerated interstice and the aerated remainder of the surface leading to a 'galvanic' corrosion process between these two zones.	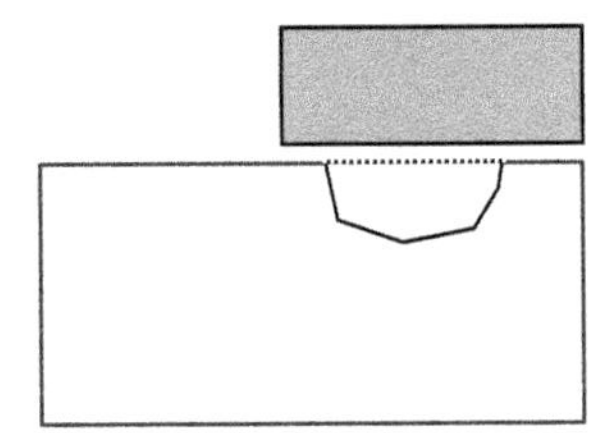

Pitting corrosion	Pitting corrosion leads to highly insidious attacks in the form of localized holes that can penetrate inwards extremely rapidly, while the rest of the surface remains intact. Pitting corrosion occurs mainly in solutions containing chloride, bromide or hypochlorite ions. Stainless steels are particularly sensitive to pitting corrosion, but other metals, such as passive iron, chromium, cobalt, aluminium, copper and their alloys are also prone to this form of damage.	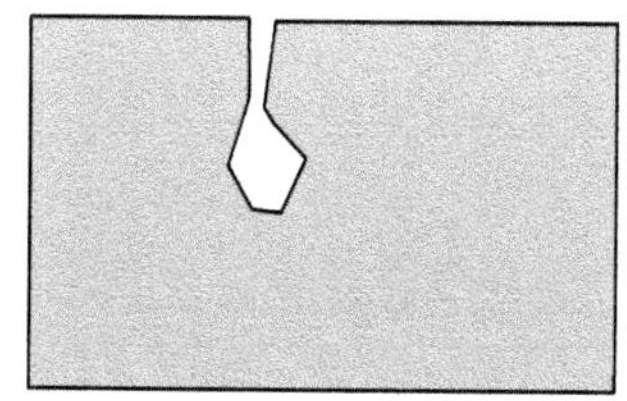
Intergranular corrosion	In most cases of corrosion, including uniform corrosion, the grain boundaries behave in essentially the same way as the grains themselves. However, in certain conditions, the grain boundaries can undergo marked localized attack while the rest of the material remains unaffected. The alloy disintegrates and loses its mechanical properties. This type of corrosion is due either to the presence of impurities in the boundaries, or to local enrichment or depletion of one or more alloying elements.	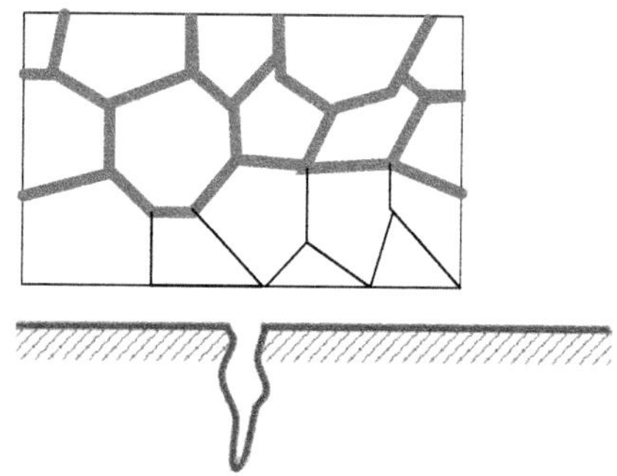

Table 7.1 *(Continued)*

Selective leaching (or dealloying)	This process involves the preferential dissolution of one of the elements in a single phase alloy or one of the phases in a multiphase alloy, leading to a porous residue on the surface. Common forms of selective leaching are decarburization, decobaltification, dezincification, denickelification and graphitic corrosion.	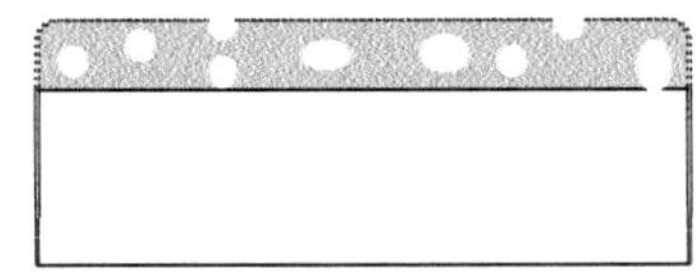
Erosion-corrosion	Erosion-corrosion is the cumulative damage induced by electrochemical corrosion reactions and mechanical effects from relative motion between the electrolyte and the corroding surface or accelerated degradation in the presence of this relative motion caused by high velocity, with mechanical wear and abrasion effects.	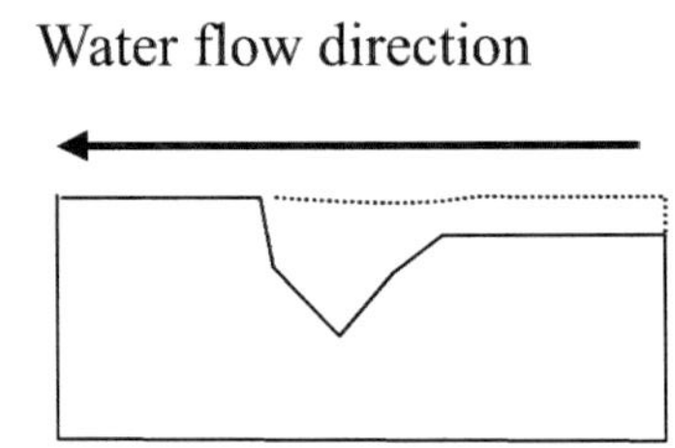
Stress corrosion cracking	Stress corrosion cracking (SCC) is a process involving the initiation of cracks and their propagation, possibly up to complete failure of a component, due to the combined action of tensile mechanical loading and a corrosive medium. In point of fact it is the presence of tensile stresses that is dangerous, as compressive stresses exert a protective influence.	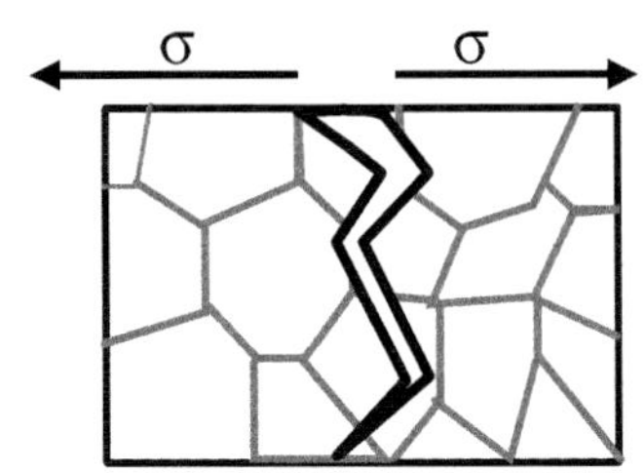

Table 7.2 *ASTM classification [7].*

General corrosion (GC)	Localized corrosion	Metallurgically influenced corrosion	Mechanically assisted degradation	Environmentally induced cracking
Corrosive attack dominated by uniform thinning	High rates of metal penetration at specific sites	Affected by alloy chemistry and heat treatment	Corrosion with a mechanical component	Cracking produced by corrosion, in the presence of stress.
• Atmospheric corrosion • Galvanic corrosion • Stray-current corrosion • General biological corrosion • Molten salt corrosion • Corrosion in liquid metals • High-temperature corrosion	• Crevice corrosion • Filiform corrosion • Pitting corrosion • Localized biological corrosion	• Intergranular corrosion • Dealloying corrosion	• Erosion corrosion • Fretting corrosion • Cavitation corrosion • Corrosion fatigue	• Stress-corrosion cracking (SCC) • Hydrogen embrittlement • Liquid metal embrittlement • Solid metal induced embrittlement

Bubble collapse can occur in a symmetric manner in the liquid bulk or in an asymmetric manner at a solid interface. During the collapse a large pulse of stress is generated followed by a high velocity fluid jet [17]. These microjets, whose average velocity, in the last μs, is 400 to 800 m s^{-1}, act like microscopic 'bombs' resulting in surface impingements, erosion, mass transfer enhancement, electron transfer activation and electrolyte chemical composition modification: see Figure 7.2.

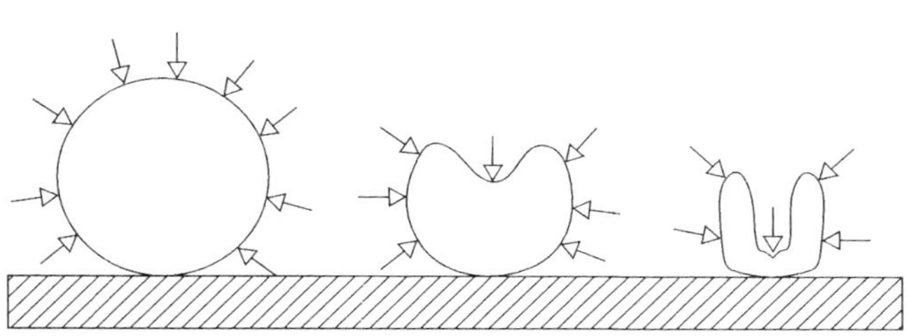

Figure 7.2 *Diagram showing formation and implosion of a cavitation bubble toward a surface.*

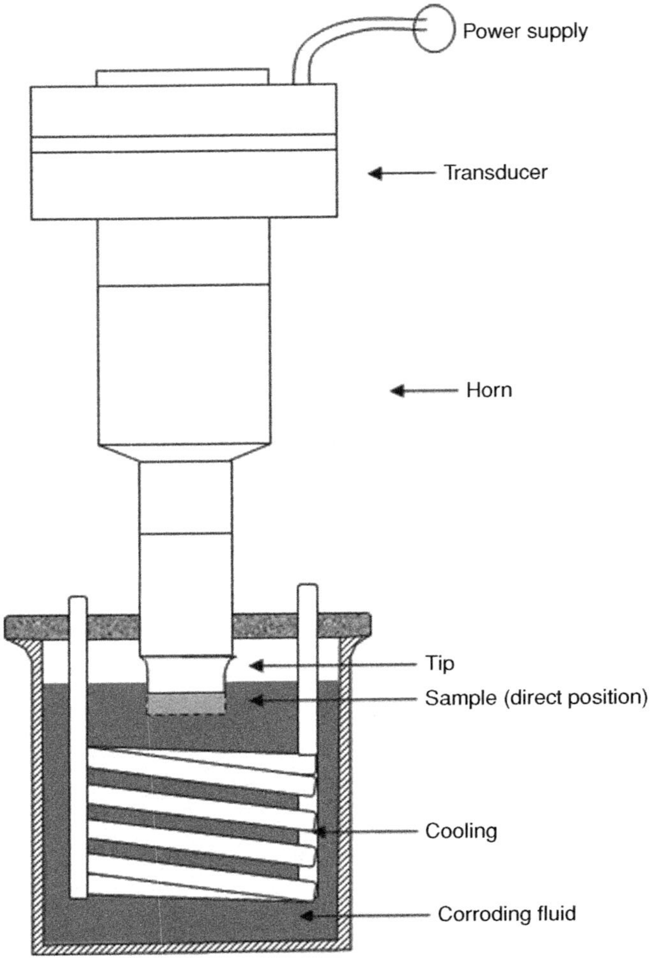

Figure 7.3 *Design of the ultrasonic setup used in ASTM G32-92 standard testing.*

7.1.10.2 Experimental Study of Corrosion-Cavitation

7.1.10.2.1 Experimental Setup To perform cavitation-corrosion testing, the ASTM G32-10 [18] standard recommends use of a 20 kHz ultrasonic apparatus either in the direct or indirect configuration (see Figure 7.3).

Although some other systems exist (rotating disc, cavitating tunnel and cavitating jet cell [13]), this is one of the most common setups used for this purpose. In the direct configuration (Figure 7.3) the specimen tip is screwed onto the probe and then immersed in the test liquid at 25 °C (usually deionized water). The probe vibrates at 20 kHz at a specific peak-to-peak amplitude (usually between 30 and 50 μm) for a defined duration leading to production of a large amount of cavitation bubbles. As the bubbles collapse, temperature and pressure increase, and shock waves and microjets are generated leading to impingements of the test specimen and, furthermore, to its erosion (i.e. material loss). This configuration is often used, and the duration of the test (continuous or intermittent) ranges from several minutes to several hours.

According to Diodati *et al.* who studied cavitation damage in water of aluminium samples fixed on the tip of the horn, at the very beginning of the exposure, cavitation damage consists of almost regular circular micro-holes, the distribution of which appears totally random [11]. After 10 minutes of exposure, many particles of eroded material are in the solution, and material loss is roughly 0.05 g (brought to a sample weight of 0.5 g). Surface morphology now exhibits deep craters and protrusions (see Figure 7.4). These authors also pointed out that in such experimental configurations, a border effect appears on the surface of the samples, leading to serious cavitation damage in the central part, while in the circular and external border almost no corrosion is present [11].

The indirect configuration consists in placing the sample in front of and parallel to the horn in its close vicinity (stationary specimen). The distance between the tip and the specimen has to be carefully adjusted and can vary from one study to another (usually between 1 and 10 mm to obtain maximum erosion) [19, 20]. This configuration is mainly used by researchers studying acoustic corrosion thanks to electrochemical methods, as in this case cavitation damage is weaker. It is also used with brittle specimens that cannot be fixed on the tip.

7.1.10.2.2 Mechanisms of Corrosion-Cavitation and Characteristic Parameters The easiest way to characterize corrosion-erosion rate is, as a first approach, to determine the mass loss. The sample is weighed before and after the test to determine its mass loss versus time of exposure to cavitation.

It is well known that the cavitation-erosion progress is divided into several stages [13,21,22]:

(i) an incubation period during which there is no damage
(ii) a transient period (composed of an acceleration stage sometimes followed by a deceleration stage)
(iii) a steady state stage where the corrosion-erosion rate is constant.

In the incubation stage, plastic deformation takes place on the material without mass loss. Mass loss then starts and proceeds at an increasing rate with erosion until it reaches a maximum rate. Finally a deceleration stage may be observed (sometimes it is not), and the erosion rate remains constant over a period of time. The transient stage includes both the acceleration stage and the deceleration stage. The final constant rate period is termed the steady-state stage. Figure 7.5 illustrates this mechanism in diagram form.

According to Soyama *et al.* [21], incubation time is defined as the intercept of the tangent of the acceleration period with the time axis on the mass loss versus time diagram. In water, it can be very short for weak materials such as pure copper and pure aluminium, but can last for nearly 20 minutes for stainless steel. For a dedicated material, the longer the incubation, the better the resistance of the material to cavitation is assumed to be.

An inverse measure of cavitation-erosion resistance is mass loss at the steady-state stage. Some other parameters can be considered to quantify damage, such as the mean and maximum depth of erosion [13]. Kwok and Man [14, 23, 24] performed extensive studies

Figure 7.4 *Appearance of aluminium plates after oscillating 10 minutes at 20 kHz in water. Illustration of the border effect. In the reported scale, the distance between each segment is 0.25 mm (Reprinted with permission from [11] Copyright (2001) Elsevier Ltd).*

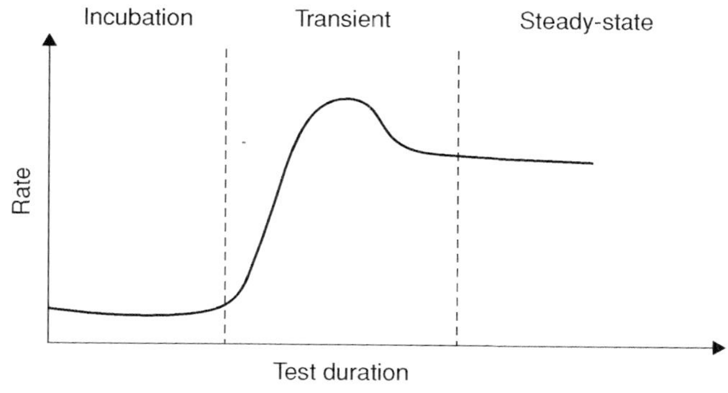

Figure 7.5 *Change in corrosion-cavitation rate as a function of test length (diagram) [10].*

on various metals and alloys to evaluate the synergetic effect of cavitation erosion and corrosion using the mean depth of penetration (MDP) and mean depth of penetration rate (MDPR) to express damage. These two magnitudes were calculated using the loss rate of materials during exposure to cavitation (using the ASTM Standard G32-10 facilities) as below:

$$\mathrm{MDP}(\mu\mathrm{m}) = \Delta\mathrm{W}/10\rho\mathrm{A} \tag{7.9}$$

and

$$\mathrm{MDPR}\ (\mu\mathrm{mh}^{-1}) = \Delta W/(10\rho A\Delta t) \tag{7.10}$$

where ΔW is the weight loss at each time interval (in mg), Δt is the time interval in hours, A is the specimen area in cm^2 and ρ the density of the specimen in g cm^{-3}.

The MDPR parameter was first measured in distilled water to obtain the cavitation-erosion resistance [$R_e = (\mathrm{MDPR})_e^{-1}$], then in 3.5% NaCl to obtain the cavitation-corrosion resistance (R_{ec}) using the same calculation. Alongside these measurements, the pure corrosion resistance R_c (in silent conditions) was measured in 3.5% NaCl using an electrochemical method.

It appears that the relative significance of corrosion, erosion by cavitation and the interaction between them depend on the material/environment system. In the case of metallic materials, the interaction can be synergetic or inhibitive. The synergetic nature of erosion-corrosion progress is determined by some major factors such as the passivity of the metal surface, the stability and adherence of the corrosion products, the electrolyte composition, the presence of dissolved oxygen and the intensity of cavitation. According to these authors, for copper-based alloys, corrosion only plays a minor role in the overall cavitation erosion-corrosion process. On the contrary, for steels and cast iron, corrosion/erosion synergism is very significant, accounting for nearly 80% of overall mass loss. Both corrosion and corrosion/erosion synergism are negligible in the case of stainless steel.

7.2 Influence of Ultrasound on the Corrosion Mechanisms of Metals

Some of the earlier works, devoted to the effect of ultrasound on metal corrosion, are due to Alkire and Perusich [25] and Tomlinson and Chapman [12], in the 1980s. The first work investigates the effect of focused ultrasound (20 kHz) on passivity of iron in sulfuric acid, while the second study deals with generalized corrosion of copper induced by 20 kHz ultrasound. These two approaches are significant in the literature on this subject. A few papers concern general corrosion enhancement due to sonication and most of them describe localized corrosion and passivity breakdown. In this section, we focus on the specific influence of sonication on corrosion.

7.2.1 Influence of Ultrasound on General Corrosion

7.2.1.1 Resistance of Various Metals to Cavitation-Erosion in Water

The earlier work by Tomlinson and Chapman [12], often cited in the literature, actually refers to cavitation-erosion testing (rather than cavitation-corrosion) of copper in 20 °C distilled water. Erosion takes place by means of a 20 kHz ultrasonic horn at various peak-to-peak amplitudes (from 13.5 to 67.5 μm) for 60 minutes. Erosion rates are determined by mass loss as a function of time at several vibration amplitudes. It appears that there is no incubation period on copper and that erosion rate is linear with time and with tip vibration amplitudes.

Similarly, Madigan *et al.* performed sonication (20 kHz) of several noble metal electrodes in deionized water for 300 s [19]. A separation distance of 1 mm between the ultrasonic probe tip and each electrode was used, and the electrical input power was 475 W corresponding to a vibrational amplitude of 125 μm. The authors studied the surface modification induced by ultrasound of gold, platinum, palladium and tungsten electrodes. The Cottrell law was used to determine the 'apparent' area of the sample after sonication and compared to SEM examinations of the surfaces. They concluded that, in the test conditions for a time range between 50 ms and 300 s, sonication does not induce any measurable surface roughening on these materials. Other studies performed on copper, silver, aluminium and zinc in similar conditions (i.e. 20 kHz, 40 W cm^{-2} (200 W), 4 to 10 minutes in distilled water, for a probe to specimen separation of 0.5 mm) showed that surface roughness of the samples gradually increases with reaction time [26]. This phenomenon appears after one minute of exposure on aluminium, and roughness attains 30 mm after 4 minutes of exposure. Copper has a longer incubation time, and requires a reaction time of 10 minutes to obtain the same roughness as on aluminium plates. Silver plates exhibited the same surface roughness as copper after 5 minutes exposure. Zinc, because of its hexagonal structure, is the hardest metal studied. Sonication time must exceed 10 minutes in order to observe surface degradation. This investigation also underlines that ultrasound induces a structural change in the metallic surface corresponding to crystallographic texture modification.

7.2.1.2 Effect of Cavitation on the Electrochemical Behaviour of Various Alloys in NaCl and KCl Solutions

Al-Hachem *et al.* carried out extensive laboratory studies on the cavitation-corrosion behaviour of some cast alloys used to manufacture seawater valves and pumps [27–30]. The first published paper [27] deals more specifically with nickel aluminium bronze (NAB), the behaviour of which was investigated using a 20 kHz ultrasonic vibrator at 25 μm vibration amplitude. In these experiments, the specimen faces the horn at a distance of 12.5 mm. The cavitation tests were conducted in natural sea water from the Arabian Gulf (pH = 8).

This paper is one of the first to associate cavitation-erosion testing with electrochemical measurements. It was shown that the presence of cavitation shifts the 'free' potential of the material in the active direction by −70 mV. A 1.6-fold increase in corrosion current density was observed in sonicated samples, and the overall rate of mass loss was 187 times greater, that is, 1.5 mg hr^{-1} cm^{-2} instead of 0.008 mg hr^{-1} cm^{-2}. This observation was associated to be due to corrosion product layer destruction, which switches the material to an active electrochemical dissolution state. When cathodic protection is applied (as it is practically the case for devices in sea water), the rate of damage by cavitation-corrosion is 47% less than that at OCP (open circuit potential). This reduction is ascribed to the cushioning of bubbles collapsed by hydrogen evolution as well as by the elimination of the oxidation process. SEM examinations showed that cavitation makes the surface of the material very rough, with large cavities, and in turns inducing grain boundary attack [30]. Cast iron underwent an opposite shift in a 'free' potential of +100 mV, when subjected to cavitation in NaCl. This would indicate that corrosion rate decreases in the presence of ultrasound, probably due to stabilization of the iron oxide corrosion product layer [28].

Another extensive study, already mentioned (see Section 7.1.5.3), was published by Man and Kwok [14,23,24], who investigated the cavitation-corrosion synergism of various alloys as well as laser-fabricated alloyed surfaces on steel and aluminium in 3.5% NaCl. The experimental setup consisted of a 20 kHz ultrasonic device vibrating with an amplitude of 30 μm on which the sample was sealed. In addition, 'free' corrosion potentials of specimens in 3.5% NaCl solution was measured under quiescent and cavitation conditions. The results of this study are presented in Figure 7.6 for five alloys, namely: mid-steel 1050, brass, stainless steel (SS) 304, SS 316 L, SS Zeron 100.

The results are consistent with those obtained by Al-Hashen *et al.* [27]: a positive shift in 'free' potential is observed on steel (as on cast iron), while a negative shift (active dissolution direction) is recorded on brass and other stainless steels. Kwok and Man concluded that the effect of cavitation on corrosion behaviour and corrosion potential depends on two competitive effects of cavitation, namely, corrosion product detachment and mass transport increase. Concerning stainless steel, the mechanisms will be discussed later in this chapter, but it can already be observed that this active shift in corrosion potential is not followed by an increase in corrosion rates as metal repassivates very rapidly.

Finally, De Morais and Brett [20] also used power ultrasound operating at 20 kHz (from 22 to 44 W cm^{-2}) to investigate the influence of ultrasonic irradiation on several metals immersed in KCl solutions at various concentrations (0.01 to 1 M). The metals included AISI M2 high speed steel and 99.99% pure aluminium. The materials were chosen due to their different rate-limiting steps in the corrosion process. In these experiments, the ultrasonic probe was placed directly above the metal surface, and the effect of varying the probe-electrode distances ('face-on' geometry) and ultrasonic power was investigated. Three electrochemical techniques were used to evaluate the corrosion process and its rate: (i) measurements at open circuit potentials (OCPs), (ii) polarization curves, and (iii) electrochemical impedance spectroscopy, before, during and after insonation in the presence of dissolved oxygen. The corroded surfaces were examined by SEM after ultrasonic irradiation. In the case of aluminium, where an oxide film is easily formed and the cathodic reaction is rate limiting, ultrasound removes the oxide film (at least partially) and leads to formation of deep pits. Nevertheless, the surface is capable of recovering from this action and reforming an oxide layer afterwards, providing that the Cl^- concentration in solution

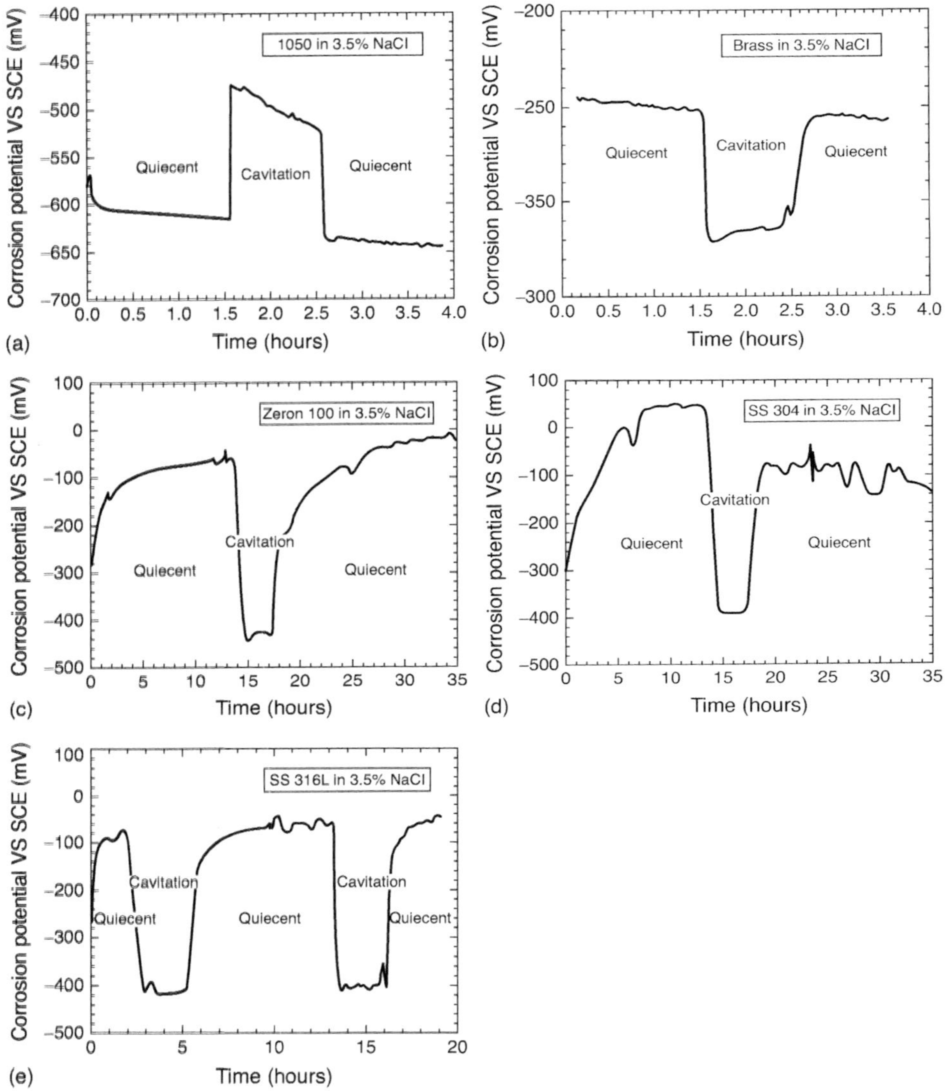

Figure 7.6 *Free corrosion potential as a function of time for various alloys in 3.5% NaCl solution under quiescent and cavitating conditions (Reprinted with permission from [14] Copyright (2000) Elsevier Ltd).*

is sufficiently low. For high chloride concentrations, ultrasound has very little effect in comparison to aggressive chloride ions which prevent formation of the passive film.

For high speed steel, where the metal oxidation reaction is rate dominating, ultrasound erodes the surface, and pits are formed since, in chloride solution, the oxide layer is not as robust as in the case of Al. Sonication leads to an enhancement of pitting and of the corrosion process, which was limited by access of the cathodic reagent, in this case, oxygen.

For steel, insonation leads to permanent effects: once sonication had ceased, some pitting corrosion continued, and reformation of a partially protective oxide layer was observed.

7.2.1.3 *Effect of Cavitation on the Electrochemical Behaviour of Steel in Electrolyte without Chloride*

Up until the work of Jiang and Zheng, who conducted their experiments in Na_2SO_4 and $NaHCO_3$ [31], most of the former studies on cavitation-corrosion were carried out in chloride-containing solutions. As SO_4^{2-} and HCO_3^- are dominant anions in river water, the authors initiated their work with the objective of elucidating corrosion mechanisms of low alloy steel in solutions containing these ions (3% Na_2SO_4 solution and 3% $NaHCO_3$) compared with traditional 3% NaCl solutions. Cavitation occurred using a 20 kHz horn vibrating with an amplitude of 60 μm. Mass loss data were generated according to the ASTM G32-85 standard by screwing the sample into the horn using the 'face-on' geometry (specimen-to-horn distance: 15 mm). Again, three electrochemical techniques were used to evaluate the corrosion process and its rate: (i) measurements at open circuit potentials (OCPs), (ii) polarization curves, and (iii) electrochemical impedance spectroscopy, before, during and after insonation. The damaged surfaces were observed by SEM. Using a comparable methodology to Kwok and Man [14], the authors also described the total cavitation-erosion-corrosion rate (W_T) as the sum of effects:

$$W_T = W_C + W_E + W_{CE} + W_{EC} \tag{7.11}$$

where W_C is the component due to corrosion in quiescent solution, W_E is the erosion component obtained in distilled water, W_{CE} is the corrosion-induced erosion and W_{EC} the erosion-induced corrosion also calculated from corrosion current density in sonicated solutions. Concerning the electrochemical behaviour of 20SiMN low alloy steel in various solutions, it appears once again that cavitation shifts the corrosion potential (E_{corr}) of steel from −670 to −520 mV in 3% Na_2SO_4 and from −730 to −660 mV in 3% $NaHCO_3$. Once cavitation stops, E_{corr} rapidly shifts back to its quiescent value. A positive shift of E_{corr} (from −570 to −470) is also recorded in 3% NaCl solution for the same steel. These observations in OCP shifts are easily accounted for by the polarization curves obtained with and without ultrasound. Cavitation strongly affects the polarization behaviour of 20SiMn low alloy steel in Na_2SO_4 by drastically enhancing the cathodic current which is diffusion controlled (reduction of dissolved oxygen). This also induces an important increase in i_{corr} (from 1.5 to 520 μA cm^{-2}). These results are in very good agreement with those obtained in NaCl solutions by the same authors and others [14, 20]. In contrast, the process is controlled by both the anodic and cathodic reactions in $NaHCO_3$, where a narrow passivation zone (possibly due to a $FeCO_3$ corrosion product layer) is obtained in quiescent conditions, which disappears in the presence of ultrasound. This leads to an increase in OCP and i_{corr}.

Ashabi-Sorkhabi *et al.* [32] investigated the effect of ultrasound on steel behaviour in concentrated sulfuric acid, particularly the influence of 24 kHz ultrasound (5 mm gap between horn and specimen) on corrosion of carbon steel in 3 N H_2SO_4 at 25 °C in the absence and presence of polyethylene glycols (PEGs) as corrosion inhibitors. The polarization resistance and the impedance spectroscopy methods showed the effectiveness of PEGs on the cavitation-corrosion inhibition of carbon steel in concentrated H_2SO_4. It was concluded that ultrasound simultaneously increases mass transfer and removes the inhibitor.

It is therefore the balance between these two competitive phenomena that determines the final corrosion rate.

7.2.1.4 Generalized Corrosion of Zinc

Two papers by Doche *et al.* focused on the cavitation effects upon the electrochemical behaviour of zinc in NaOH and Na_2SO_4 solutions [33, 34]. The investigations showed that cavitation-corrosion mechanisms of zinc, either used as a sacrificial material in corrosion protection or as an anode material in batteries, could offer potential development in the field of accelerated corrosion tests and power generation. In these two studies, a 'face-on' geometry configuration was used (but with the ultrasonic horn placed at the bottom of the sonoreactor). The authors observed that this setup is more convenient for electrochemical measurements and allows the use of a rotating disc working electrode (RDE). The experimental setup is shown in Figure 7.7.

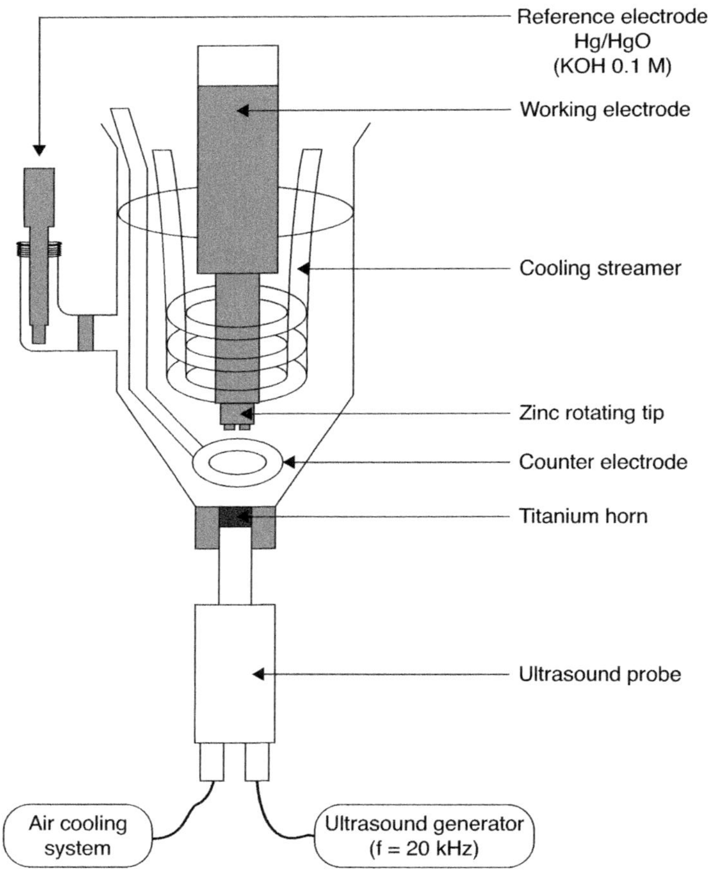

Figure 7.7 *Experimental setup for the study of zinc corrosion under ultrasound in NaOH and Na_2SO_4 electrolytes.*

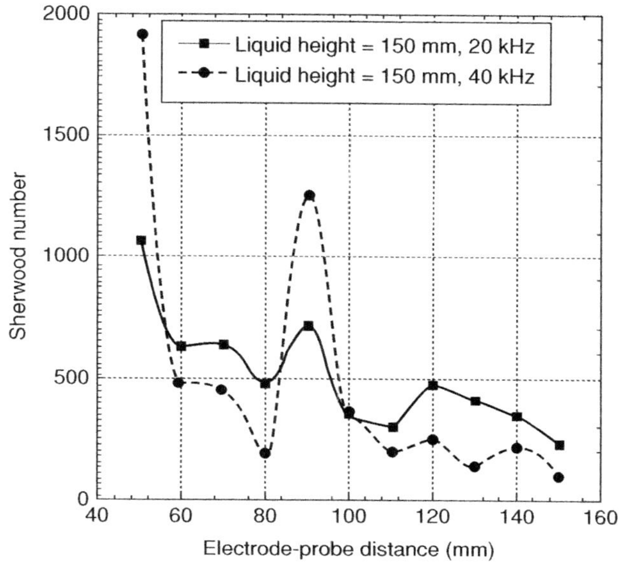

Figure 7.8 *Influence of ultrasonic frequency on mass transfer variation as a function of distance between the working electrode and the probe for a constant liquid height of 150 mm and a transmitted power of 70 W.*

An innovative and interesting approach in these two papers is that the authors carefully quantify the ultrasonic power transmitted to the solution as a function of the generator input power and frequency (20 and 40 kHz), but also take into account liquid height in the reactor. Transmitted power is the only representative data allowing the determination of ultrasonic intensity (= power/area) compared to electrical power and vibrational amplitude. Prior to corrosion experiments, a series of experiments were also carried out to measure mass-transfer coefficient (quoted in Sherwood number) distribution along the transducer axis during sonication. Experiments performed at a constant transmitted power of 70 W, and for 20 and 40 kHz, show similar trends as shown in Figure 7.8.

It can be observed that the Sherwood number decreases as the electrode is moved away from the ultrasonic probe. However, this decrease exhibits a pseudo-sigmoid shape related to maximum stirring points alternating with minimum ones and separated by a distance of one quarter of a wave length ($\lambda/4$). Such observations confirm establishment of a steady-state wave system as the result of complete reflection at the air/electrolyte interface.

The same authors also found that 0.5 M Na_2SO_4 saline media induces a relatively corrosive condition [34]. In this work, the corrosion progress of zinc was followed by means of electrochemical methods (mainly polarization curves). They observed that as the zinc corrosion electrolyte is aerated, the mechanism is under the control of the limiting step of oxygen diffusion and the limiting oxygen diffusion current was taken to be the corrosion current. Zinc corrosion results in Na_2SO_4 are presented in Figure 7.9.

At 20 kHz, corrosion current fluctuates according to the electrode position and reaches a maximum value (seven to eight times greater than in silent conditions) for a distance

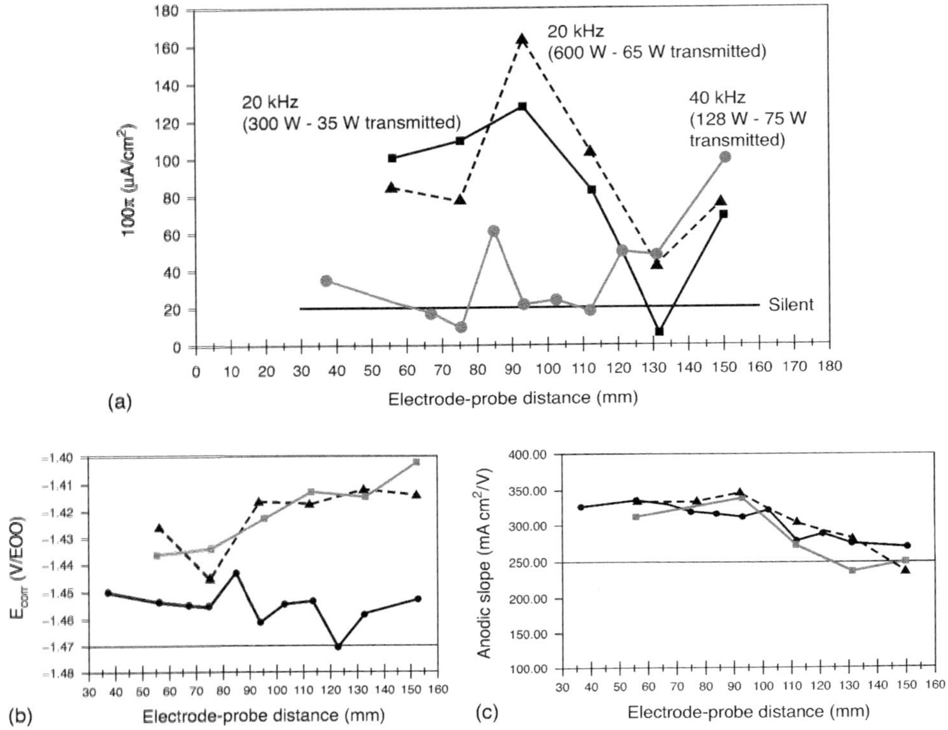

Figure 7.9 *Zinc corrosion parameters in Na_2SO_4 0.5 M as a function of the distance between the electrode and the horn for different powers and frequencies: (a) corrosion current density, (b) corrosion potential, (c) anodic oxidation rate (Reprinted with permission from [34] Copyright (2003) Elsevier Ltd).*

of nearly 94 mm ($\lambda + \lambda/4$). This observation is consistent with previous experiments as this position was found to be crucial for high mass-transfer. It was also observed that when the ultrasonic power is increased, the corrosion rate slightly increases and at 40 kHz, the corrosion rate is far lower than at 20 kHz. Furthermore, a specific behaviour was observed when the electrode was positioned at the liquid interface where corrosion rates were increased by a factor of 3–5. They found that (i) the observed corrosion potentials (at $I = 0$ A) are the result of two competitive reactions, that is, the oxygen reduction and zinc oxidation, and (ii) a positive shift in corrosion potential was observed under ultrasound (+60 mV at 20 kHz). This confirms an enhancement of oxygen reduction current due to high stirring as already observed on steel in Cl^- solutions (see previous sections). However a greater value would have been expected and suggests that sonication may also modify the rate of zinc oxidation partial reaction.

The behaviour of zinc in NaOH electrolytes (0.1 and 1 M) was also studied using similar experimental procedures but at a fixed ultrasonic frequency of 20 kHz. It was found that when the electrode was placed at a distance greater than 1.9 cm (for a liquid height of 9.4 cm), no significant ultrasonic effect on the polarization characteristics of zinc was observed.

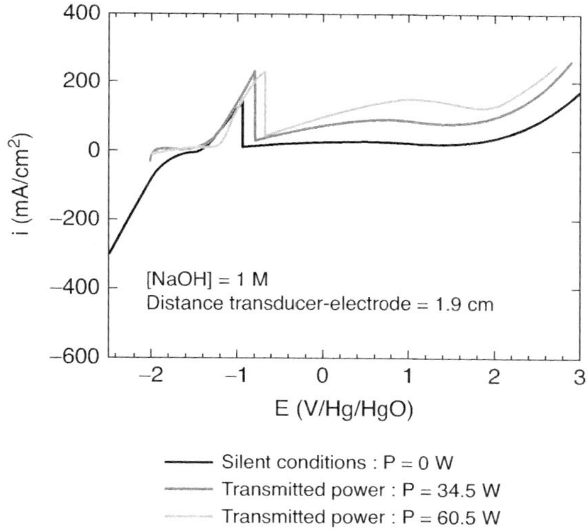

Figure 7.10 *Polarization curves obtained on a zinc electrode immersed in a sonicated 1 M NaOH electrolyte as a function of transmitted power (Reprinted with permission from [33] Copyright (2001) Elsevier Ltd).*

Figure 7.10 shows steady-state polarization curves performed under both silent and sonicated conditions. In the absence of ultrasound, even at pH = 14, the current-potential curves exhibit typical passivation behaviour. In the potential range below −1.4 V versus Hg/HgO, the net current is mainly due to the cathodic contribution of water and dissolved oxygen reduction.

At $E > -1.4$ V versus Hg/HgO, the current starts to become more positive as the anodic oxidation process becomes predominant, and a passivation peak is visible followed by a passivity plateau ranging from −0.9 to 1.8 V versus Hg/HgO. According coulometric analyses, it is more likely that a porous hydrated hydroxide layer was formed on the zinc surface in 1 M NaOH.

It was also found that as the ultrasonic power was increased, a slight negative shift in OCP was observed (and the passivation potential shifted from +200 to +400 mV in the positive direction). In their conditions, they observed that the passivation peak areas doubled, indicating that passivation was delayed and the passivating film had difficulties in forming under sonication. However, no complete passivation breakdown was observed in 0.1 M NaOH solutions; the probably far more stable passivating film appeared to be barely affected by sonication.

7.2.1.5 Sonoelectrochemical Etching of Aluminium

Aluminium etching is another type of ultrasonic application relating to corrosion. This process was recently developed by Kang and Tak [35, 36]. Aluminium etching is an anodizing process consisting in generating a high density of pits of controlled size at the aluminium surface with the aim of increasing its surface and hence its capacitance. In the referred papers, etching was performed in mixed solutions of 1 M HCl + 3 M H_2SO_4 at 60 °C.

The aluminium foil was mounted onto a holder in front of and parallel to the horn with a gap of 100 mm in an ammonium borate solution. In order to induce etching, a current density of 300 mA cm^{-2} was applied to the sample. Two ultrasonic frequencies (28 and 68 kHz) were used during these experiments. It was found that (i) the use of ultrasound during etching inhibits oxide film formation and favours the formation of corrosion pits, and (ii) pit densities are increased and a more uniform tunnel length distribution is obtained. These effects are further enhanced at higher ultrasonic frequencies and powers as a consequence of intense cavitation activities. Kang and Tak [37, 38] also observed that cavitation contributes to pit creation at early stages of etching, and then to growth in length of initial cubic etches by enhancing transport of $AlCl_3$ corrosion products outside the pits.

7.2.2 Influence of Ultrasound on Passivity of Metals

In 1983, Alkire and Perusich [25] were the first to study the effects of ultrasound on the passivation of iron. They carefully demonstrated that cavitation influences the behaviour and integrity of the passive film on the electrode surface. A later study by the same authors proposes a mathematical model to determine the reaction(s) and transport between a liquid microjet and the solid surface [39]. Prior to this work, focused ultrasound for initiation of metal depassivation had never been employed. This work was also innovative due to the fact that focused ultrasound (1.58 MHz, focal region of 3.5 mm diameter) was used, the ancestor of the current HIFU. The system chosen for this study was iron in 2 N H_2SO_4. First, it was shown that focused ultrasound inhibits metal repassivation when depassivation is electrochemically generated. Secondly, local depassivation of iron was achieved by the use of high frequency ultrasound only if the focal intensity exceeded a critical value of 1.5 kW cm^{-2}. For lower ultrasonic intensities, continuous exposure led to a sudden depassivation after an incubation period.

Whillock and Harvey extensively studied the ultrasonically enhanced corrosion of 304L stainless steel in nitric/chloride solutions [40–42]. Preliminary investigations were performed in 2 M nitric acid containing various chloride ion concentrations (0.3, 0.5 or 2 g l^{-1}) at 50 °C. This system was chosen since the three parameters: nitric acid strength, chloride ion concentration and temperature, normally allow fine adjustment of solution 'corrosiveness'. A microprobe operating at 55 kHz was used as the ultrasonic source, and the transmitted power was determined calorimetrically. At that time, this procedure was innovative and is currently used by many workers in the field. In their experimental setup, the sample was mounted using the 'face-on' configuration (1.1 mm from the horn) and immersed in the corrosive solution at non-atmospheric hydrostatic pressure [39]. The most spectacular effect recorded by the authors is shown in Figure 7.11 where the OCP variations are presented as a function of ultrasonic pulses. It appears that, in certain corrosive conditions, ultrasound can switch metal from a passive to an active state. This operates in reverse when ultrasound is stopped, and is referred to as the 'switch on/off effect'. The corrosion rate determined by EIS indicates that sonication promotes passivation breakdown and enhances active corrosion of stainless steel by a factor of circa 3 to 6. Corrosion rates were also determined in various operating conditions: temperatures, hydrostatic pressures, ultrasonic powers and frequencies, and source to sample distances [39, 40]. It was shown that (i) 50 °C corresponds to the 'ideal' temperature for maximum corrosion rates and (ii) increasing the hydrostatic pressure in the range (0.5–4 bar) causes increase in corrosion

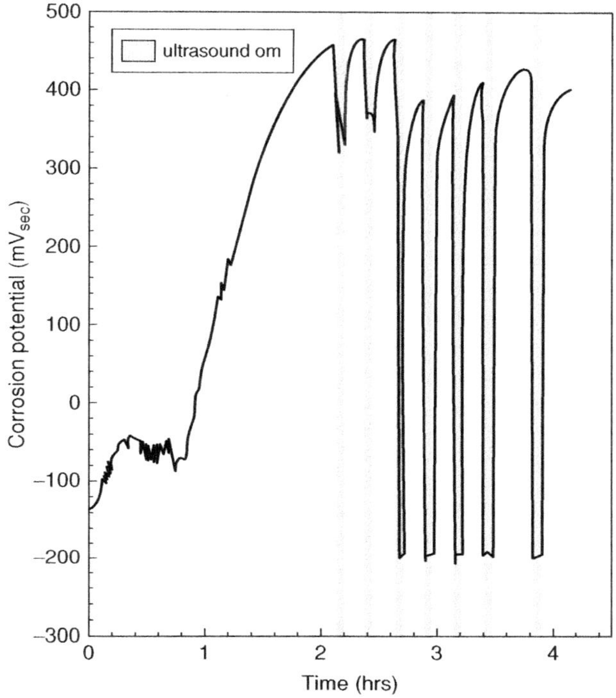

Figure 7.11 *Effect of sonication on the corrosion potential of 304 L stainless steel in 2 M nitric acid + 0.3 g L^{-1} Cl^- at 50 °C [40] (Reprinted with permission from [40] Copyright (1996) Elsevier Ltd).*

rates. Finally, the authors concluded that ultrasound is capable of enhancing the corrosion rate of stainless steel if, and only if, the solution is inherently sufficiently corrosive to initiate, even transiently, the first attack. This was demonstrated by sonicating the samples in the absence of Cl^- ions in solutions. In this case, whatever the ultrasonic frequency, power or sample-horn distance, no visible attack occurs after several hours of sonication.

Recently some studies conducted by a Chinese team have dealt with this topic [43, 44]. For example, Wang and Zhu [43] studied the electrochemical behaviour of stainless steel 0Cr13Ni5Mo passive film in HCl solutions under static state (quiescence) and ultrasonic conditions (20 kHz, 1 mm horn-specimen distance) by means of polarization curves, electrochemical impedance spectroscopy (EIS) and capacitance potential measurements. The results indicated that the passive film shows a multi layer structure distribution, and presents a *p*-type semiconductor property under static state conditions. An extremely remarkable result is that ultrasound, apart from decreasing the stability of the passive film, changes the semi-conducting property of the film to an *n*-type semi-conductor in the presence of cavitation. This is explained by the fact that thermo-excitation of electrons in the valence band caused by ultrasonic cavitation resulted in an increase in electron densities in the passive film, and led to the conversion of the semiconducting property of the passive film. The amount of exciting electrons from valence band depended on the Fermi level height of doping semiconductor and the ultrasonic intensity. Nevertheless, only a few details are given on the operating conditions such as the transmitted power and the surface of the transmitter. Finally, the most recent works are those of Wang and Kido in Japan [45–47], who investigated the influence of ultrasound on pitting and crevice corrosion of SUS304 (ASI 304) stainless steel. The methodology used here are largely inspired by other workers such as Whillock *et al.* [38] and Doche *et al.* [33, 34]. The experimental setup includes a corrosion cell filled with 3.5% NaCl (+ tap water being the coupling media) solution immersed in an ultrasonic bath ($f = 42$ kHz [45, 46] or 19.5 kHz [47]). Here, the distance from the horn to the specimen was changed from 19 to 96 mm, bearing in mind that the bottom of the cell acts as an obstacle between the horn and the specimen. It was found that the best configuration for ultrasonic wave transmission is a specimen-to-horn distance corresponding to an ultrasonic wavelength of 76 mm. In this study, pits or crevices were initiated on stainless steel by applying simultaneously a pitting current density and ultrasound. It was observed that both crevice and pitting corrosion of 304 SS can be suppressed by ultrasound which is mainly attributed to a decrease or suppression of enrichment of metallic ions, hydrogen ions and chloride ions in the crevice by the high stirring effects of sonication. These results are somewhat contradictory with those obtained by other authors [38, 41, 42], in that they highlight the recovering effect of ultrasound (probably at low ultrasonic powers) on passive layers.

To conclude this section, it is also worth noting the work of Birkin [22], whose approach differs from other studies in that this researcher uses the corrosion phenomenon as a sensor to map or/and characterize cavitation events (see Chapter 2).

7.3 Ultrasound as a Tool to Develop Accelerated Corrosion Testing

De Morais *et al.* [20] are amongst the first workers to study and to point out the advantages of ultrasound-assisted corrosion as a rapid diagnostic test for corrosion resistance. In the same way, ultrasonic tests were used by Ashabi-Sorkhabi [32] to test the efficiency of PEG addition as cavitation-corrosion inhibitors in 3 N H_2SO_4. However, papers dealing with this specific application of ultrasound remain unusual.

7.3.1 Atmospheric Corrosion of Zinc Plated Steel

The Besançon group [48] were the first to study the accelerated atmospheric corrosion testing of zinc under ultrasound. Corrosion resistance of new zinc coatings is commonly evaluated by long-term atmospheric exposure or by accelerated corrosion tests. Atmospheric metal corrosion can be considered as the result of chemical action by atmospheric pollutants [49] and the more physical action of surface erosion by natural precipitation. This mechanical action must be added to the abrasive action of solid particles hitting the metal surfaces due to wind action. When dealing with electrolytes, an important method of inducing mechanical and even chemical changes is the use of an ultrasonic source. Simulation of atmospheric corrosion at laboratory scale remains particularly difficult due to the wide variety of natural atmospheres: marine, rural, urban and industrial atmospheres. A large part of Ligier's work was to formulate electrolytes based on $[Cl^-][SO_4^{2-}][HCO_3^-][H_2O_2]$ species, able to simulate all types of natural environments from marine to industrial [49].

Two ultrasonic frequencies (20 and 500 kHz) were used in views of accelerating the corrosion process. Acoustic transmitted power measurement was performed and, for the first time in corrosion studies, the author also followed the rate of H_2O_2 production during sonication. While accelerated corrosion tests performed in specially designed electrolytes yield satisfactory results with respect to corrosion rate enhancement, the nature of zinc corrosion products and the corrosion sequences involved, surface morphology changing with ultrasound is not totally satisfactory for accurately reproducing surface damage after a natural exposure [49].

7.3.2 Accelerated Corrosion Test for Stainless Steel Used in Exhaust Systems

The Besançon group also developed an accelerated corrosion test using low frequency ultrasound to simulate corrosion conditions in the cold end of an automotive exhaust system made of middle range 1.4512 (AISI 409) stainless steel [50]. The automotive industry currently uses the conventional dip dry test (DDT) to qualify new grades of materials in terms of corrosion resistance. The DDT uses a sequence of immersion phases in the electrolyte followed by a drying and a heating stage. In the case of exhaust material testing, the test is performed in a synthetic condensate (which almost reproduces the composition of gas condensates in a cool part of the exhaust line), and wet sequences alternating with dry sequences simulate the thermal and humidity cycles to which the metal is subjected in driving conditions. Alternatively, Doche *et al.* developed a new accelerated test (the so-called UST) by introducing a 40 kHz ultrasonic stirring during the wet sequence of the cycling test [50]. Corrosion tests were performed in two different media to simulate internal corrosion due to exhaust gas condensates and external cosmetic corrosion, greatly enhanced by road salt during winter. This concerns, respectively, a synthetic gas condensate, the composition of which is derived from what is obtained from motor gasoline combustion, and NaCl solution. In both electrolytes, stainless steel suffers from pitting corrosion. As expected, the use of ultrasound allows pit growth right from the beginning of exposure time, so that the maximum pit depth recorded after 180 hours of immersion is twice that of the classical dip dry test. It was found that pit density increases by a factor of 4 under sonication and ultrasound does not modify the pit initiation mechanism but only increases growth kinetics. This new test appeared efficient in reducing the diagnostic time of the corrosion test by the combined action of the chemical corrosion process and of mechanical degradation.

7.3.3 Accelerated Corrosion Test for Evaluating Oilfield Corrosion Inhibitors

Internal corrosion of oil pipelines is a serious problem for oil producers, which is often treated by corrosion inhibitors. Studies by Petersen *et al.* describe an ultrasonic based test bench to measure the effectiveness of corrosion inhibitors on carbon steel when submitted to cavitation [51, 52]. It is assumed that cavitation bubbles produced by ultrasound behave in like manner to cavitation encountered in slug flow, a type of flow common in multiphase pipelines characterized by a series of liquid plugs (slugs) separated by a relatively large gas pockets. In vertical flow, the bubble is an axially symmetrical bullet shape that occupies almost the entire cross-sectional area of the tubing. The resulting flow alternates between high-liquid and high-gas composition.

In this case, ultrasound is used as an accelerating 'agent' and also it simulates material service use. A 20 kHz horn was used and positioned in the corrosion cell horizontally in the flow directly opposite to the working electrode (distance = 1 mm). The corrosive solution was a 1 M NaCl solution conditioned by bubbling with CO_2 for at least 1 h (as it is corrosion by dissolved CO_2 that the authors intended to measure). Three corrosion inhibitors were used (commercial and homemade), and corrosion resistances were measured by linear polarization resistance. It was observed that in the absence of inhibitors, the collapse of cavitation bubbles generate very high corrosion rates. The ultrasonic test also appears efficient in selecting and differentiating corrosion inhibitors that display undistinguishable behaviour in gently stirred electrolytes as in classical testing. In the presence of inhibitors [52], different processes occur at the electrode surface. It was concluded that the inhibitor film forms a barrier between the surface and the solution, preventing enhanced corrosion.

7.3.4 Accelerated Corrosion Test for Surgical Implant Materials in Body Fluids

Lesniewics *et al.* [53] used a very simple ultrasonic assisted corrosion test to evaluate corrosion of eight implant materials, namely niobium and niobium alloys, nickel and nickel alloys, titanium and alloys, as well as two types of austenitic stainless steel (316 L type) in three synthetic body fluids, namely a phosphate-buffered solution known as PBS, Hank's solution (a nutritive salt solution used for culturing animal cells, made of $CaCl_2$, $MgSO_4$, KCl, KH_2PO_4, $NaHCO_3$, NaCl and Na_2HPO_4) and saliva. During the tests, material samples were simply immersed into body fluid flasks and then subjected to ultrasonic irradiation (50 kHz, ultrasonic bath). Concentrations of released metals was measured by ICP-OES after 30, 50, 100 and 150 hours of immersion, and compared to those obtained under silent conditions and gently mechanically stirred solutions. Implant material surfaces were also examined using the X-ray diffraction method (XRD). The results showed that the application of ultrasound increases implant material bio-deterioration by a factor of 20 to 40 depending on the material used. This allows rapid evaluation (just a few days) of the 'corrosiveness' of some body fluids. XRD studies also showed that several salts present in body fluids were deposited on the metal surface thus protecting the implants against further degradation.

Douglas *et al.* investigated the effect of ultrasound on 316 L stainless steel (a common grade in fracture fixation implant material) in the lactated Ringer's solution to simulate corrosive body fluid [54]. This study was dictated by the fact that high frequency (1.5 MHz), low intensity pulsed ultrasound is recommended as a fracture healing adjuvant. The authors speculated that such ultrasound irradiation can have a negative effect on the passive film protecting 316 L fracture fixation implants against corrosion. Polarization curves were generated to measure the corrosion parameters of the passive materials, that is, passivation potentials, breakdown potentials and residual current densities. The results, as expected, confirmed that high frequency ultrasound does not promote passivation breakdown of 316 L SS in biological media. It should be emphasized that high frequency ultrasound has not been reported to be a corrosion accelerator, contrary to low frequency ultrasound.

References

General References on Corrosion

(1) http://www.iso.org (accessed on May 2011).
(2) http://www.nace.org (accessed on May 2011).

(3) Koch, G.H., Brongers, M.P.H., Thompson, N.G., Virmani, Y.P. and Payer J.H. (2002) *Corrosion Costs and Preventative Strategies in the United States, a Supplement to Materials Performance*, NACE International, Houston TX.
(4) http://www.corrosion-doctors.org/ (accessed on May 2011)
(5) Barbara A. Shaw and Robert G. Kelly (2006) What is corrosion? *The Electrochemical Society Interface*, Spring.
(6) Guthrie, J., Battat, B. and Grethlein, C. (2002) Accelerated corrosion testing. *The AMPTIAC Quarterly*, **6** (3), Fall.
(7) *Corrosion Tests and Standards: Application and Interpretation*, 2nd edn (ed. Robert Baboian), ASTM, 2004.
(8) Roberge, Pierre R. (2000) *Handbook of Corrosion Engineering*, McGraw-Hill.
(9) Audisio, S. (2011) *Multimedia Corrosión Guide*, 2nd edn, http://www.cdcorrosion.com (accessed on May 2011).
(10) Landolt, Dieter (1997) *Traité des Matériaux Vol 12, Corrosion et Chimie de Surface des Métaux*, Presses Polytechniques et universitaires Romandes, Lausanne.

Specialized References

(11) Diodati, P. and Giannini, G. (2001) Cavitation damage on metallic plate surfaces oscillating at 20 kHz. *Ultrasonics Sonochemistry*, **8**, 49–53.
(12) Tomlinson, W.J. and Chapman, D. (1989) Effect of amplitude on the erosion and corrosion of copper by 20 kHz ultrasound. *Wear*, **130** (2), 381–384.
(13) Steller, J. (1999) International cavitation-erosion test and quantitative assessment of material resistance to cavitation. *Wear*, **233–235**, 51–64.
(14) Kwok, C.T., Cheng, F.T. and Man, H.C. (2000) Synergetic effect of cavitation erosion and corrosion of various engineering alloys in 3.5% NaCl solutions. *Materials Science & Engineering A*, **290**, 145–454.
(15) Wagner, C. and Traud, W. (1938) Uber die deutung von korrosionsvorgängen durch überlagerung von elektrochemischen teillvorängen und über die potentialbildung an mischelektroden. *Zeitschrift Electrochemistry*, **44**, 391–454.
(16) Stern, M. and Geary, A.L. (1957) Electrochemical polarization: I. A theoretical analysis of the shape of polarization curves. *Journal of the Electrochemical Society*, **104**, 56.
(17) Lauterborn, W. and Ohl, C.D. (1997) Cavitation bubble dynamics. *Ultrasonics Sonochemistry*, **4**, 65–75.
(18) ASTM G32 – 10, *Standard Test Method for Cavitation Erosion Using Vibratory Apparatus*, http://www.astm.org (accessed on May 2011).
(19) Madigan, N.A., Hagan, C.R.S., Zhang, H. and Coury, L.A. (1996) Effects of sonication on electrode surfaces and metal particles. *Ultrasonics Sonochemistry*, **3**, S239–S247.
(20) De Morais, N. and Brett, C.M.A. (2002) Influence of power ultrasound on the corrosion of aluminium and high speed steel. *Journal of Applied Electrochemistry*, **32** (6), 653–660.
(21) H. Soyama, M. Futakawa (2004) Estimation of incubation time of cavitation erosion for various cavitating conditions. *Tribology Letters*, **17**, 1.
(22) Birkin, P.R., O'Connor, R., Rapple, C. and Martinez, S.S. (1998) Electrochemical measurement of erosion from individual cavitation events generated from continuous ultrasound. *Journal of the Chemical Society, Faraday Transactions*, **94** (22), 3365–3371.
(23) Man, H.C., Kwok, C.T. and Yue, T.M. (2000) Cavitation erosion and corrosion behaviour of laser surface alloyed MMC of SiC and Si3N4 on Al alloy AA6061. *Surface and Coatings Technology*, **132**, 11–20.
(24) Kwok, C.T., Cheng, F.T. and Man, H.C. (2001) Laser-fabicated Fe-Ni-Co-Cr-B austenitic alloy on steels. Part II : Corrosion behaviour and corrosion-erosion synergism. *Surface and Coatings Technology*, **145**, 206–214.

(25) Alkire, R.C. and Perusich, S. (1983) The effect of focused ultrasound on the electrochemical passivity of iron in sulfuric acid. *Corrosion Science*, **23** (10), 1121–1132.
(26) Verdan, S., Burato, G., Comet, M., Reinert, L. and Fuzellier, H. (2003) Structural changes of metallic surfaces induced by ultrasound. *Ultrasonics Sonochemistry*, **10**, 291–295.
(27) Al-Hachem, A., Caceres, P.G., Riad, W.T. and Shalaby, H.M. (1995) Cavitation corrosion behavior of cast nickel-aluminium bronze in seawater. *Corrosion*, **51** (5), 331–343.
(28) Al-Hachem, A., Abdullaah, A. and Riad, W. (2001) Cavitation corrosion of nodular cast iron (NCI) in seawater: microstructural effects. *Materials Characterization*, **47**, 383–388.
(29) Al-Hachem, A. and Riad, W.T. (2001) The effect of duplex steel microstructure on its cavitation morphology in seawater. *Materials Characterization*, **47**, 389–395.
(30) Al-Hachem, A. and Riad, W.T. (2002) The role of microstructure of nickel-aluminum-bronze alloy on its cavitation corrosion behavior in natural seawater. *Materials Characterization*, **48**, 37–41.
(31) Jiang, S.L., Zheng, Y.G. and Yao, Z.M. (2006) Cavitation erosion behavior of 20 SiMn low alloy steel in Na_2SO_4 and $NaHCO_3$. *Corrosion Science*, **48**, 2614–2632.
(32) Ashassi-Sorkhabi, H. and Ghalebsaz-Jeddi, N. (2006) Effect of ultrasonically induced cavitation on inhibition behavior of polyethylene glycol on carbon steel corrosion. *Ultrasonics Sonochemistry*, **13**, 180–188.
(33) Doche, M.-L., Hihn, J.-Y., Touyeras, F., Lorimer, J.P., Mason, T.J. and Plattes, M. (2001) Electrochemical behaviour of zinc in 20 kHz sonicated NaOH electrolytes. *Ultrasonics Sonochemistry*, **8**, 291–298.
(34) Doche, M.-L., Hihn, J.-Y., Mandroyan, A., Viennet, R. and Touyeras, F. (2003) Influence of ultrasound power and frequency upon corrosion kinetics of zinc in saline media. *Ultrasonics Sonochemistry*, **10**, 357–362.
(35) Kang, J., Shin, Y. and Tak, Y. (2005) Growth of etch pits formed during sonoelectrochemical etching of aluminium. *Electrochimica Acta*, **51**, 1012–1016.
(36) Kang, J., Ko, E. and Tak, Y. (2006) Influence of ultrasound power on the formation of submicron etch pits during sonoelectrochemical etching of aluminium. *Key Engineering Materials*, **326–328**, 433–436.
(37) Kang, J., Shin, Y. and Tak, Y. (2005) Growth of etch pits formed during sonoelectrochemical etching of aluminium. *Electrochimica Acta*, **51**, 1012–1016.
(38) Kang, J., Ko, E. and Tak, Y. (2006) Influence of ultrasound power on the formation of submicron etch pits during sonoelectrochemical etching of aluminium. *Key Engineering Materials*, **326–328**, 433–436.
(39) Perusich, S.A. and Alkire, R.C. (1983) Ultrasonically induced cavitation studies of electrochemical passivity and transport mechanisms. *Journal of the Electrochemical Society*, **138** (3), 700–707.
(40) Whillock, G.O.H. and Harvey, B.F. (1996) Preliminary investigation on the ultrasonically enhanced corrosion of stainless steel in the nitric acid(chloride system. *Ultrasonics Sonochemistry*, **3**, S111-S118.
(41) Whillock, G.O.H. and Harvey, B.F. (1997) Ultrasonically enhanced corrosion of 304L stainless steel I: The effect of temperature and hydrostatic pressure. *Ultrasonics Sonochemistry*, **4**, 23–31.
(42) Whillock, G.O.H. and Harvey, B.F. (1997) Ultrasonically enhanced corrosion of 304L stainless steel II: The effect of frequency, acoustic power and horn to specimen distance. *Ultrasonics Sonochemistry*, **4**, 33–38.
(43) Bao-Cheng, Wang and Jin-hua, Zhu (2008) Influence of ultrasonic cavitation on passive film of stainless steel. *Ultrasonics Sonochemistry*, **15**, 239–243.
(44) Zhu, Li-qun and Wang, Hui (2007) Electrochemical behavior of oxide films of stainless steel in 40 kHz sonicated sulfate electrolytes. *Chinese Journal of Aeronautics*, **20**, 148–152.

(45) Wang, R. (2008) Influence of ultrasound on pitting corrosion and crevice corrosion of SUS304 stainless steel in chloride sodium aqueous solution. *Corrosion Science*, **50** (2), 325–328.
(46) Wang, R.G. and Kido, M. (2008) Influence of ultrasound on corrosion behavior of SUS304 Stainless steel with crevice. *Materials Transactions*, **49** (8), 1806–1811.
(47) Wang, R.G. and Kido, M. (2009) Influence of the input power to vibrator and vibrator-to-specimen distance of ultrasound on pitting corrosion of SUS304 Stainless steel in 3.5% chloride sodium aqueous solution. *Corrosion Science*, **51** (8), 1604–1610.
(48) Ligier, V., Hihn, J.Y., Wery, M. and Tachez, M. (2001) The effects of 20 kHz and 500 kHz ultrasound on the corrosion of zinc precoated steels in [Cl^-][SO_4^{2-}][HCO_3^-][H_2O_2] electrolytes. *Journal of Applied Electrochemistry*, **31** (2), 213–222.
(49) Ligier, V., Wéry, M., Hihn, J.-Y., Faucheu, J. and Tachez, M. (1999) Formation of the main atmospheric zinc end products: $NaZn_4Cl(OH)_6SO_4$ $6H_2O$, $Zn_4SO_4(OH)_6$ nH_2O and $Zn_4Cl_2(OH)_4SO_4$ $5H_2O$ in [Cl^-] [SO_4^{2-}] [HCO_3^-] [H_2O_2] electrolytes. *Corrosion Science*, **41** (6) 1139–1164.
(50) Doche, M.L., Hihn, J.Y., Mandroyan, A., Maurice, C., Hervieux, O. and Roizard, X. (2006) A novel accelerated corrosion test for exhaust systems by means of power ultrasound. *Corrosion Science*, **48** (12), 4080–4093.
(51) Petersen, A.G., Klenerman, D., Hedges, W.M. and Harris, M.L. (2002) Effect of cavitation on carbon dioxide corrosion and the development of a test for evaluating inhibitors. *Corrosion*, **58** (3), 216–224.
(52) Petersen, A.G., Klenerman, D. and Hedges, W.M. (2004) Evaluation of the effect of carbon dioxide corrosion inhibitors on cavitation damage caused during the ultrasound test. *Corrosion*, **60** (2), 187–194.
(53) Lesniewicz, A., Gackiewcz, L. and Zyrnicki, W. (2010) Biodegradation of metallic surgical implants investigated using an ultrasound-assisted process combined with ICP-OES and XRD. *International Biodeterioration & Biodegradation*, **64** (1), 81–85.
(54) Pittner, D.E., Levin, L. and Archdeacon, M.T. (2008) In vitro corrosion analysis in low-intensity pulsed ultrasound. *The American Journal of Orthopedics*, **37** (2), E32–E37.

8

Sonoelectropolymerisation

Fabrice Lallemand, Jean-Yves Hihn, Mahito Atobe and Abdeslam Et Taouil

8.1 Introduction to Electropolymerisation

In the previous chapters, the necessity of surface treatments to protect bulk material against corrosion or to impart to the surfaces required properties in terms of friction coefficient or conductive ability have been extensively described in the case of metallic coatings. However, while standard plating has considerably contributed to industrial applications, a challenger to this supremacy has emerged with the development of chemistry. A new class of polymers exhibiting high electronic conductivity in the partially oxidised state was discovered in the 1970s [1, 2]. Their conductivity is appreciably identical to that of metals, and this emergence of a new material has resulted in a change in our way of thinking. New paths have been opened up in chemistry and physics, and numerous studies have begun worldwide in the field of π-conjugated systems! Three researchers: Heeger, MacDiarmid and Shirakawa, played a major role in the development of these conducting materials, and were awarded the Nobel Prize for Chemistry in 2000. After preparation of polymers such as polyacetylene and the discovery of the large increase in its conductivity after 'doping', more stable polyaromatic systems such as polyaniline, polypyrrole and polythiophene have been developed (Figure 8.1).

Their very interesting properties enable them to be used as potential materials in a wide range of promising applications such as light-emitting diodes, solar cells, electrochromic devices, batteries, fuel cells, protection of metals, biosensors, to name but a few. Their good temperature resistance means they can be used for electromagnetic shielding or antistatic coatings. Moreover, after several years of research, the fundamental nature of charge propagation has been explained by the transfer of charge carriers via electronic conjugation inside a same polymer chain; and charge transfer reaction (i.e. Π-dimerization)

Power Ultrasound in Electrochemistry: From Versatile Laboratory Tool to Engineering Solution, First Edition. Edited by Bruno G. Pollet.

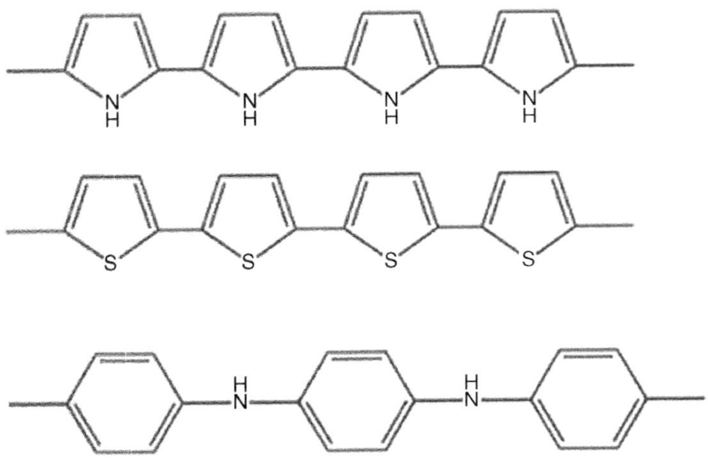

Figure 8.1 Most commonly studied aromatic conducting polymers (polyaniline PANI, polypyrrole PPy, polythiophene PT).

between neighbouring chains. Even if the intellectual curiosity of scientists has been focused on understanding the behaviour of these systems, a significant number of investigations have been dedicated to improvements of the properties of conducting films in order to integrate these films in several devices [3]. These films can be fabricated in a number of ways. Electrochemical synthesis is certainly the most common method due to its facility, the rapidity of the reaction and the possibility of perfectly controlling the experimental conditions.

Classically, as for example in the case of polypyrrole, a first step consists in forming radical cations by pyrrole oxidation at the anode. Secondly, two radical cations join up (oxidative coupling) to form a dimer. Thirdly, the dimer undergoes a re-aromatisation by the loss of two protons. Thus the dimer and then the trimer becomes oxidised, and so on (Figure 8.2).

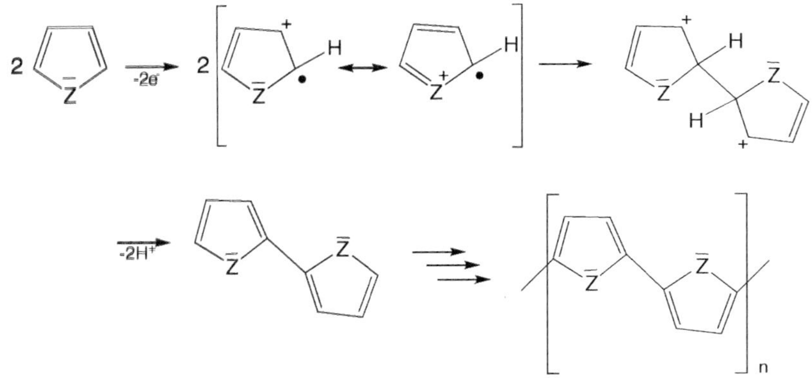

Figure 8.2 Electropolymerisation mechanisms – example of heteroaromatic compounds.

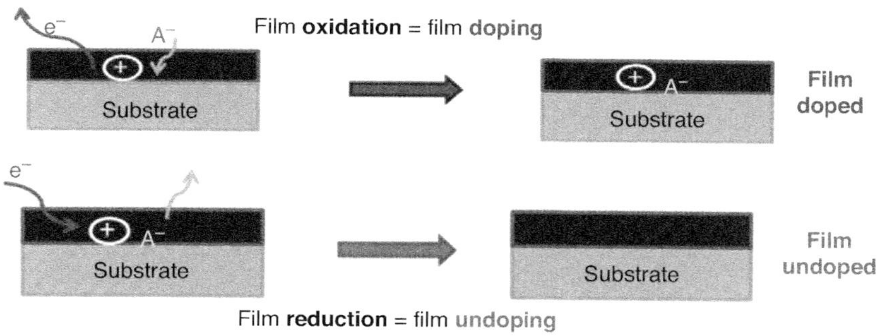

Figure 8.3 *Example of doping undoping processes of conducting polymers.*

In addition, in almost every case, chemical and physical properties of electrochemically deposited conducting polymers have been shown to be strongly dependent upon several parameters: the nature of the solvent, the supporting electrolyte, the temperature and the deposition potential/current. Doping and undoping process, which assumes oxidation and reduction of polymers as well as incorporation and expulsion of counter ions are also of prime importance. Figure 8.3 shows the example of positive doping by oxidation. Electron extraction of the polymer chains leads to the insertion of anions into the layer to preserve electro neutrality. Polymer structure is altered, and conducting properties appear by delocalisation of positive charges along polymer chains. If the system is once again applied at low potentials, a reduction reaction is noticeable, which corresponds to the reinjection of electrons into the chains and to the reject of anions imbedded during the doping phase. The polymer layer is restored to its initial neutral and insulating state.

Owing to the diversity and complexity of these materials, the development of new functionalised monomers raises new problems to solve and, consequently, more studies are needed to optimise the various parameters. Recently, an increase in interest has been observed in improving conducting film properties using original techniques such as the use of magnetic field, centrifugation or ultrasound. For this reason, this chapter is dedicated to this innovative technique named sonoelectropolymerisation and to its contribution to the improvement of conducting polymer properties.

8.2 Innovative Processes for Electrode Activation

Many original techniques have proved advantageous in terms of conducting polymer properties if they are performed during electrosynthesis. For example, the use of magnetic field [4] and centrifugation [5] leads to beneficial modifications in structural and electrochemical properties of the polymer deposits. It is generally accepted that an applied magnetic field influences the molecular or/and electronic spin state of materials, and the chemical reactivity of radical pairs in solution may be influenced by nuclear spins [6], meaning that polymers can be oriented and chemical or electrochemical reactions can be affected. In particular, morphological changes are observed under magnetic field due to the large anisotropy of their magnetic susceptibility. Moreover, the formation of special orientation of conjugated polymers would enhance their charge transfer capability and therefore lead

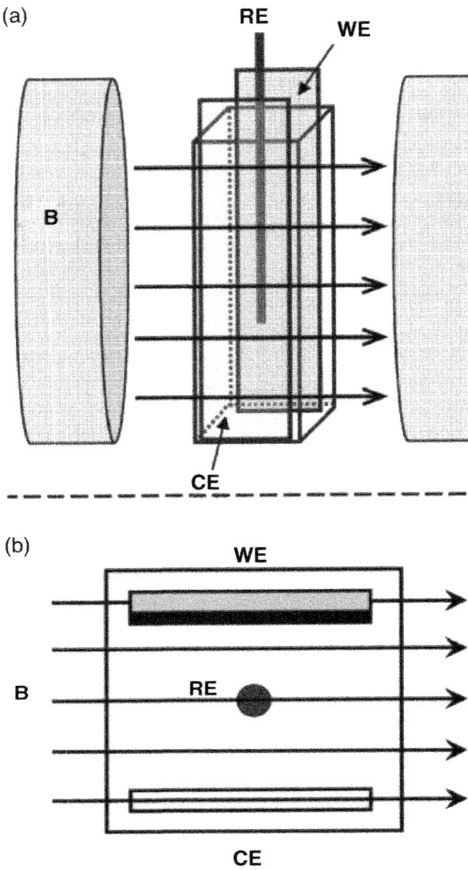

Figure 8.4 Schematic illustration of the polymerisation cell in a magnetic field (a) side view and (b) top-view. WE = working electrode, RE = Ag/AgCl reference electrode, CE = platinum counter electrode (Reprinted with permission from [4] Copyright (2008) Elsevier Ltd).

finally to a very high level of conductibility together with good mechanical properties. Cho *et al*. have tested poly(3,4-ethylenedioxythiophene) (PEDOT) synthesis on ITO glass under an applied magnetic field of 1.17 T at room temperature [4]. Water was degassed by nitrogen bubbling and polymerisation was carried out at a constant potential of 2.0 V. The experimental setup is shown in Figure 8.4; and the EDOT behaviour was followed by cyclic voltammetry, Raman spectroscopy and scanning electron microscopy (SEM).

It is clearly evident that current density is much higher in the presence of the magnetic field (more than one third), and the voltammogram confirms a 'speeding up' of polymerisation. The Raman spectrum indicated that the peaks of the stretching vibration of C=C shift to a lower wavenumber, which might be due to the formation of a more expanded chain structure. Finally, SEM pictures show marked changes in polymer morphology, with sharper images of PEDOT, with complicated needle shape. The authors accounted for these changes by an increase in nucleating centres, allowing PEDOT growth in the current direction [4].

Another specific method used to influence polymerisation tested with conducting polymers consists in performing synthesis under centrifugal acceleration ranging from 1 to 300 g [5]. Microgravity fields have been applied to broad areas in science and technology, particularly to material processing, with the purpose of producing unique and improved materials that cannot be prepared under normal earth conditions [6, 7]. Microgravity fields can be generated by means of free fall, parabolic flight and artificial satellite. Experiments in such fields are costly, but the use of small but powerful centrifuge equipment offers an excellent alternative.

Atobe *et al.* [5] reported successful fabrication of a centrifuge by improving commercially available centrifuge equipment with a 17-cm arm to generate a stable acceleration of roughly 300 g during electrochemical deposition of polyaniline (PANI), polypyrrole (PPy) and polythiophene (PT). The experimental setup consisted of a cylindrical electrolytic cell made of polytetrafluoroethylene resin, 7 mm long and 14 mm in diameter. Contacts between the classical three-electrode cell and the potentiostat/galvanostat were possible through silver rotating rings and carbon brushes. The entire system was suspended from the lid of a centrifuge tube with a polyethylene line, as shown in Figure 8.5.

Typical cyclic voltammograms were recorded during electropolymerisation of aniline at various centrifugal accelerations. Although their shape seems to be similar, oxidation and reduction peak currents are increased at 315 g (electrode A) and decreased at 290 g (electrode B – with a reversed direction force), compared to electropolymerisation performed under 1 g. Using the anodic peak current of circa 0.5 V as an indicator, the polymerisation rate appeared to be greatly affected. Higher rates of polymer deposition at electrodes were observed for each monomer. This enhancement of the formation rate also leads to a higher electrochemical capacity on the electrode situated opposite the centrifugal force. Scanning electron microscopy pictures revealed a high degree of sensitivity of the deposit morphology, which appears to be aggregated into large bundles. Chemical bonding and, finally, conductivity of the obtained polymers were considerably affected by the gravity fields. The clear anisotropy of the effects led the authors to propose mechanisms for electropolymerisation under centrifugal forces based on the basic differences between the solvent and macromolecules, shown concretely by different inertial behaviours. The accumulation of the oligomers produced at the electrode opposite to the centrifugal force resulted in an increase in nucleation and precipitation at the surface. As a final result, the high density of oligomers closer to the surface led to denser polymer films.

In the case of precipitation and under sonication, a very similar theory has been proposed and quantified. The basic mechanism is that, at the end of the bubble collapse, the fast inward motion of the mixture is stopped violently by gas re-compression in the bubble, yielding a huge outward acceleration. If the clusters are denser than the surrounding liquid, which is generally the case for crystallisation in solutions, they undergo an inward drift motion relative to the liquid, and become over-concentrated near the bubble wall for a very short time (typically 1 ns). This forced diffusion of two species of different densities in a pressure gradient physics is somewhat similar to that underlying ultra-centrifugation, and is known as 'pressure diffusion' [8].

It is a fact that propagation of a power ultrasonic wave through a liquid medium induces strong mechanical, physical and chemical changes leading to activation of chemical or

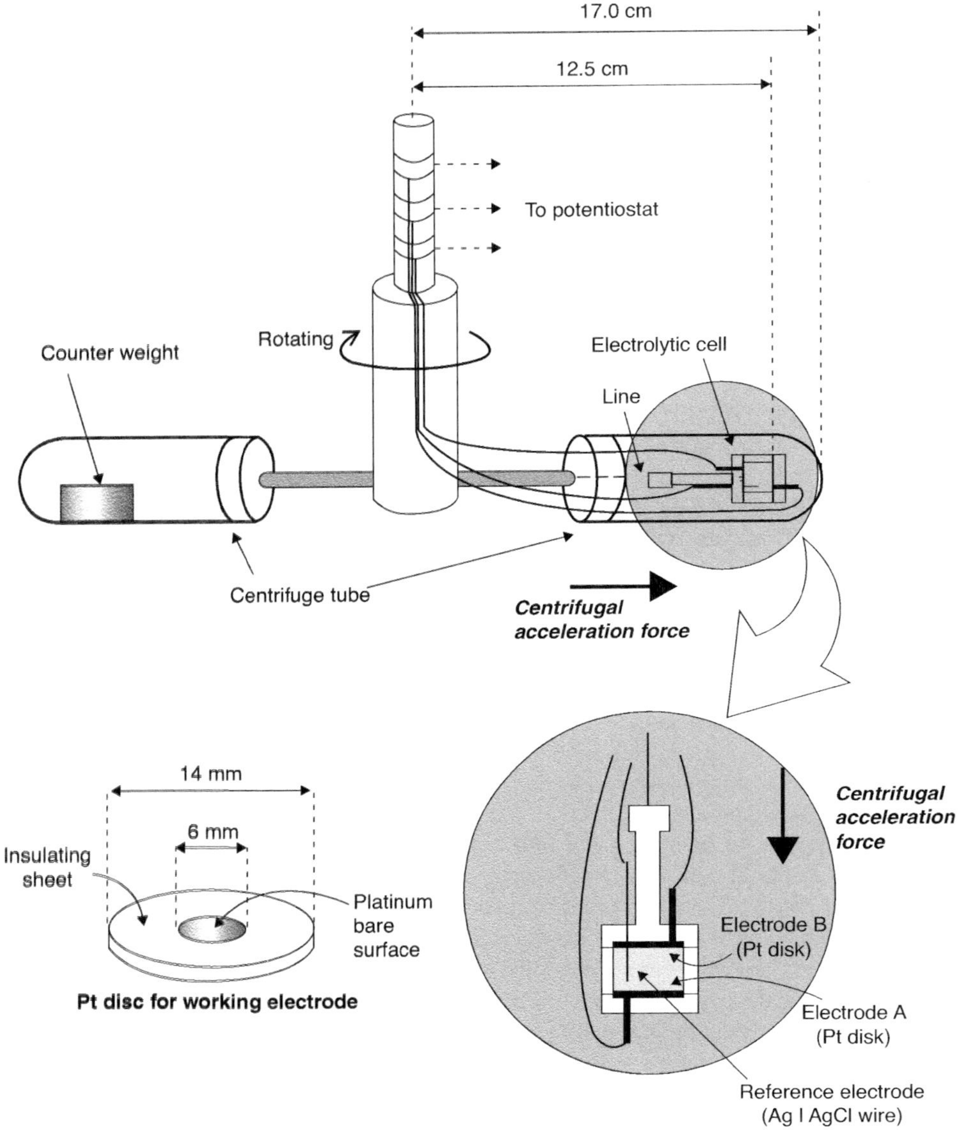

Figure 8.5 *Centrifugal setup equipped with an electrolytic cell (Reprinted with permission from [5] Copyright (2004) Elsevier Ltd).*

electrochemical reactions. Several researchers in the field of sonoelectroorganic chemistry, such as Luche and Walton [9, 10], have shown that ultrasound improves current efficiencies and product yields, and activates the electrode surface continuously (see Chapter 5). Luche showed that these enhancements are mainly due to improved mass transfer of the electroactive species, the formation of radical species and improved 'single electron transfer' – SET. However, it should be acknowledged that these process improvements may only be due to intense stirring and increase in bulk temperature, but not solely to sonication. In the late 1920s, it was observed that ultrasound increases emulsification of oil and water [11]. This observation prompted many investigations into the possible mechanisms for emulsification. For example, Fox *et al.* [12] reported that organic initiators could be replaced by ultrasound due to its source of free radicals in the emulsion polymerisation of styrene [12]. More recently, Walton and Phull reported that sonication offers several advantages in styrene polymerisation: a reduced induction period, a faster and better emulsion and an increased oxidation rate [13]. There are several possible explanations for the increase in oxidation rate:

(i) $H^\bullet$ and $OH^\bullet$ radicals, produced by sonolysis, may recombine to form hydrogen peroxide (H_2O_2) and thus oxidise impurities, which could act as inhibitors
(ii) ultrasonic degassing may remove absorbed oxygen
(iii) the breakdown of the initiator may be accelerated by ultrasound
(iv) increased mixing of the components producing a finer emulsion.

The explanation in (iv) was investigated by Atobe *et al.* who showed that ultrasound irradiation appears to be an excellent technique for monomer dissolution, which contributes considerably to polymerisation enhancement [14, 15]. Another effect is that ultrasound minimises or, in some cases, removes the thin film of polymer blocking the electrode surface that impedes further electrochemical reactions. In this case, electroinitiation of chain polymerisation is enhanced [16–19]. Some of these studies concern sonoelectrodeposition of conducting polymers, and obvious effects of ultrasound had been highlighted, such as:

(i) altering the effective reactivity ratio of the monomers
(ii) generating an improvement in percentage conversion versus polymerisation time characteristics
(iii) improving quality of the electrodeposited materials
(iv) improving mechanical characteristics of the electrodeposits, for example, flexibility and toughness [18]
(v) providing higher yields and better quality films [18]
(vi) ensuring better polymer conductivity (e.g. >100 S cm^{-1}) [19].

The work of Atobe and co-workers was probably the first 'modern' example investigating electropolymerisation under sonication in a complete series of papers at low ultrasonic frequencies. Starting from electroorganic reactions under ultrasonic fields [20], polymerisation of aniline was studied using the electrochemical [21] and chemical routes [22, 23] as well as the nanoparticle synthesis route [24, 25].

The behaviour of polypyrrole films electropolymerised under ultrasonic irradiation was also investigated, and unique properties in the doping-undoping processes were highlighted. The authors attributed these observations to the production of highly dense film under sonication, but also deplored the degradation of the film due to high cavitational effects at 20 kHz [26]. The Besançon group also investigated the use of high frequency ultrasound (500 kHz, 25 W) for the electropolymerisation of 3,4-ethylenedioxythiophene (EDOT) or polypyrrole in aqueous media in order to investigate its effects on conducting polymer properties. They showed that: (i) mass transfer enhancement induced by sonication improves electropolymerisation, and that (ii) stirring effect is not the only phenomenon induced by ultrasound during electrodeposition [27, 28].

8.3 Solubilisation of Monomers with Ultrasound

Water is the most ideal medium for synthetic chemistry due to its low operational costs, safety and environmental aspects, although many organic materials are water-insoluble or only sparingly soluble. To overcome the issue of solubility in water, 'artificial' processes employing surfactant-stabilised emulsions and suspensions have been developed. Nevertheless, the presence of surfactants may cause other problems, such as difficulty in separating them from the reaction mixtures and the production of toxic wastes. Therefore, new methods are being investigated to conduct emulsion synthetic processes in the absence of surfactants. The most obvious method is mechanical agitation; indeed, mechanically stirred emulsions of immiscible liquid materials in water produce two-phase systems, which can be electrolysed. But, as a rule, it is not possible to obtain high electron transfer rates between the droplet and the electrode since the size of the droplets is usually larger than the micrometre range. Thus in this case, the use of ultrasound seems to be a very promising technique as it is known to provide stable emulsions without the need for surfactants, simply by the action of mechanical forces which arise at the liquid/liquid phase boundaries.

A test was conducted with 3,4-ethylenedioxythiophene (EDOT) in aqueous media containing 1.0 M $LiClO_4$ at 25 °C, with the view of comparing the stability of acoustic emulsification obtained at 20 kHz with mechanical agitation at 1500 rpm [14]. As shown in Figure 8.6, 'de-emulsification' was observed immediately after emulsification by mechanical stirring, while turbidity still remained for up to 60 min after ultrasonic irradiation. In addition, using the dynamic light-scattering method, the mean size of EDOT droplets formed by acoustic emulsification was found to be circa 211 nm. Here, electropolymerisation was performed with untreated solutions and both acoustically and mechanically emulsified solutions. If polymerisation occurred in untreated solutions, oxidation and reduction peaks in mechanically treated solutions reached average levels, but decreased versus time as EDOT droplets were unstable. On the contrary, electropolymerisation occurred in

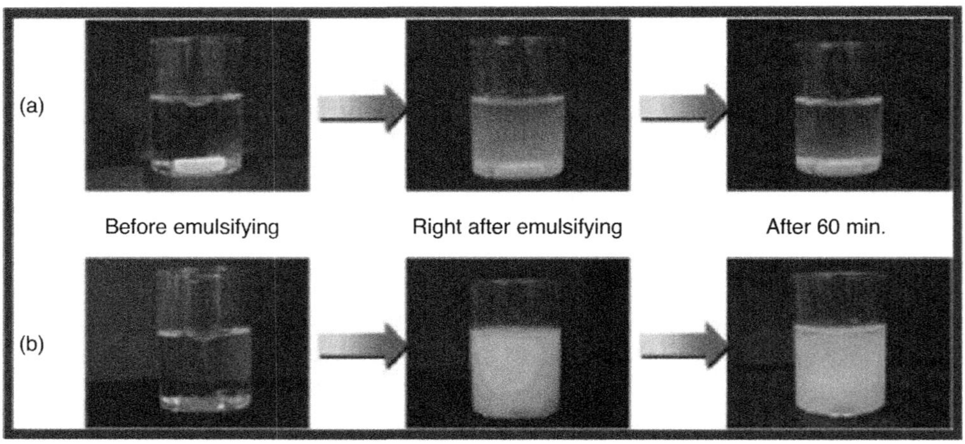

Figure 8.6 Photographic images of emulsification and 'demulsification' of EDOT (7.5 mmol) in 1.0 M $LiClO_4$ aqueous solution. (a) emulsification by mechanical stirring (1500 rpm) (b) by ultrasound (20 kHz) (Reprinted with permission from [14] Copyright (2005) American Chemistry Society).

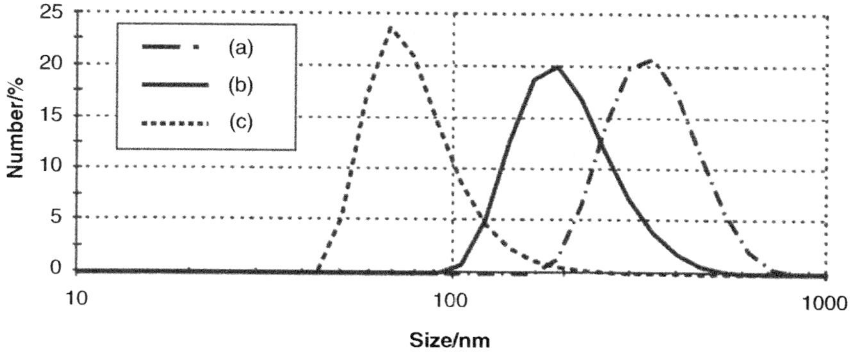

Figure 8.7 *Size distribution of EDOT droplets (0.02 M EDOT, 1.0 M $LiClO_4$) in aqueous solutions exposed to: (a) 20 kHz, 5 min, (b) 20 kHz, 5 min then 1.6 MHz, 5 min, (c) 20 kHz, 5 min then 1.6 MHz, 5 min, and finally (d) 2.4 MHz, 5 min (Reprinted with permission from [29] Copyright (2011) Royal Society of Chemistry).*

acoustically emulsified solutions even after 50 cycles, and the PEDOT film formed showed properties comparable to those obtained by conventional methods [14].

The same authors also showed the dependency of the mean droplet size on the ultrasonic frequency [29]. From mixtures containing water and EDOT monomers, the solutions containing immiscible EDOT droplets were prepared sequentially at 20 kHz, 1.6 MHz and 2.4 MHz, which they referred to as 'tandem acoustic emulsification processing'. In other words, after preparation of stable emulsion at 20 kHz, a second exposure to 1.6 MHz ultrasound for 5 mins leads to a clearer emulsification. If this solution is then exposed to 2.4 MHz ultrasound for 5 mins, the solution becomes extremely clear and transparent [29]. This is sufficiently stable to keep the clear and transparent appearance for 12 months, even under 'surfactant-free' conditions. They attributed these observations to the reduction in mean droplet sizes when increasing the ultrasonic frequency, as shown in Figure 8.7.

When using emulsions to prepare films by electropolymerisation, the obtained morphology is directly linked to mean droplet sizes, that is, the surface roughness is increasingly reduced by an increase in frequency [29]. Moreover, films obtained with clear and transparent emulsions are transparent too, with high conductivities.

Thus in this case, the formation of stable monomer droplets can be formed in aqueous electrolytes under ultrasonic irradiation without adding any surfactants. In other words, a very smooth electropolymerisation in aqueous electrolytes via direct electron transfer between the electrode surface and the immiscible monomer droplets with good conductivity after doping of the polymer films formed is observed.

8.4 Chemical Polymerisation

In addition to dispersion by emulsification, ultrasound has been used to improve polymerisation processes using conducting polymers in the absence of 'electrochemistry'. This concerns rare studies (mainly in Japan) on chemical polymerisations [22, 23] or on nanoparticle syntheses [24, 25]. Indeed, the preparation of metallic nanoparticles is well known, and many examples still exist in the literature where hydrogen radicals formed

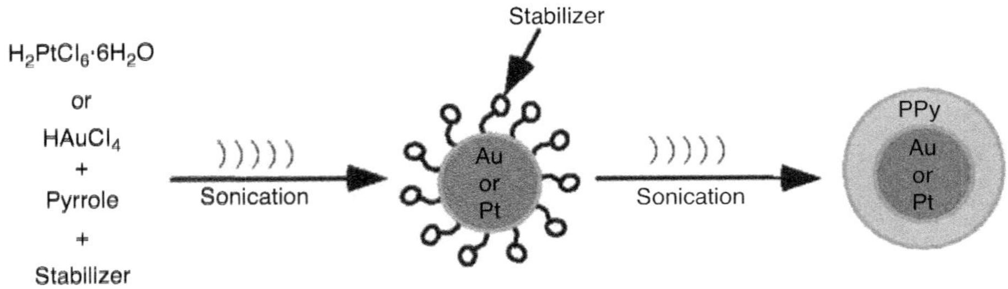

Figure 8.8 Formation of gold nanoparticles coated with a polypyrrole shell (Reprinted with permission from Electrochimica Acta, Sonochemical synthesis of conducting polymer-metal nanoparticles nanocomposite by J-E Park, M. Atobe and T. Fuchigami, 51, 5, 849–854 Copyright (2005) Elsevier Ltd).

during water sonolysis act as a reducing agent (see Chapters 9 and 10). The literature also reports a possible chemical oxidation of aniline by hydrogen peroxide [30], meaning that the possibility of using H_2O_2 formed by recombination of hydroxyl radicals obtained by sonication may be an attractive option.

Studies on gold nanoparticle production is particularly representative [23]. Colloidal dispersions of hybrid nanocomposites composed of polypyrrole (PPy) and gold nanoparticles (Au-NPs) or platinum nanoparticles (Pt-NPs) were prepared by a sonochemical method, in which Au ions and pyrrole monomers in aqueous solutions were respectively reduced and oxidised by ultrasonic irradiation in the presence of sodium dodecyl sulfate (SDS) in an one-step synthesis [24]. The net reaction is shown in Figure 8.8.

In the presence of ultrasound, the net reaction occurs via two competitive steps (Figure 8.9): one step consists of the reduction of $AuCl_4^-$ or $PtCl_6^{2-}$ by hydroxyl radicals formed by sonication, while the other step consists of the chemical oxidation of pyrrole by hydrogen peroxide formed by continuous recombination of hydroxyl radicals during sonolysis. The authors noticed that both reactions can be conducted under silent conditions, but the formation of the nanomaterial is clearly accelerated by ultrasound.

The formation of colloidal dispersion of Au nanoparticles/PPy nanocomposite clusters was confirmed by UV-Vis spectra changes, cyclic voltammetry and TEM pictures, as shown in Figure 8.10 [24].

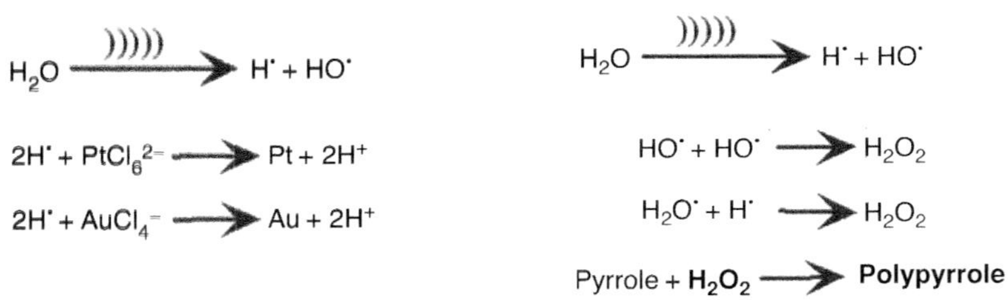

Figure 8.9 Mechanisms involved in the formation of Me-NPs (Au, Pt)/PPy nanocomposite (Reprinted with permission from Electrochimica Acta, Sonochemical synthesis of conducting polymer-metal nanoparticles nanocomposite by J-E Park, M. Atobe and T. Fuchigami, 51, 5, 849–854 Copyright (2005) Elsevier Ltd).

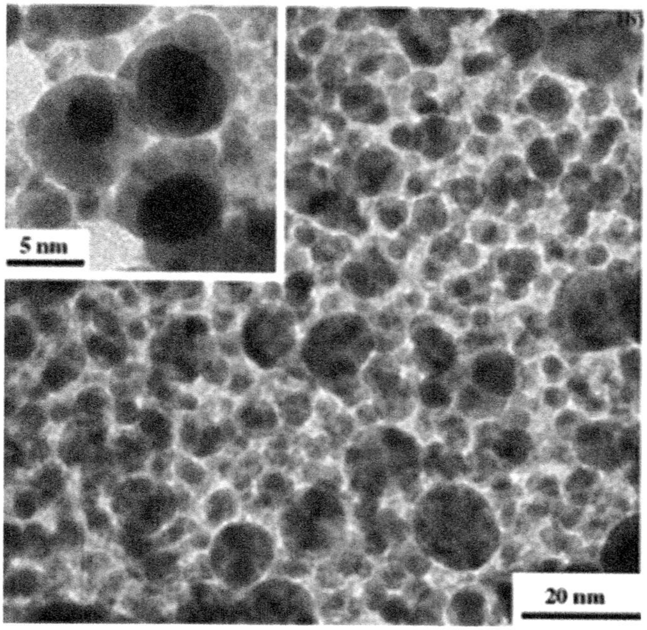

Figure 8.10 *TEM pictures of Au/PPy nanocomposite core-shell prepared by ultrasonic irradiation of 4 hours (Reprinted with permission from Electrochimica Acta, Sonochemical synthesis of conducting polymer-metal nanoparticles nanocomposite by J-E Park, M. Atobe and T. Fuchigami, 51, 5, 849–854 Copyright (2005) Elsevier Ltd).*

8.5 Electropolymerisation under Ultrasonic Irradiation

Pioneer investigations were performed by Osawa *et al.* at the end of the 1980s whereby polythiophene films were prepared by electrochemical oxidation under a 20 kHz ultrasonic field [18]. They observed that under silent conditions and for upper current density values, the films became brittle and under sonication the films' properties were improved, exhibiting increased conductivities (130–150 S cm^{-1}) and excellent mechanical behaviours. Moreover, the films obtained by this method exhibited a higher diffusion rate of dopants during oxidation process [18]. In a sister paper [19], an electrochemical study from the same authors showed that polymerisation is greatly affected by reaction potentials. It was shown that sonication suppressed the overall overpotential by reducing the diffusion layer thickness due to the strong stirring induced by ultrasound. Yang *et al.* [31] succeeded in the electropolymerisation of pyrrole under a 20 kHz ultrasonic field in a series of propylene carbonate solutions at various ultrasonic powers and in the presence of various background electrolytes. They noticed that over the cavitation threshold, agitation at the electrode surface adversely affected polymerisation, and the polymer was degraded. Under this threshold, that is, at moderate ultrasonic intensities, an enhancement of conductivity and the tensile strength of the polypyrrole films were observed [31]. On the contrary, Atobe *et al.* showed that the electrooxidation of aniline in acidic solutions occurred over the cavitation threshold at 20 kHz [21]. Nevertheless, a decrease in current densities was observed over an ultrasonic power of 10 W, when cavitation occurred. In those experimental conditions, highly uniform and dense films were formed, and Atobe *et al.* reported for the first time an alteration of the films on a macroscopic scale [21]. This was confirmed by Kojima *et al.* for styrene monomers [32], and by Walton *et al.* for pyrrole and azulene monomers [10]. They found that 20 kHz sonication interfered with the film formation at the electrode surface, and thus recommended using a higher frequency, for example, a 850 kHz ultrasonic bath, to obtain better yields and films properties. These films exhibited finer structural features with denser morphologies than silent ones. More recently, Galal [33] prepared poly(3-methylthiophene) films under 20 kHz ultrasonic irradiation under potentiostatic conditions at various electrode-to-horn distances. The level of doping of the resulting polymer films varied according to the electrode-to-horn distances during the synthesis. They found that the shorter the distance; the higher the level of doping and, consequently, the increase in film surface conductivities were observed. While FTIR spectra showed no alteration in aromatic ring integrity for polymer films produced under ultrasonic conditions, SEM pictures showed films that are more compact and less porous compared to silent conditions. These changes in morphology were accounted for by the reduction of structural defects. Indeed, five-member heterocycles such as thiophenes or pyrrole electropolymerise in highly ordered chains, provided that co-planarity of units between chains is respected, a fact favoured by sonication. Nevertheless, they observed a lower average doping of the films, attributed to less ionic diffusion into the films induced by film compactness. EIS (impedance) data showed that restriction of diffusion and decrease in doping level affected the rate of charge transfer at the electrolyte/polymer interface and the diffusion parameters of the film, exhibiting the same behaviour as over-oxidised films (at a potential of circa +1.5 V). Atobe *et al.* [26] investigated using electrochemical methods the 'doping-undoping' properties of polypyrrole films fabricated under ultrasound, and observed, for example that *p*-toluenesulfonate-doped polypyrrole films exhibited a more negative cathodic peak than under silent conditions. This voltammetric change at the reduction process was accounted for as a change from anion undoping to cation doping. They also observed unique electrochemical properties in the 'doping-undoping' processes under sonication attributed to the highly dense structure of the films formed [26].

Finally, specific applications of films elaborated by sonoelectropolymerisation emerged in the late 2000s with the work carried out by Reyman *et al.* [34]. They studied the effect of 530 kHz ultrasonic irradiation on the electrodeposition of polythiophene in acetonitrile using lithium perchlorate as a background electrolyte. The originality of their work was based upon the use of the fluorophore BCCM (9H-pyrido(3,4-*b*)indole-3-carboxylic acid methyl ester, which can be embedded in the polymer when produced under ultrasound at potentials of 2.2, 2.5 and 2.8 V. This was confirmed by the fluorescence properties of the film, not present under silent conditions [34]. Another application consists of steel protection against corrosion by polypyrrole films fabricated under 28 kHz ultrasound in acid phosphate solutions containing molybdophosphate ($PMo_{12}O_{40}{}^{3-}$) ions [35]. Kowalski *et al.* concluded that:

(i) corrosion resistance of steel coated by the PPy film was improved by ultrasonic irradiation during the electropolymerisation of PPy. The time required to maintain the passive state for the steel coated by PPy film prepared under ultrasonic irradiation is one and a half times as long as that for the steel coated by PPy film in the absence of ultrasonic irradiation.

(ii) surface morphology of PPy films was changed with imposition of ultrasound in the electropolymerisation process. Under irradiation, a dense and compact PPy layer was formed, which was assumed to result from a change in the nucleation-growth mechanism under sonication, and increased mixing of the components.

(iii) the doping level of $PMo_{12}O_{40}^{3-}$ ions was doubled by the ultrasonic irradiation during electropolymerisation.

In recent years the Besançon group (France) undertook several studies focused both on sonopolymerisation mechanism(s) [27, 28] and on their applications at high ultrasonic frequencies. One may cite pH sensors [36], protection against corrosion [37] and biological cell proliferation [38]. The development of new sonoreactor equipment in the team [39] led the Besançon group to an original use of high intensity focused ultrasound (HIFU) as a selective electrodeposition technique [40] (see previous chapters).

Reactors used in these studies consist mainly of specially designed cells built to accommodate the ultrasonic source, that is, a titanium horn of various diameters. The transducers are embedded in the reactor bottom, so that ultrasound is emitted parallel to the liquid surface, opposite the working electrode surface. For high ultrasonic frequencies, two experimental setups may be used: a flat transducer consisting of a piezoelectric ceramic fixed on a circular glass plate placed at the bottom of the reactor, and a high intensity focused ultrasound transducer.

Figure 8.11 shows two types of equipment used in the Besançon group. Experiments conducted with the flat transducer were carried out at a constant ultrasonic frequency of 500 kHz in a 750 mL cell. An input power of 25 W was chosen, as the generated calorimetric measurement data were linear and reproducible. Here, ultrasound is emitted

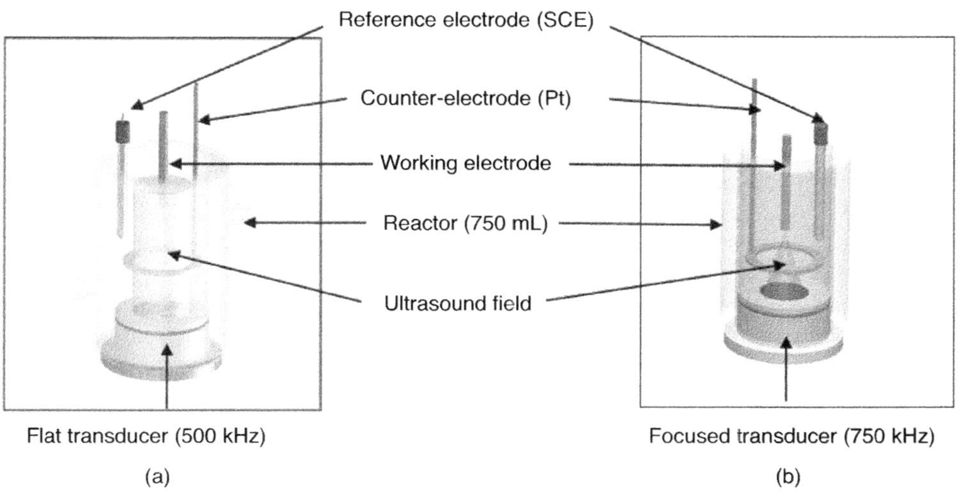

Figure 8.11 *Diagram of (a) flat transducer and (b) focused transducer-based sonoreactor setup (Reprinted with permission from [40] Copyright (2010) Elsevier Ltd).*

parallel to the liquid surface, opposite the working electrode surface (Figure 8.11(a)). The high ultrasonic frequency flat transducer shown in the figure was developed by the Laboratory of Molecular and Environmental Chemistry (University of Savoie, Chambéry, France). The circular transducer consists of a piezoelectric ceramic fixed on a circular glass plate (56 mm). An electronic circuit consisting of a variable inductance and capacitor allows adjustment between generator impedance and the highly capacitive transducer impedance. Experiments conducted with thehigh intensity focused ultrasound (HIFU) transducer were carried out at 750 kHz in the same 750 mL cell with 25 W as input power. The transducer was manufactured by the French company IMASONIC (France) (Figure 8.11(b)).

8.6 Effects of Ultrasound on Film Properties

As previously discussed in this chapter, many film properties can be greatly improved by electropolymerisation in the presence of ultrasound. In particular, beneficial effects can be observed on the production of polymeric films on: (i) mass-transfer control, (ii) morphological properties, (iii) doping properties, and, finally, (iv) local control of the surface.

8.6.1 Mass-Transfer Effect

As previously mentioned in early chapters of this book, one well-known effect of ultrasonic irradiation is the generation and subsequent collapse of cavitation bubbles in an electrochemical cell. This phenomenon can be considerable within the electrolyte bulk as well as close to the electrode surface. The electrode surface is affected by induced asymmetrical collapse of a bubble, which in turn leads to the formation of a high velocity jet of liquid directed towards its surface. This jet, together with microstreaming, is thought to lead to the enhancement of mass transfer at the electrode. This is experimentally observed by an increase in current in all studies reported in this book, and is the origin of the increase in polymerisation rate, whatever the ultrasonic frequency and power employed.

First, to quantify the contribution of mass transfer to the enhancement in polymerisation rates, a 'calibration' is required prior to film elaboration. For example, Pollet *et al.* showed that mass-transfer characterisations of the sonoelectrochemical setup are necessary prior to sonoelectrochemical work. In their studies, mass-transfer experiments were performed in aqueous solutions of 5 mM potassium ferrocyanide ($K_4Fe(CN)_6$) in 0.2 M sodium hydroxide. Current potential polarisation curves (sonovoltammograms) of the $Fe(CN)_6{}^{4-}/Fe(CN)_6{}^{3-}$ quasi-reversible couple were recorded at a 2 mV s^{-1} scan rate [41]. Note that mass-transfer limited currents under ultrasonic irradiation are usually made of a steady-state and time-dependent components (oscillating around the average plateau current value). In their conditions, only steady-state current values were considered, and mass-transfer limited currents were determined by a statistical study [41]. The mass-transfer coefficient k_d (m s^{-1}) was calculated from the experimental mass-transport limited current density (A m^{-2}) values j_{lim}, the number of exchanged electrons n, the Faraday constant (C mol^{-1}) F and the concentration in electroactive species (mol m^{-3}) C in the solution, according to Equation 8.1:

$$k_d = \frac{j_{lim}}{nFC} \tag{8.1}$$

The average mass-transport limited current density was found to be equal to 2.2 mA cm^{-2} (Figure 8.12) [28].

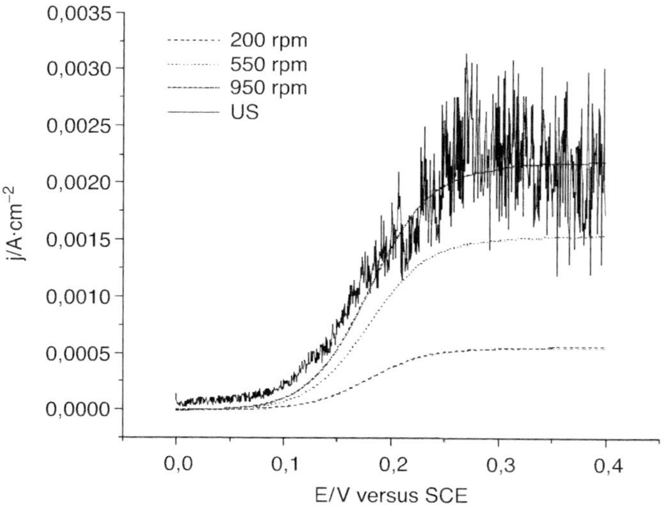

Figure 8.12 *Rotating disc and sono-voltammograms (500 kHz, 25 W) of 5 mmol dm^{-3} $Fe(CN)_6^{4-}/Fe(CN)_6^{3-}$ in 0.2 M NaOH on a platinum electrode at 2 mV s^{-1} scan rate and at 298 K (Reprinted with permission from [28] Copyright (2011) Elsevier Ltd).*

In order to work at the same stirring level both under silent and ultrasonic conditions, a calibration was carried out by measuring the mass-transfer coefficients at various rotating speeds (Figure 8.12). A comparison between all linear sweep voltammograms indicated that a rotation of 950 rpm led to the same transport-limited current density found at 25 W ultrasound (500 kHz). A corresponding mass-transfer coefficient was found to be 4.6×10^{-5} m s^{-1} [28]. For further information on mass-transfer characterisations, the reader is invited to refer to the Appendix section in this book entitled: 'Ultrasound Effect on Electrode Kinetics'.

PEDOT conducting polymer films were then elaborated on FTO substrates for 100 s and chronoamperometric curves were generated under silent, ultrasonic conditions or with an electrode rotating at 950 rpm (in the absence of sonication), as shown in Figure 8.13.

Under silent conditions, the decrease in current density is attributed to the growth of the polymer film at the electrode surface, whose conductivity is lower than that of the substrate. The chronoamperometric curves in Figure 8.13 show that under ultrasound, current values are much higher than those obtained under silent conditions, thus an enhancement in electropolymerisation of conducting films is observed. After a few seconds, the current reaches its average value corresponding to the nucleation step, that is, the creation of the first active centres at the electrode interface (occurring faster) leading to polymer growth. Thereafter, the current does not decrease but remains constant, in other words the monomers are brought to the electrode interface more efficiently due to the high ultrasonic stirring effects at higher current density, leading to better electropolymerisation. In this study, a deposit was also formed under potentiostatic conditions and at a rotation speed of 950 rpm in the absence of ultrasound (in order to work in similar agitation conditions, that is, same mass transfer coefficient k_d as when ultrasound is used). The data presented a clear increase in current density, showing a dependency of electrodeposition on the

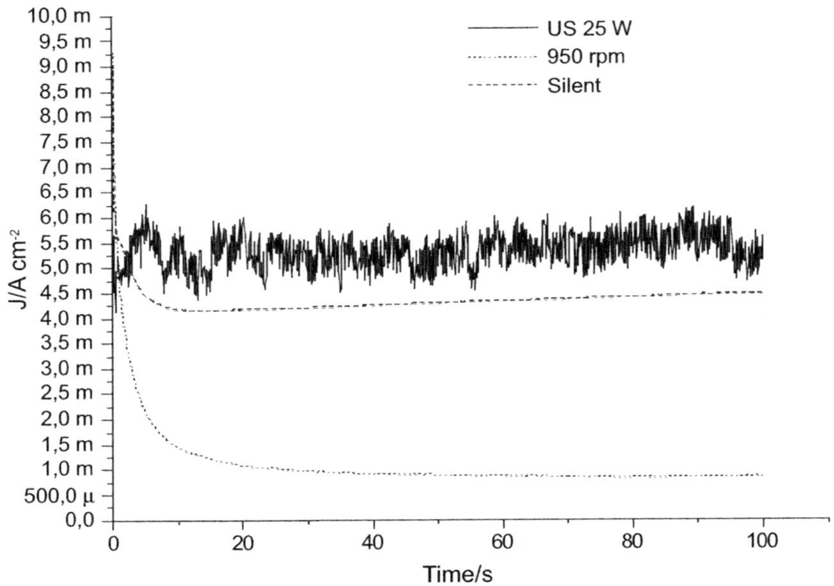

Figure 8.13 *Chronoamperometric curves of PEDOT electrosynthesis in aqueous solution (5×10^{-3} M EDOT + 0.1 M $LiClO_4$) in the absence and presence of ultrasound (500 kHz, 25 W) – $\omega = 950$ rpm and $E_{app} = 1$ V versus SCE for 100 s (Reprinted with permission from [28] Copyright (2011) Elsevier Ltd).*

stirring conditions. Nevertheless, when comparing the two chronoamperometric curves in Figure 8.13 (US 25 W and RDE recorded for the same $k_d = 4.6 \times 10^{-5}$ m s^{-1}), it is obvious that the electrosynthesis current density is much higher under ultrasonic conditions. Indeed, ultrasound does not only have a stirring effect at the interface, but other phenomena that improve electropolymerisation need to be considered [28]. For example, the energy losses during wave propagation in the liquid medium create a global convection, thus increasing the global mass transfer of the monomers in solution to the electrode surface, just like any other agitation methods (e.g. RDE). Nevertheless, when the ultrasonic activity occurs close to the interface, the shocks waves as well as the asymmetric collapse of cavitation bubbles, which generates micro-jets, provide an additional effect to mass transfer to the electrode surface, allowing its activation by impingement. This continuous but fugitive local energy supply is certainly the key to nuclei site multiplication.

8.6.2 Morphology Effect

Since the early work carried out by Osawa *et al.* [18], the influence of ultrasound on surface morphology has been frequently observed. For example, one of the first SEM pictures reported in the literature of polythiophene surfaces obtained at 5 mA cm^{-2} showed a marked change in morphology in the presence of ultrasound, as shown in Figure 8.14.

It was found that the surface of the films elaborated under sonication was more homogeneous and smoother compared to those obtained under silent conditions. This observation was attributed to cavitational events that were able to alter film solidity [10, 31]. In

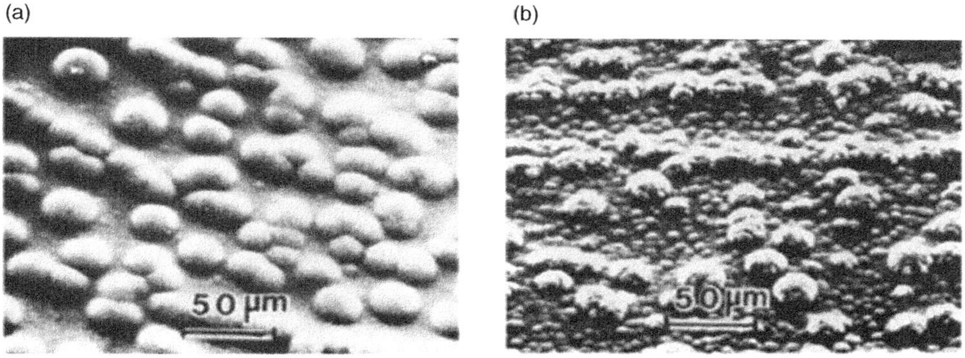

Figure 8.14 *SEM pictures of film surface (a) in presence of 20 kHz ultrasound and (b) in silent conditions (Reprinted with permission from [18] Copyright (1987) Elsevier Ltd).*

the work of Et Taouil *et al.* [42], conducting polymer films (essentially PEDOT and PPy) were evaluated by complementary methods such as mechanical sensors or optical and electronic techniques as shown in Figure 8.15. The polymer surface was characterised by scanning electron microscopy (SEM, JEOL 5600), atomic force microscopy (AFM PicoSPM from Molecular Imaging, USA), contact mode with gold-coated Si_3N_4 tip (200-μm long triangular cantilever and 0.1 N m^{-1} force constant), stylus-based mechanical probe profiler (Alpha-Step IQ, KLA Tencor) and optical surface profiler (Alicona® Infinite Focus Intralux® dc-1100) to measure film roughness. They observed that, by using the AFM imaging indicated, (i) the PPy films exhibited a high roughness level under silent conditions, with a growth in a cauliflower-like morphology, and (ii) a marked refining of the surface under ultrasonic irradiation was observed. Such surface refining for polymer films electrosynthesised under high frequency ultrasound has already been observed and discussed in previous works [40] where it was stipulated that thinner surfaces could be promoted for corrosion protection and sensing applications where surface morphology plays an important role.

SEM pictures of PPy films obtained in the absence and presence of ultrasound showed in both cases a typical nodular structure. The film obtained under sonication appeared to have a thinner and more homogeneous surface structure. The globules of the film elaborated under silent conditions were of 1.2 ± 0.2 μm diameters, while the sample obtained under sonication exhibited thinner microspherical grains with a diameter of circa 0.5 ± 0.2 μm. This observation was once again attributed to the asymmetric collapse of cavitation bubbles, leading to a larger number of nuclei sites during the first electropolymerisation step and therefore to a change in the nucleation-growth mechanism. Optical profiling led to similar conclusions with roughness values of 9.4 ± 0.5 μm under silent conditions compared to 5.0 ± 0.5 μm under sonication.

8.6.2.1 Biological Applications

Most biological systems offer many advantages as they possess specific binding interactions, and recognise foreign objects by their physical shape, topography and chemical properties right down to molecular dimensions. This allows the creation of useful materials for a large

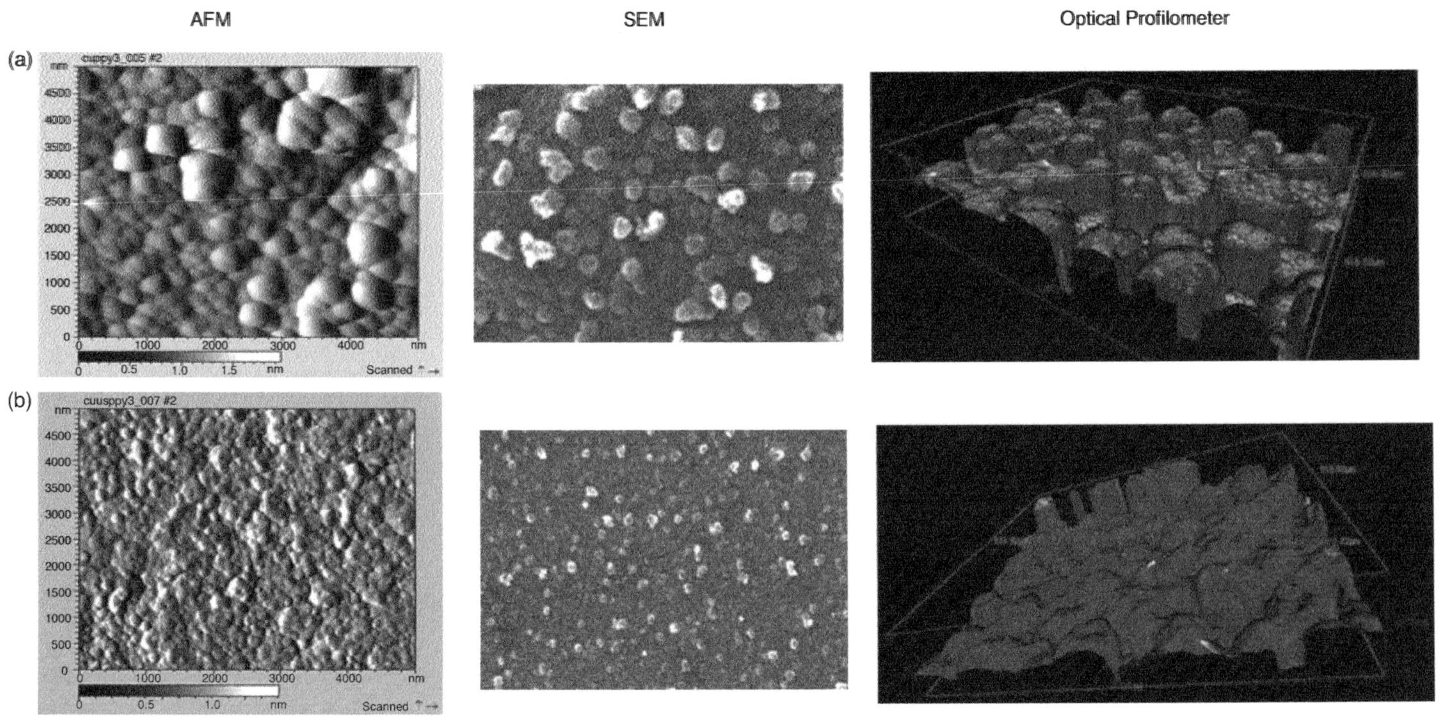

Figure 8.15 *AFM, SEM and optical profilometry pictures of polymer films obtained under (a) silent conditions and (b) ultrasonic conditions (500 kHz).*

number of biotechnology applications, where interfaces with enhanced specific binding properties are required. Examples of applications include medical implants that replace or improve lost or impaired vital body functions in humans [1,43], biosensors for biomedical diagnostics [44, 45], non-biofouling surfaces for the food industry or for device implants, and polymers for controlled drug release [46, 47]. Consequently, for all these applications, it is important to study the adhesion of biological cells on various materials, since adhesion plays an important role in several key cellular processes.

Recently the use of conductive polymers in biological applications have received a lot of attention as they offer:

(i) good biocompatibility, that is, implants in animals and humans [48]
(ii) controlled porosity since conducting polymers have an intrinsic property that can be varied according to the preparation methods [49]
(iii) easy control of location, thickness or morphology by applying current or potential/cell voltage
(iv) physical, chemical and electrical control by varying the doping agents [26].

Moreover, adhesion and growth of olfactory neuronal cells on electrosynthesised polypyrrole confirm the ability of these polymers to be used as cell culture substrates [50]. Therefore, polypyrrole is one of the most interesting materials since it is: (i) easily deposited from aqueous and non-aqueous media, (ii) extremely adherent to many types of substrate, (iii) a good conductor and (iv) very stable. However, even if surface chemistry is known to play a major role in cell adhesion and proliferation mechanisms, the effects of polypyrrole film morphology on cell cultures have not yet been studied in any great depth. For this reason the authors of this chapter decided to produce polypyrrole films in the absence and presence of ultrasound in order to evaluate the effects of surface smoothness on cellular adhesion and growth [51].

The human osteosarcoma cell line Saos-2 (ATCC-HTB85) was used to study, qualitatively and quantitatively, human bone cell adhesion on polymer films, following a procedure described in the literature [51]. After 24 hours, the cells displayed a round or/and polygonal morphology on all the samples tested. The actin cytoskeleton was essentially represented by actin fibres at the periphery of the cells (Figure 8.16). After 3 days the cells proliferated and displayed polygonal or/and spindle-shape morphologies. Their cytoskeleton was still essentially at the periphery of the cells but in some cells stress fibres were formed and oriented in the *y*-axis direction (Figure 8.16).

After 6 days, the actin cytoskeleton was formed essentially by stress fibres parallel to the longer axis of cells, and cells reached confluence on all the samples (Figure 8.16). It was concluded that the Saos-2 cells adhered and proliferated normally on all the polymers tested, and their morphology was the same as on FTO glass control, thus proving the non-toxicity of polypyrrole for osteoblastic cells. Furthermore, no quantitative difference was observed for treated and untreated irradiated polypyrrole samples proving that osteoblastic cells reacted similarly to both polymeric samples [51].

Then, in this particular study, tests were performed using microbial cells. *Escherichia coli* PHL628 was chosen as it is known to be able to reproduce curli and exo-cellularpolymeric substances. Cells were cultivated following a procedure described in the literature [51]. As shown in Figure 8.17, the number of planktonic bacteria in growth medium after 24 hours

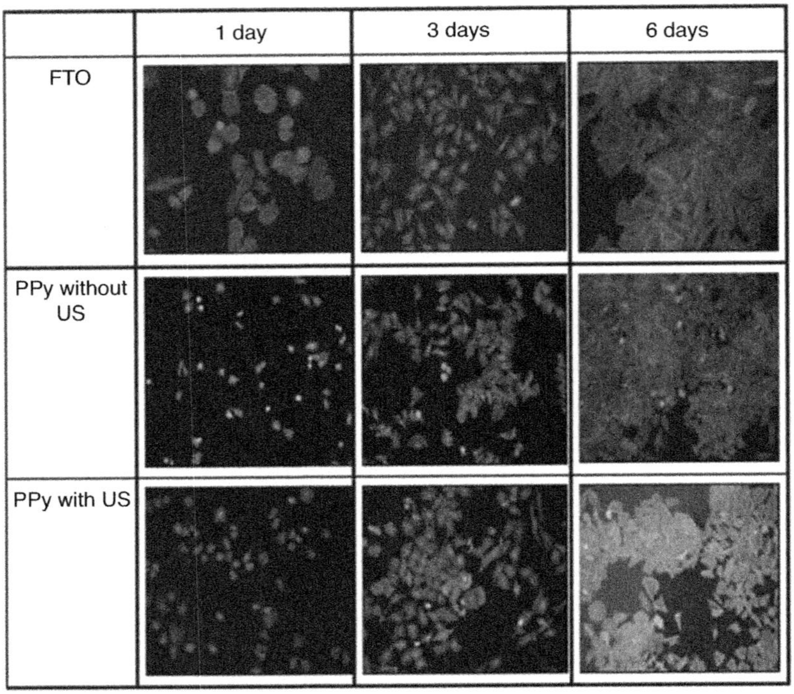

Figure 8.16 *Immunostaining of FITC-phalloidin on osteoblast cells SaOs-2 cultured on FTO and on polypyrrole elaborated in presence or absence of ultrasound (Reprinted with permission from [51] Copyright (2009) Elsevier Ltd).*

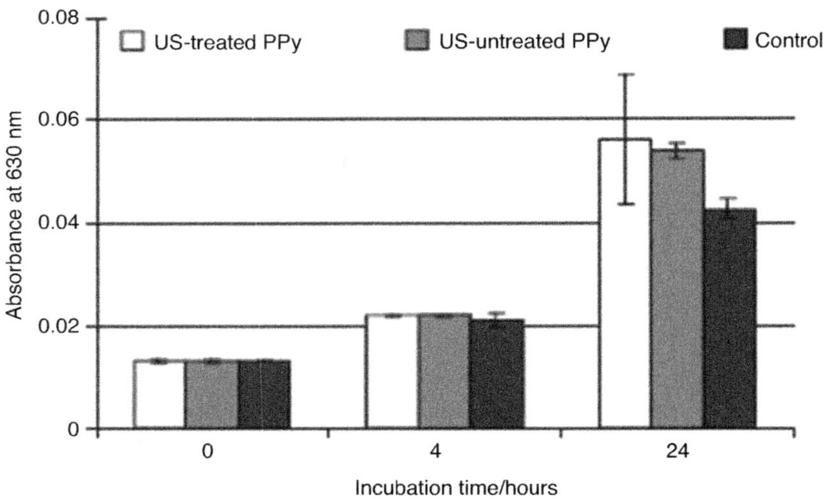

Figure 8.17 *Amount of planktonic bacteria on various substrates through A630 measurements after re-suspension of cells in NaCl solution (9 g/L in water) (Reprinted with permission from [51] Copyright (2009) Elsevier Ltd).*

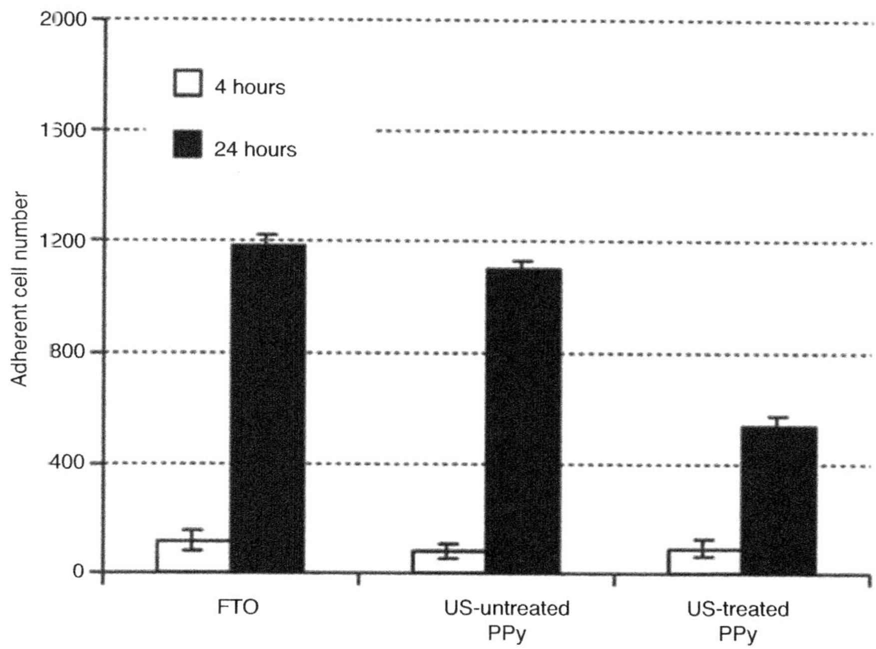

Figure 8.18 *Number of adherent bacteria on various substrates (Reprinted with permission from [51] Copyright (2009) Elsevier Ltd).*

incubation was not significantly different in the presence of polymer in the samples (FTO with polypyrrole films), compared to the control medium (bacterial culture in Petri in the absence of the polymer). This result clearly demonstrated the non-toxicity of the polymers (polypyrrole films elaborated in the absence and presence of ultrasound) for *Escherichia coli*.

With regard to bacterial adhesion and adherent bacterial growth, Figure 8.18 evidently shows that there are no significant differences between samples after 4 hours incubation.

By contrast, the number of cells was smaller on polypyrrole surfaces compared to FTO surfaces after 24 hours incubation. Furthermore, bacterial colonisation on ultrasonically treated polypyrrole samples were smaller than those elaborated under silent conditions. This lower level of bacterial colonisation can be related to the porous structure of the polymer films, which hide parts of the adherent bacteria. In accordance with the weak colonisation level obtained on sonicated films compared to non-irradiated ones, polymer structures formed under sonication exhibited more pores with a rougher surface [51]. Figure 8.17 proves that all polypyrrole samples are non-toxic since the number of cells on FTO and polymers is similar. However, since there are less cells adhering to sonicated films, the use of ultrasound could be a promising method to prevent surfaces from bacterial biofouling.

8.6.2.2 Protection against Corrosion Applications

The ability of polymer films to protect metals against corrosion has already been reported [35, 37] and described in this book (Chapters 6 and 7), and one of the many possibilities to enhance film properties is to increase their compactness by sonication. Therefore, the use of high frequency ultrasound irradiation, known to modify polypyrrole film morphology, has been used to produce compact films on zinc-coated steel samples (here, the films are usually compared to inorganic passivation, e.g. of Cr^{III} and Cr^{VI} used as reference materials). Tsuchiya *et al.* [52] and Kowalski [35] studied the protection role of PPy against corrosion on various metals. In their studies, sodium tartrate was chosen as supporting salt for the electrosynthesis of PPy in which 5 mM of molybdate were added to improve corrosion protection properties. Glow discharge optical emission spectroscopy (GD-OES) measurements showed that Mo elements were incorporated into the film in the absence and presence of ultrasound [37].

In their investigation, the first test was to evaluate protection against corrosion on samples dipped in 3.5% sodium chloride solutions. The evolution of the open circuit potential (OCP) was monitored as a function of time for zinc-plated steel covered with films prepared in the absence and presence of ultrasonic irradiation (Figure 8.19). The figure shows that polypyrrole films prepared under silent conditions exhibit potentials between −570 and −550 mV versus SCE for the first 250 s of immersion. The potential then decreases progressively due to film breaking until it reaches the active potential of zinc at 1500 s. However, PPy films generated under sonication exhibit higher potentials in the range of −500 and −410 mV versus SCE with large fluctuations. The origin of these fluctuations is not clear, but it is speculated that they are possibly due to the PPy film being attacked by corrosive species. The figure also shows that after circa 1500 s of experiment, the potential decreases suddenly to reach the active potential of zinc at 2000 s. It was concluded that the PPy films obtained under ultrasound maintain zinc in the passive state longer than film produced under silent conditions (> 500 s). This observation can be explained by the fact that under sonication the surface structure of the produced

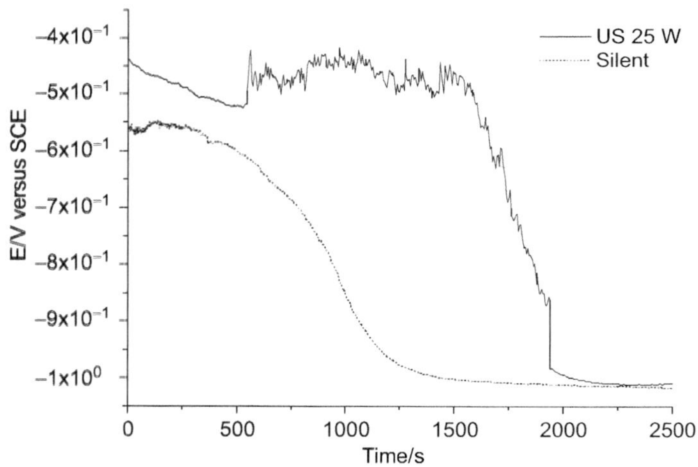

Figure 8.19 *Open circuit potential (OCP) as a function of immersion time in 3.5 wt.% NaCl aqueous solution for Zn-coated steel covered with polypyrrole electrosynthesised at 3.0 V versus SCE at 6 C cm^{-2} in the absence and presence of ultrasonic irradiation.*

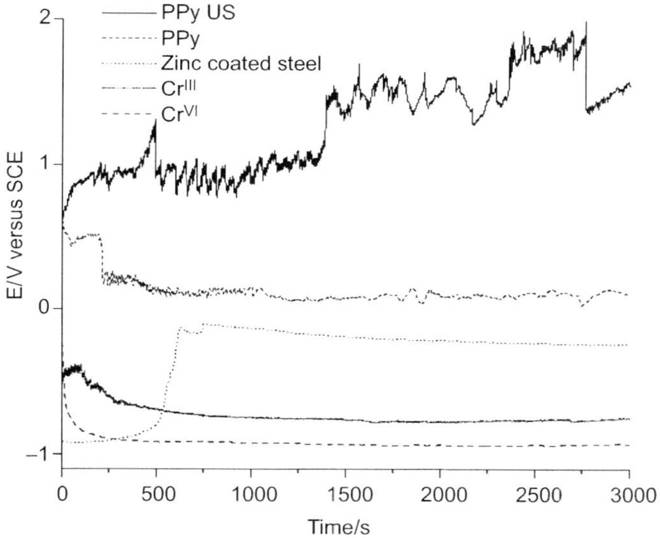

Figure 8.20 *Evolution of potential with immersion time in 3.5 wt.% NaCl aqueous solution at 10 mA cm^{-2} applied current density for bare Zn-coated steel, Cr^{III}/Cr^{VI} passivated Zn-coated steel and silent/sonicated PPy modified Zn-coated steel.*

film is more compact and dense, forming a more efficient barrier against the attacking environment [37].

The second test is more severe as it consists of applying a constant current density of 10 mA cm^{-2} on the polypyrrole-coated zinc samples in 3.5% sodium chloride solution and then recording chronopotentiometric curves, as shown in Figure 8.20 [37].

For bare zinc-coated steel samples, the potential remains around −900 mV versus SCE for the first 500 s before a sudden increase and stabilisation at circa −200 mV versus SCE. This increase corresponds to the destruction of the Zn layer and the attainment of steel substrate. For Cr^{III} and Cr^{VI} passivated layers, the potential starts at −300 mV versus SCE and decreases rapidly to stabilise at −750 and −900 mV versus SCE for Cr^{III} and Cr^{VI} layers, respectively [37]. These low potential values confirm the fact that chromated layers do not homogeneously cover the Zn surface. Polypyrrole film behaviour is complex and no clear corrosive attack of the film is evident. However, the potentials remain relatively stable and far higher than those observed for inorganic protection.

Finally, in order to obtain quantitative data, Tafel plots from DC polarisation curves in 3.5% sodium chloride solutions were generated, and the values are quoted as corrosion potentials (E_{corr}) and corrosion current densities (j_{corr}) as shown in Table 8.1. The table shows that all coatings seem to prevent corrosion because, in all cases, j_{corr} strongly decreases compared to the bare zinc surface.

As expected, inorganic coatings, and particularly Cr^{III}, show the best passivation behaviour as they lead to the lowest j_{corr} values (14 μA cm^{-2} compared to 116 μA cm^{-2} for Cr^{VI}). For Cr^{VI} passivation, an obvious shift of E_{corr} towards more anodic potentials is also observed. For polypyrrole films, a strong decrease in j_{corr} is observed irrespective

Table 8.1 Corrosion current densities (J_{corr}) and potentials (E_{corr}) from DC polarisation curves obtained in 3.5 wt.% NaCl solution at 2 mV s^{-1} scan rate [37].

Sample	E_{corr} (mV versus SCE)	J_{corr} (μA cm^{-2})
Bare Zn	−1088 ± 50	1006 ± 100
Cr^{III}/Zn	−1038 ± 30	14 ± 3
Cr^{VI}/Zn	−713 ± 20	116 ± 5
PPy/Zn	−804 ± 90	191 ± 10
PPy US/Zn	−455 ± 40	11 ± 5

of the use of ultrasound. However, sonication of polypyrrole leads to far better corrosion protection; indeed j_{corr} decreases drastically under ultrasonic irradiation (11 μA cm^{-2}) by circa 20-fold compared to silent conditions (191 μA cm^{-2}). It is interesting to note that j_{corr} values are similar for Cr^{VI} and polypyrrole coatings on the one hand, and for Cr^{III} and sonicated polypyrrole on the other, whereby a great increase in E_{corr} was also observed.

8.6.3 Doping Effect

As highlighted earlier, the common mechanisms behind rechargeable batteries, electrocatalysis, and so on are 'doping' and 'undoping' processes, which assume oxidation and reduction of conducting polymers and the movements of ions to preserve electroneutrality. In the pioneer work of Osawa *et al.* [18], it has been shown that conductivity is normally influenced by the doping level. In their study, they determined by elemental microanalysis the doping levels of the polymerised films obtained in the absence and presence of ultrasound, and plotted the generated values against several current densities, as shown in Figure 8.21. The figure shows that with increasing current density, the level of doping increases, and this effect is more pronounced under sonication. In all cases, the levels of doping obtained under ultrasonic irradiation and at a given current density, are always higher than those obtained in silent conditions.

When plotting film conductivity values versus the doping level, Osawa *et al.* [18] found an increasing linear relationship [18], indicating that conductivity of the polymerised films is primarily controlled by the doping level.

More recently, Atobe *et al.* studied the effects of ultrasound at low frequencies upon 'doping-undoping' properties of polypyrrole films [26]. They observed that the reduction process of the doped polypyrrole film accompanies the undoping of anions, as shown in Figure 8.22(a). But when mobility of the dopant anion within the film is so small, the reduction process leads to cation doping instead of undoping of anions for charge compensation, as shown in Figure 8.22(b).

To highlight changes in 'doping-undoping' path, Atobe *et al.* used various ions of different sizes. For example, they used TsO^- as it is a relatively large anion and its mobility within PPy film is not excessively high. Therefore, as shown in Figure 8.23, the cyclic voltammograms of PPy/TsO films show two distinctive reduction cathodic peaks. For example, the cathodic peak (I_c) is significantly larger at −0.62 V versus SCE compared to the one observed at −0.73 V versus SCE (II_c, distinct peak) which is smaller, corresponding to 'undoping' of TsO^- (path A) and doping of Na^+ (path B), respectively [26]. In sharp

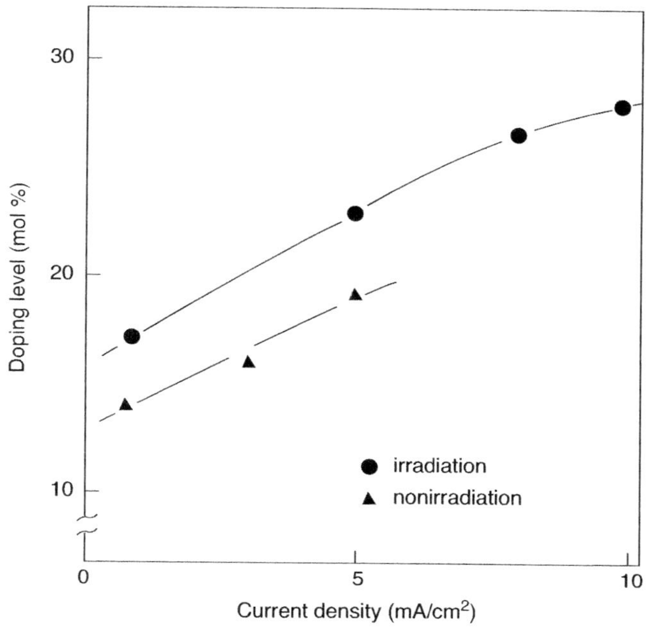

Figure 8.21 Plots of doping level of polymerised samples versus current density for irradiated and silent samples (Reprinted with permission from [18] Copyright (1987) Elsevier Ltd).

contrast, as shown in Figure 8.23(b), the −0.62 V versus SCE peak cannot be distinguished in the voltammogram of the film prepared under ultrasound while the −0.73 V versus SCE peak is significantly larger compared to the one prepared under silent conditions. This interesting observation suggests that Na^+ doping (path B) predominantly occurs in the reduction process of the polypyrrole film prepared under ultrasonic irradiation.

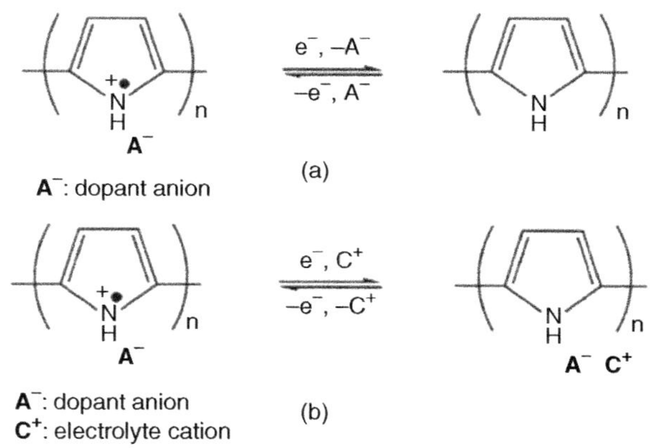

Figure 8.22 Diagram (a) anion undoping of a polypyrrole film during the reduction process (path A) and (b) cation doping during the reduction process (path B) (Reprinted with permission from [26] Copyright (2006) The Electrochemical Society Inc).

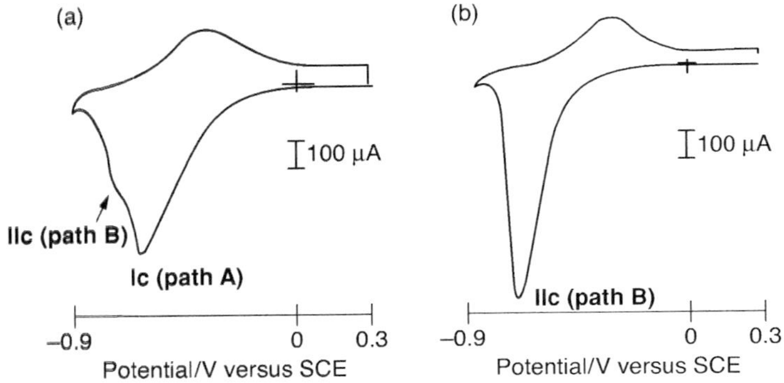

Figure 8.23 Cyclic voltammograms of polypyrrole/TsO films prepared (a) in the absence and (b) presence of ultrasound (Reprinted with permission from [26] Copyright (2006) The Electrochemical Society Inc).

The authors also attributed this observation to the increase in compactness as observed in previous investigations with polyaniline, polythiophene and polypyrrole, where smooth surfaces were obtained [15, 20, 26].

In order to investigate ion-sieving into electrosynthesised polymers, the films obtained in the absence and presence of ultrasound were oxidised to dope the film with anions (by dipping) under polarisation conditions using various electrolytes containing respectively: 0.1 M $NaClO_4$, 0.1 M NaCl, 0.1 M $EtSO_3Na$, 0.1 TsONa and 0.1 M NSNa aqueous solutions. The results shown in Figure 8.24 are plots of the anodic peak currents (I_{pa}) obtained by LSV (linear sweep voltammetry) versus anion diameters under silent (a) and sonicated solutions (b).

It can be noted that the doping rate (directly correlated to the anodic peak current) is greatly affected by the anion diameter. Because polypyrrole films were fabricated in the presence of $NaClO_4$, both films prepared under silent and sonicated conditions were permeable to ClO_4^- and smaller anions such as Cl^-, whereas the films depressed penetration of anions with larger diameters such as ethanesulfonate ($EtSO_3^-$), *p*-toluenesulfonate (TsO^-) and 1-naphthalenesulfonate (NS^-). Nevertheless, the ion-sieving behaviour of the films prepared under sonication were superior, as the films more effectively prevented larger ion penetration into the matrix [26].

In the same manner, the Besançon group studied the polymerisation of 5×10^{-3} M in EDOT and 0.1 M in $LiClO_4$ by cyclic voltammetry studies (100 mV s^{-1}) [28]. A potential of 1 V versus SCE was applied to avoid EDOT aromatic cycle overoxidation, and the solutions were kept under an argon flow during the experiments. For successive cycles, the oxidation of the PEDOT film was observed at potentials of circa +0.7 V versus SCE. It is clearly evident that high frequency ultrasound leads to more marked oxidation peaks, corresponding to higher current densities. Thus, during film oxidation, perchlorate counterions are brought more efficiently to the working electrode surface and are better incorporated into the film structure, in turn improving the film oxidation process [28]. Doping level was evaluated by XPS measurements, and the recorded spectra presented exactly the same peaks at similar binding energy levels, in other words, the chemical form of species followed by this method is always the same [28]. PEDOT films electrodeposited under ultrasonic irradiation always provide higher doping levels than under silent conditions. Nevertheless, this observation

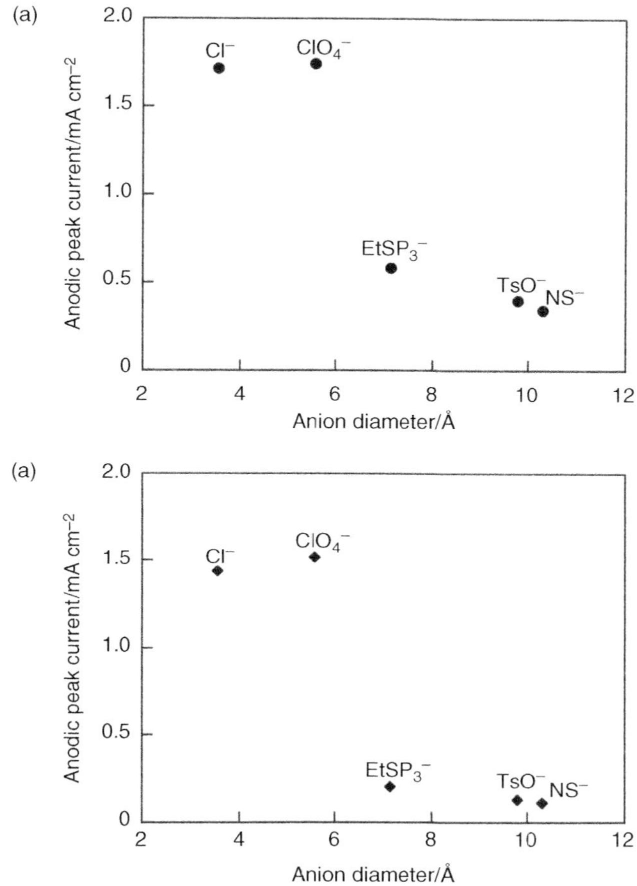

Figure 8.24 *Anodic peak current of linear sweep voltammograms for various anion diameters (a) without and (b) in presence of ultrasound (Reprinted with permission from [26] Copyright (2006) The Electrochemical Society Inc).*

seems to be influenced by other parameters, ignored up to now, for example charge density brought during electrosynthesis as shown in Figure 8.25.

The difference in doping levels between PEDOT films elaborated with and without ultrasound as a function of charge density becomes significant and stabilises at 33% only when the charge density reaches 76 mC cm^{-2}. Before this threshold, doping levels are almost the same regardless of the mode of convection. For low film thickness, perchlorate counterions easily penetrate even without mass-transfer improvement. Thus, the doping level remains nearly the same in the absence and presence of insonation. After a certain film thickness is reached, counterions are incorporated in the film with more difficulty during the electrosynthesis in silent conditions, that is, a 7% decrease in doping level is observed from 15 to 150 mC cm^{-2} on synthesised films. On the contrary, for larger thicknesses, ultrasound facilitates counterion penetration in the film leading to the observed 33% increase in doping level. In these less favourable conditions, that is, larger thicknesses,

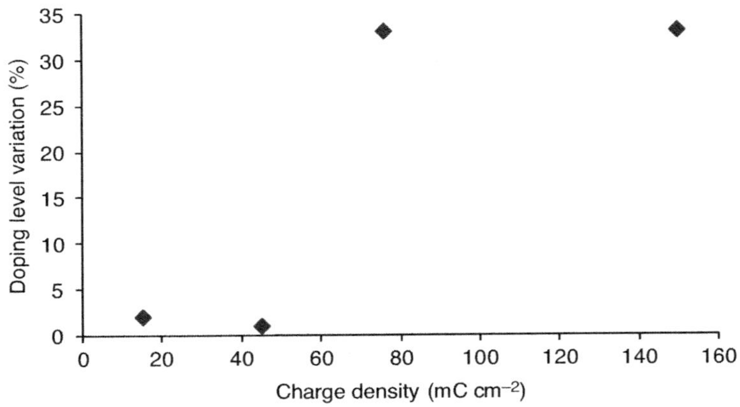

Figure 8.25 *Percentage of variation in doping levels between PEDOT films electrosynthesised in the absence and presence of ultrasound in function of charge density (Reprinted with permission from [27] Copyright (2011) Elsevier Ltd).*

an increase in mass transfer induced by sonication introduces counterions more efficiently into the film structure. An additional effect could be attributed to the small wavelength at high ultrasonic frequency, that is, for a sufficient thickness, the system (film + substrate) will not be completely reflecting, and part of the acoustic energy will be absorbed by the film. In this case, diffusion coefficients of perchlorate ions in the film will increase with film temperature, leading to better penetration into the film matrix [28]. It also seems that radical species (i.e. $OH^{\bullet}$) are responsible for film better oxidation under sonication [53].

Investigations are in progress at the Besançon group concerning the behaviour of 'doping-undoping' properties with various ions, and their mobility and distribution in the film matrix using the powerful technique GD-OES.

8.6.4 Effect on Local Control of Surfaces

Control of a defined area on the surface, that is, the possibility to influence in one direction or in another a chemical reaction in a small zone with a precise location, is the main attribute of the high intensity focused ultrasound sonoreactor (described earlier in this chapter and elsewhere in this book). In previous studies, acoustic power distribution was investigated [54, 55] whereby the very high intensities available in the focal zone were used to interact with an organic film [56]. For example, while controlling the operating parameters, ablation of the organic film in a chosen zone was possible, allowing selective plating [57].

Thus, why not use this high quantity of energy concentrated in a small volume to act on reaction selectivity? The first step, consisting of choice of electrolyte, proved to be crucial for the electrodeposition of polypyrrole on copper due to possible substrate oxidation. Three types of additional salts were tested to ensure the formation of an oxide layer on the metallic surface, condition *sine qua none* to obtain a polymer film [40]. The first type (sodium perchlorate and tartrate) always makes polypyrrole deposition impossible. The second type (sodium salicylate) enables polymerisation whatever the experimental conditions, and the last type (sodium oxalate) allows polymerisation in silent conditions (impossible under sonication). The latter offers a very interesting prospect of 'switch on-off' effects, depending

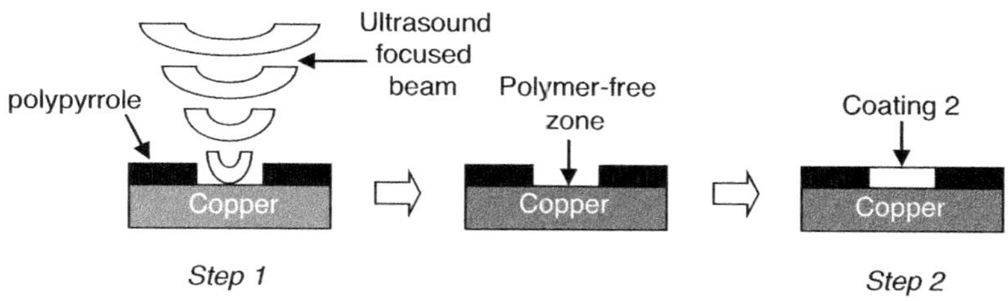

Figure 8.26 *Diagram showing the acoustic masking process (Reprinted with permission from [40] Copyright (2010) Elsevier Ltd).*

on the presence of ultrasound. Nevertheless, flat transducers do not offer the power density capacity (commonly limited to 5 W cm^{-2}) required for complete polymerisation inhibition. Focalisation of ultrasound appears to be very promising for enhancing these effects and locating them on a specific defined area. The recent experimental setups described in Figure 8.11(b) use concave-shaped transducers, allowing wave focalisation and a considerable gain in acoustic energy in a small zone known as the 'acoustic focal area'.

Different transducers of various ultrasonic frequencies are available, but already for 750 kHz, acoustic intensity is 110 times higher, and the gain in pressure is up to 10.5 in the focal zone. These values are only reliable for exposed zones reduced to the minimum (limited to the focal), but if a larger zone of up to a few cm is taken into account, a dedicated calculation throughout the average value of the gain function on the electrode distance has to be considered. A region of the copper substrate can be placed at this location to be exclusively irradiated by the ultrasonic wave (Figure 8.26).

Therefore, in a medium such as sodium oxalate in which electropolymerisation cannot occur under high powered ultrasound, the irradiated zone of the metallic substrate is protected from polymer deposition, making possible selective electropolymerisation of polypyrrole only on chosen parts of the substrate. Ultrasound beam irradiation on the selected zone of the copper surface occurs as soon as the oxidation current for electropolymerisation is applied. During the experiment, polypyrrole deposition is clearly visible all around the protected areas which are circular shaped. Conditions for polymer deposition are the same as before (1.0 V versus SCE for 100 s) [40]. A picture of the substrate obtained is shown in Figure 8.27.

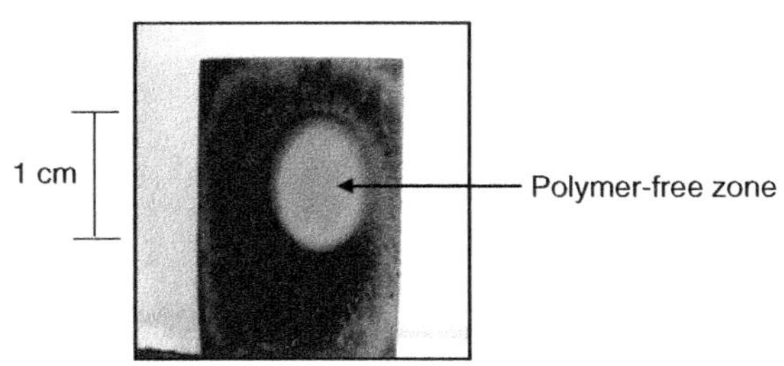

Figure 8.27 *Picture of the copper substrate obtained after ultrasound masking against polymer deposition. The central area which was irradiated by the focused ultrasound beam (750 kHz, 25 W) during electropolymerisation is polymer free (Reprinted with permission from [40] Copyright (2010) Elsevier Ltd).*

Finally in this study, the sample was used to produce organic layers exhibiting various structural properties from the surrounding polypyrrole. For example, the Besançon group succeeded in grafting in the polymer-free area a dodecanthiol self-assembled monolayer [40]. Nevertheless, many other treatments of this copper polymer-free area are possible, for example, from organic to inorganic deposits.

This remarkable novel 'acoustic masking process' offers considerable advantages such as simplicity and rapidity as it is only a one-step process. Furthermore, the size of the selected area that can be obtained by this technique varies from 1 mm to a few cm. It can also be proposed as an alternative to well-known masking techniques such as lithography, which normally is a multiple and more restricting process [58].

References

(1) Shirakawa, H., Louis, E.J., MacDiarmid, A.G., Chiang, C.K. and Heeger, A.J. (1977) Synthesis of electrically conducting organic polymers – halogen derivatives of polyacetylene, (CH)X. *Journal of the Chemical Society, Chemical Communications*, 578.

(2) Chiang, C.K., Fincher, C.R., Park, Y.W., Heeger, A.J., Shirakawa, H., Louis, E.J., Gua, S.C. and MacDiarmid, A.G. (1977) Electrical-conductivity in doped polyacetylene. *Physical Review Letters*, **39**, 1098.

(3) Inzelt, G. (2008) *Conducting Polymers: a New Era in Electrochemistry*, Springer Verlag.

(4) Cho, M.S., Yun, Y.Y., Nam, J.D., Son, Y. and Lee, Y. (2008) Effect of magnetic field on electrochemical polymerization of EDOT. *Synthetic Metals*, **158** (21–24), 1043–1046.

(5) Atobe, M., Murotani, A., Hitose, S., Suda, Y., Sekido, M., Fuchigami, T., Chowdhury, A.N. and Nonaka, T. (2004) Anodic polymerization of aromatic compounds in centrifugal fields. *Electrochimica Acta*, **50** (4), 977–984.

(6) Maret, G. and Dransfeld, K. (1985) *Strong and Ultrastrong Magnetic Fields and their Applications*, Springer, Berlin.

(7) Ximei, W., Liqun, Z., Huicong, L. and Weiping, L. (2008) Influence of surface pretreatment on the anodizing film of Mg alloy and the mechanism of the ultrasound during the pretreatment. *Surface and Coating Technology*, **202**, 4210–4217.

(8) Harzali, H., Baillon, F., Louisnard, O., Espitalier, F. and Mgaidi, A. (2011) Experimental study of sono-crystallisation of $ZnSO_4{\cdot}7H_2O$, and interpretation by the segregation theory. *Ultrasonics Sonochemistry*, **18** (5), 1097–1106.

(9) Luche, J.L. (1993) Sonochemistry from experiment to theoretical considerations, in *Advances in Sonochemistry*, Vol. **3** (ed. T.J. Mason), JAI Press.

(10) Walton, D.J., Iniesta, J., Plattes, M., Mason, T.J., Lorimer, J.P., Ryley, S., Phull, S.S., Chyla, A., Heptinstall, J., Thiemann, T., Fuji, H., Mataka, S. and Tanaka, Y. (2003) Sonoelectrochemical effects in electro-organic systems. *Ultrasonics Sonochemistry*, **10**, 209–216.

(11) Wood, R.W. and Loomis, A.L. (1927) *Philosophical Magazine*, **4**, 417.

(12) Fox, R., Yeager, E. and Havovka, F. (1960) *Journal of the Acoustical Society of America*, **32**, 1499–1502.

(13) Walton, D.J. and Phull, S.S. (1996) *Sonoelectrochemistry, Advances in Sonochemistry*, **4**, 205–284.

(14) Asami, R., Atobe, M., and Fuchigami, T. (2005) Electropolymerization of an immiscible monomer in aqueous electrolytes using acoustic emulsification. *Journal of the American Chemical Society*, **127**, 13160–13161.

(15) Atobe, M., Chowdhury, A.N., Fuchigami, T. and Nonaka, T. (2003) Preparation of conducting polyaniline colloids under ultrasonication. *Ultrasonics Sonochemistry*, **10** (2), 77–80.

(16) Walton, D.J. (1991) *Electronic Materials – From Silicon to Organics* (eds L.S. Miller and J.B. Mullin), Plenum Press.
(17) Walton, D.J., Hall, C.E. and Chyla, A. (1992) *The Analyst*, **117** 1305–1308.
(18) Osawa, S., Ito, M., Tanaka, K. and Kuwano, J. (1987) Electrochemical polymerization of thiophene under ultrasonic field. *Synthetic Metals*, **18**, 145–150.
(19) Osawa, S., Ito, M., Tanaka, K. and Kuwano, J. (1992) The application of ultrasonic waves to the electrochemical polymerization of thiophene. *Journal of Polymer Science – Part B Polymer Physics*, **30**, 19–24.
(20) Atobe, M. and Nonaka, T. (1998) New developments of sonoelectrochemistry – Electroorganic reactions under ultrasonic fields. *Nippon Kagaku Kaishi*, (4), 219–230.
(21) Atobe, M., Kaburagi, T. and Nonaka, T. (1999) Ultrasonic effects on electroorganic processes. XIII. Role of ultrasonic cavitation in electrooxidative polymerization of aniline. *Electrochemistry*, **67** (12), 1114–1116.
(22) Atobe, M., Chowdhury, A.N., Fuchigami, T. and Nonaka, T. (2003) Preparation of conducting polyaniline colloids under ultrasonication. *Ultrasonics Sonochemistry*, **10** (2), 77–80.
(23) Chowdhury, A.N., Atobe, M. and Nonaka, T. (2004) Studies on solution and solution-cast film of polyaniline colloids prepared in the absence and presence of ultrasonic irradiation. *Ultrasonics Sonochemistry*, **11** (2), 77–82.
(24) Park, J.E., Atobe, M., and Fuchigami, T. (2005) Sonochemical synthesis of conducting polymer-metal nanoparticles nanocomposite. *Electrochimica Acta*, **51** (5), 849–854.
(25) Park, J.E., Atobe, M. and Fuchigami, T. (2005) Sonochemical synthesis of inorganic-organic hybrid nanocomposite based on gold nanoparticles and polypyrrole. *Chemistry Letters*, **34** (1), 96–97.
(26) Atobe, M., Tsuji, H., Asami, R. and Fuchigami, T. (2006) A study on doping-undoping properties of polypyrrole films electropolymerized under ultrasonication. *Journal of the Electrochemical Society*, **153** (1), D10–D13.
(27) Et Taouil, A., Lallemand, F., Hihn, J.-Y. and Blondeau-Patissier, V. (2011) Electrosynthesis and characterization of conducting polypyrrole elaborated under high frequency ultrasound irradiation. *Ultrasonics Sonochemistry*, **18** (4), 907–910.
(28) Et Taouil, A., Lallemand, F., Hihn, J.-Y., Melot, J.-M., Blondeau-Patissier, V. and Lakard, B. (2011) Doping properties of PEDOT films electrosynthesized under high frequency ultrasound irradiation. *Ultrasonics Sonochemistry*, **18** (1), 140–148.
(29) Nakabayashi, K., Amemiya, F., Fuchigami, T., Machida, K., Takeda, S., Tamamitsu, K. and Atobe, M. (2011) Highly clear and transparent nanoemulsion preparation under surfactant-free conditions using tandem acoustic emulsification. *Chemistry Communication*, **47**, 5765–5767.
(30) Sarma, T.K., Chowdhury, D., Paul, A. and Chattopadhyay, A. (2002) Synthesis of Au nanoparticle–conductive polyaniline composite using H_2O_2 as oxidising as well as reducing agent. *Chemistry Communication*, **10**, 1048–1049.
(31) Yang, J. and Li, Y.F. (1996) Effect of ultrasonic wave on the electropolymerization of pyrrole. *Chinese Journal of Polymer Science*, **14** (3), 270–276.
(32) Kojima, Y., Koda, S. and Nomura, H. (2001) Effect of ultrasonic frequency on polymerization of styrene under sonication. *Ultrasonics Sonochemistry*, **8** (2), 75–79.
(33) Galal, A. (2006) Characterization of conducting poly(3-methylthiophene) films prepared under sono-electrochemical conditions. *Journal of Applied Polymer Science*, **102** (3), 2416–2425.
(34) Reyman, D., Guereca, E. and Herrasti, P. (2007) Electrodeposition of polythlophene assisted by sonochemistry and incorporation of fluorophores in the polymeric matrix. *Ultrasonics Sonochemistry*, **14** (5), 653–660.
(35) Kowalski, D., Ueda, M. and Ohtsuka, T. (2008) The effect of ultrasonic irradiation during electropolymerization of polypyrrole on corrosion prevention of the coated steel. *Corrosion Science*, **50**, 286–291.

(36) Et Taouil, A., Lallemand, F., Melot, J.-M., Husson, J., Hihn, J.-Y. and Lakard, B. (2010) Effects of polypyrrole modified electrode functionalization on potentiometric pH responses. *Synthetic Metals*, **160** (9–10) 1073–1080.
(37) Et Taouil, A., Mahmoud, M.M., Lallemand, F., Lallemand, S. and Gigandet, M.-P. (2011) Corrosion protection of sonoelectrodeposited organic films on zinc coated steel. *Ultrasonics Sonochemistry* (in press).
(38) Et Taouil, A., Lallemand, F., Morand-Villeneuve, N., Melot, J.-M., Lakard, S., Hihn, J.-Y. and Lakard, B. (2011) Electrosynthesis and characterization of conducting poly(11-N-pyrrolylundecanoic) acid films on FTO substrates. Effect of functionalization on cell proliferation. *Journal of Chemistry and Chemical Engineering* **5** (12).
(39) Hihn, J.-Y., Doche, M.-L., Mandroyan, A., Hallez, L. and Pollet, B.G. (2011) Ultrasound and Better Reactor design, Chapter 23 of Handbook on Applications of Ultrasound: Sonochemistry for Sustainability, (eds D. Chen, S.K. Sharma and A. Mudhoo), CRC Press Taylor & Francis.
(40) Taouil, A.E., Lallemand, F., Hallez, L. and Hihn, J.Y. (2010) Electropolymerization of pyrrole on oxidizable metal under high frequency ultrasound irradiation. Application of focused beam to a selective masking technique. *Electrochimica Acta*, **55** (28), 9137–9145.
(41) Pollet, B.G., Hihn, J.Y, Doche, M.L., Lorimer, J.P., Mandroyan, A. and Mason, T.J. (2007) Transport limited currents close to an ultrasonic horn equivalent flow velocity determination. *Journal of Electrochemical Society*, **154** (10), E131–E138.
(42) Taouil, A.E. (2011) Effets des ultrasons haute fréquence sur l'électrosynthèse des polymères conducteurs. PhD Thesis, Université de Franche-Comté, Besançon, France.
(43) Chiang, C.K., Fincher, C.R., Park, Y.W., Heeger, A.J., Shirakawa, H., Louis, E.J., Gua, S.C. and MacDiarmid, A.G. (1977) Electrical-conductivity in doped polyacetylene. *Physical Review Letter*, **39**, 1098–1001.
(44) Dimitrakopoulos, C.D. and Malenfan, P.R.L. (2002) Organic thin film transistors for large area electronics. *Advanced Materials*, **14** (2), 99–117.
(45) Horowitz, G. (1998) Organic field-effect transistors. *Advanced Materials*, **10** (5), 365–377.
(46) Yu, G., Gao, J., Hummelen, J.C., Wudl, F. and Heeger, A.J. (1995) Polymer photovoltaic cells: enhanced efficiencies via a network of internal donor-acceptor heterojunctions. *Science*, **270**, 1789–1791.
(47) Argun, A.A., Cirpan, A. and Reynolds, J.R. (2003) The first truly all-polymer electrochromic devices. *Advanced Materials*, **15** (16), 1338–1341.
(48) Cui, X., Wiler, J., Dzaman, M., Altschuler, R.A. and Martin, D.C. (2003) In vivo studies of polypyrrole/peptide coated neural probes. *Biomaterials*, **24**, 777–787.
(49) Hallik, A., Alumaa, A., Kurig, H., Janes, A., Lust, E. and Tamm, J. (2007) On the porosity of polypyrrole films. *Synthetic Metals*, **157**, 1085–1090.
(50) Lakard, S., Morrand-Villeneuve, N., Lesniewska, E., Lakard, B., Michel, G., Herlem, G., Gharbi, T. and Fahys, B. (2007) Synthesis of polymer materials for use as cell culture substrates. *Electrochimica Acta*, **53**, 1114–1126.
(51) Lakard, B., Ploux, L., Anselme, K., Lallemand, F., Lakard, S., Nardin, M. and Hihn, J.Y. (2009) Effects of ultrasound on the electrochemical synthesis of polypyrrole, application to the adhesion and growth of biological cells. *Bioelectrochemistry*, **75**, 148–157.
(52) Tsuchiya, S., Ueda, M. and Ohtsuka, T. (2007) Polypyrrole coating on zinc for corrosion prevention of zinc-coated steels. *Isij International*, **47**, 151–156.
(53) Et Taouil, A., Lallemand, F., Hihn, J.-Y., Hallez, L., Moutarlier, V., Blondeau-Patissier, V. (2011) Relation between structure and ions mobility in polypyrrole electrosynthesized under high frequency ultrasound irradiation. *Electrochimica Acta* (in press).
(54) Hallez, L., Touyeras, F., Hihn, J.Y. and Klima, J. (2007) Energetic balance in an ultrasonic reactor using focused or flat high frequency transducers. *Ultrasonics Sonochemistry*, **14** (6), 739–749.

(55) Hallez, L., Touyeras, F., Hihn, J.-Y., Klima, J., Guey, J.-L., Spajer, M. and Bailly, Y. (2010) Characterization of HIFU transducers designed for sonochemistry application: Cavitation distribution and quantification. *Ultrasonics*, **50** (2), 310–317.
(56) Hallez, L., Touyeras, F., Hihn, J.Y. and Bailly, Y. (2010) Interactions H.I.F.U. / polymer films. *Physics Procedia*, **3** (1) 179–184.
(57) Rochon, S., Hihn, J.Y., Hallez, L. and Touyeras, F., Procédé de dépôt sélectif d'un métal précieux sur un support par ablation ultrasonore d'un élément de masquage et son dispositif. *International Patent* WO 2009/150083A1.
(58) Lallemand, F., Hihn, J.Y., Hallez, L. and Et Taouil, A., Modification de propriétés de surface de polymères conducteurs par irradiation ultrasonore focalisée. *French Patent* PCT/EP2009/067588.

9

Sonoelectrochemical Production of Nanomaterials

Jonathan P. Metters and Craig E. Banks

In this chapter we overview the application of power ultrasound applied with electrochemistry which facilitates the design and fabrication of novel nanomaterials allowing unique nanoparticle morphologies to be readily obtained that can sometimes be '*un*obtainable' using conventional approaches.

9.1 Introduction

The recent introduction of nano-sized materials has found to induce significant advantages when applied to a range of technological fields due to their size-dependent electronic, chemical [1], magnetic [2, 3] and optical [4, 5] properties. Consequently, the technology has been applied in a vast array of areas within industry, with its list of applications ever growing. Applications of nanotechnology include, for example, in the field of medicine, where gold nanoparticles are being used in order to allow targeted delivery of drugs to tumours [6]; electronic merchandise with LCD displays in television screens where nanotube-based conductive films are employed [7]; research into the application of nanomaterials for the introduction of vitamins and other nutrients into foods without affecting the taste or odour of the product through encapsulation within nanoparticles [8]; and in hydrogen and fuel cell technologies where hydrogen is generated using nanoparticle-coated electrodes allowing for improved efficiency of hydrogen generation by electrolysis [9] (see Chapter 10) to name just a few.

Power Ultrasound in Electrochemistry: From Versatile Laboratory Tool to Engineering Solution, First Edition. Edited by Bruno G. Pollet.

Commercial companies developing nanomaterials for implementation into their technology through tailoring their physical and chemical properties have found that the cost is hugely prohibitive and consequently there is an enormous research interest in the area of fabricating novel nanomaterials. A range of methodologies exist to fabricate nanomaterials, for example, vapour deposition, chemical reduction, radiation methods [10, 11] and so on. The cost of each approach varies depending upon a whole range of parameters and while methodologies may work well in a research lab, the scalability of a given technique may be cost prohibited and time consuming. Another approach available to a researcher in order to fabricate nanoparticles lies in the application of electrochemistry.

In this approach the application of a suitable electrode material held at an appropriate electrode potential immersed into a solution containing the target metal salt (and electrolyte) will result in the electrodeposition of the target metal onto the electrode surface, as described by Equation (9.1):

$$M^{n+}(aq) + ne^- \rightarrow M(electrode) \tag{9.1}$$

Typically this can be reversed, such that the electrodeposited metal is 'voltammetrically' stripped back to the corresponding ion (see Chapter 3) which forms the basis of the electroanalytical sensing of the target analyte. Focusing on the electrodeposition of target metal ions to form metallic domains (Equation 9.1), the *in situ* formed metallic domains can be used as an electrode material. Figure 9.1 demonstrates the *in situ* observation of the nucleation of metallic cobalt grown on a boron-doped diamond electrode as a function of applied deposition time which allows the growth rate of nuclei to be elegantly determined [12].

Following modification, the electrode is 'heterogeneous' consisting of two electrode materials, that is to say, the underlying electrode and the newly formed metallic domains, both exhibiting their own heterogeneous rate constant (k^o) which would exhibit towards an electrochemical active species such as $Ru(NH_3)_6{}^{3+}$ in a suitable electrolyte such as 1 M KCl, two voltammetric waves reflecting the difference seen in the k^o values from the two contributing electrode domains. However, typically in many electrochemical applications the metallic domains are only designed to be electrocatalytic towards the target analyte. The electrodeposited metallic domain is readily controlled by various parameters such as applied deposition time and potential and is also influenced by the solution and electrode composition whilst further being dependent upon the thermodynamic and kinetic parameters of the target metal ion. A key parameter is the spacing between domains and the coverage of the nanoparticles on the electrode surface. For example, it has been shown that for the electrochemical reduction of hydrogen peroxide (H_2O_2) at silver nanoparticle modified electrodes where the silver acts as a working electrode domain, the voltammetric response is greatly affected. A decrease in the size of the nanoparticles results in a shift of the voltammetric peak to a more negative potential, while increasing the coverage results in a shift to a more positive potential [13]. For a theoretical analysis, we invite the reader to examine the fundamental work by Belding *et al.* [14]. It is important to note that many applications, as mentioned earlier, require the nanomaterials to be 'free', usually in solution, rather than being immobilised and it is this methodology, sonoelectrochemistry, which permits this. Consequently in this chapter we will explore the synergy of power ultrasound with electrochemical methodologies to produce advantageous nanomaterials.

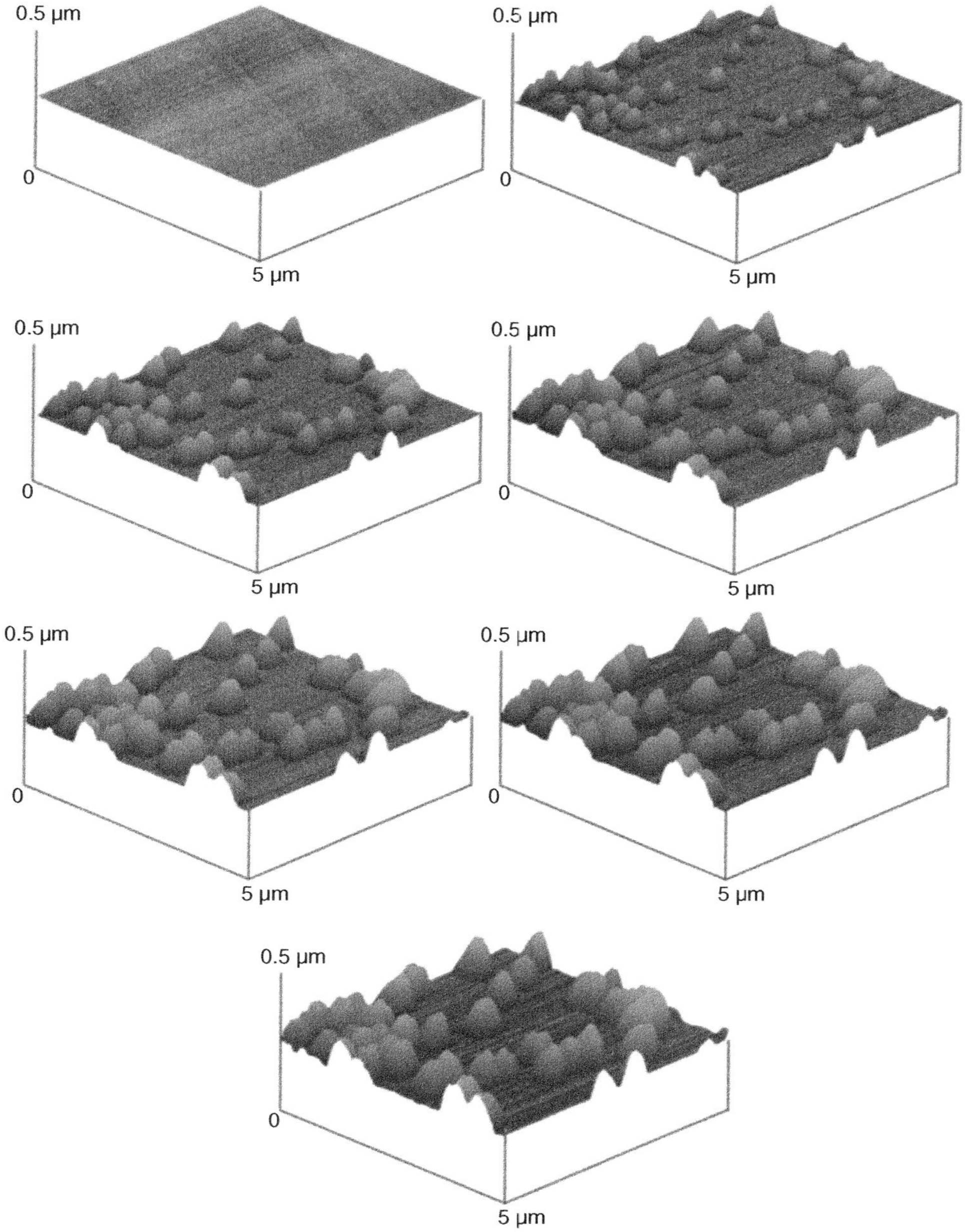

Figure 9.1 In situ *AFM images of the growth of cobalt nuclei on boron-doped diamond using a deposition potential of – 1.15 V for 10 second periods (Reprinted with permission from [12] Copyright (2006) Wiley).*

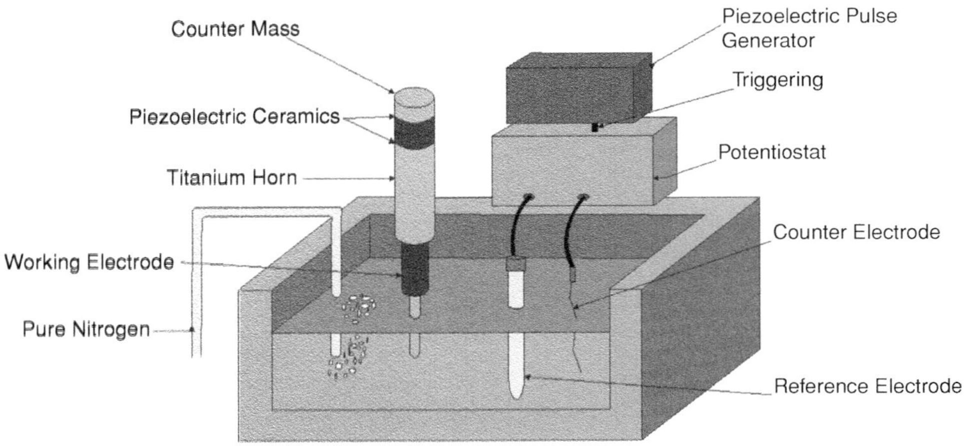

Figure 9.2 *A 3-dimensional representation of the sonoelectrochemical experimental setup used for fabricating nanomaterials.*

9.2 Experimental Configurations

The sonoelectrochemical approach for producing micro- and nano-sized materials uses a configuration originally developed by Reisse and co-workers [15], where rather than using an electrode in the vicinity of an ultrasonic source, the tip of the ultrasonic probe is also utilised as an electrode; such an approach is shown schematically in Figure 9.2.

A pulsed mode is usually used, with the time of applied ultrasound, t_{us}, typically varying between 100 and 1000 ms and where each pulse is conveniently triggered by the potentiostat. The horn tip of the ultrasonic source acts as the working electrode, which can be polarised cathodically or anodically where the potentiostat is used in the pulse mode, with an on time, t_{on}, ranging between 1 and 1000 ms and an off time, t_{off}, ranging from 100 and 1000 ms. Figure 9.3 shows a typical representation of the applied electrochemical and ultrasonic pulses used in the sonoelectrochemical synthesis of nanomaterials.

The nanomaterials are fabricated by the out of phase pulses and after each electrochemical pulse, a signal is sent from the potentiostat to the generator to start the ultrasound pulse. The duration of this ultrasonic pulse is shorter than the t_{off} period. During the electrochemical pulse, nuclei are formed on the titanium surface of the ultrasonic horn tip. It has been shown that the higher the current density or the overpotential, the smaller the nuclei and the faster the process of nucleation occurs [16]. During the applied ultrasonic pulse, mechanical shaking removes the *in situ* formed nuclei, cleaning the surface in preparation for the next nuclei to be formed and also replenishing the diffusion layer via acoustic streaming at the electrode surface [17]. It is this beneficial approach that allows nanomaterials to be continuously formed into the solution that may be collected and utilised in a plethora of areas and it is this ability that allows the approach to provide the scalable production of nanomaterials rapidly in large quantities. It is interesting to note, however, that some authors consider cavitational activity either the sole origin or at least a contribution to the detachment of the *in situ* formed nanomaterials.

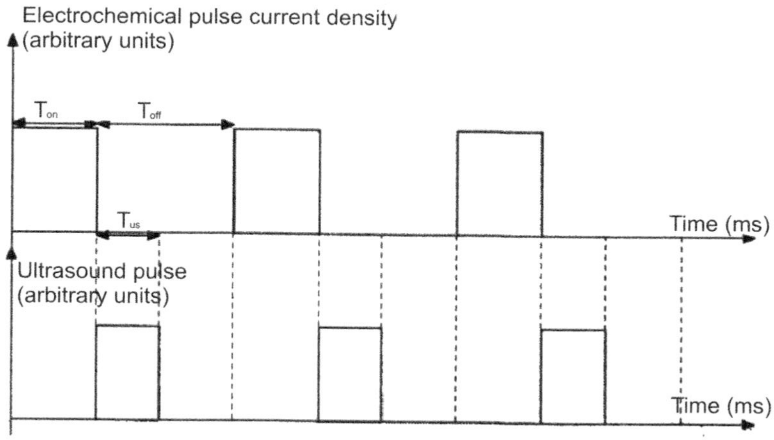

Figure 9.3 A representation of electrochemical and ultrasound pulses (Reprinted with permission from [78] Copyright (2010) Elsevier Ltd).

Related to this area is sonochemistry without the application of electrochemistry, which can be used to beneficially produce nanomaterials (see Chapter 10). Interested readers are directed to the authoritative work of Gedanken [18] who has overviewed this area, which has been shown to be superior for the preparation of amorphous products in the nanometre range and which has been shown to be useful for: the insertion of nanomaterials into mesoporous materials producing a smooth layer of the required catalyst inside the structure from applying ultrasound; immobilisation of nanoparticles onto polymeric and ceramic surfaces; and the formation of proteinaceous microspheres and nanospheres, where for example, tetracycline can be encapsulated while retaining biological activity [18].

In this chapter we focus upon the synergy of power ultrasound with electrochemistry in the configuration where the ultrasonic horn tip has a dual function of not only providing the ultrasonic source but also as the electrode, which greatly simplifies the protocol allowing some very interesting nanomaterial morphologies to be derived. Here, we overview this area with pertinent literature examples.

9.3 Pure Metals

9.3.1 Cobalt, Iron and Nickel

Delplancke and co-workers explored the fabrication of cobalt, iron, and nickel nanomaterials utilising a t_{on} of 600 ms, t_{off} of 400 ms with a t_{us} of 300 ms at a constant current density of 2000 A m^{-2} which was shown to usefully produce nickel (fcc) particles with an average diameter of 50 nm with a deposition current yield of 60% over a 3-hour duration [16].

9.3.2 Silver

Zhu *et al.* [19] have reported the sonoelectrochemical fabrication of silver nanoparticles using a silver salt in the presence of nitrilotriacetate. Using a combination of silver salt concentration with different electrosonication times ranging from 5 to 25 minutes, the silver nanomaterial morphology can be readily tailored from spherical type nanoparticles through to rods and dendritic nanostructures; clearly the sonoelectrochemical methodology allows a range of nanomaterial morphologies to be derived through careful experimental choice. Other studies have reported amorphous silver from aqueous solutions of AgBr in gelatine [20] and Socol *et al.* [21] have attempted to explore the underlying sonoelectrochemical mechanism while Tang *et al.* [22] have reported a poly(*N*-vinylpyrrolidone) (PVP) assisted approach (see Chapter 10).

9.3.3 Copper

Haas *et al.* [23] have reported the size-controlled synthesis of spherical copper nanoparticles in aqueous solutions using PVP as a stabiliser. Mono-dispersed copper nanoparticles in the range of 25–60 nm have been shown to be possible where the size of the copper particles were shown to be readily controlled by the time of reaction, current density, temperature and ultrasonic power [23]. The same authors reported separately that in the presence of poly(*N*-chloroethanediyl) (PVC) and PVP metallic copper is produced, but in the presence of CTAB (cetyltrimethylammonium bromide), a mixture of 95 wt% CuBr and 5 wt% metallic copper is produced [23], which demonstrates how changing the stabiliser may alter the particle composition. Furthermore, the authors also reported [24] that copper dendrites could be obtained using PVC. Lastly, Mancier *et al.* [25] were able to achieve copper(I) oxide nanoparticles in the range of ~8 nm from a copper(II) lactate electrolyte solution.

9.3.4 Magnesium

One notable use of sonoelectrochemistry is the formation of magnesium nanoparticles reported recently by Haas and Gedanken [26]. Magnesium attracts strong interest because of its high performance applications, with hydrogen storage materials being one of the most significant [26–33] since storage capacity is limited by material weight and consequently significant attention has been directed to lightweight materials such as magnesium. Synthetic methodologies that may be utilised for fabricating magnesium nanoparticles are: cryochemistry [34], gas-phase synthesis [35], deposition from ethereal solutions [36–38] and deposition from high-temperature molten salts [39] such as $MgCl_2$ [26]. Haas and Gedanken used a cathodic current density of 5 mA cm^{-2} with a pulse duration of 600 seconds, t_{off} = 600 ms, and t_{on} = 300 ms and an ultrasonic power of 76 W [26]. This sonoelectrochemical approach efficiently highlights the sole purpose of this methodology where species that cannot be easily reduced by chemical means can be reduced via the application of a large negative reduction potential which exceeds the standard reduction potential (SRP) plus an additional driving force in an applied overpotential [26]. In the case of preparing Mg nanoparticles, a key limitation is the choice of solvent due to the electrophilic nature where polar aprotic solvents readily react and hence Grignard solutions can only be prepared in ethers, such as tetrahydrofuran (THF) and dibutyldiglyme (DBDG). Haas and Gedanken have demonstrated that magnesium nanoparticles could be readily formed from Bu^tMgCl in a 1.45 M dibutyldiglyme solution without $AlCl_3$ [26], clearly showing aggregated particles consisting of smaller ones (Figures 9.4 and 9.5).

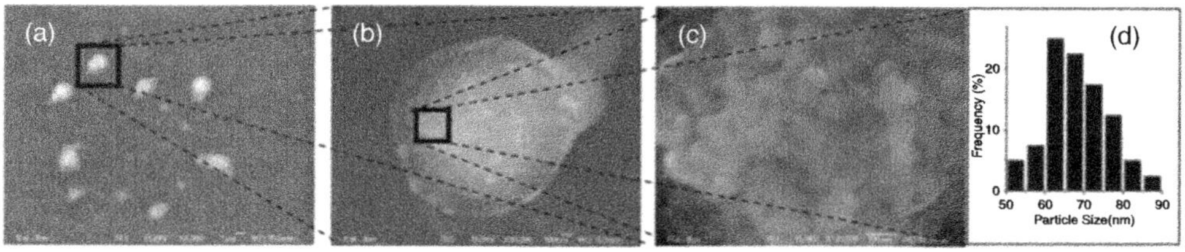

Figure 9.4 SEM images of magnesium nanoparticles produced sonoelectrochemically from BuMgCl in a 1.45 M dibutyldiglyme solution, without $AlCl_3$. (a)–(c): images are from low magnification to high magnification. (d) A histogram of the size of individual magnesium nanoparticles (image (c)) (Reprinted with permission from [26] Copyright (2008) Royal Society of Chemistry).

Further resolution with HR-TEM, as depicted in Figure 9.5, reveal that actually the Mg nanoparticles are of 4.5 ± 0.5 nm in size.

9.3.5 Aluminium

Other notable work worth highlighting has been reported by Mahendiran *et al.* where the above approach for magnesium has been further extended towards aluminium nanoparticles [40]. Al metals can be prepared electrochemically by the Hall-Heroult process which uses a temperature of ~1233 K using a carbon anode and an Al pool as the cathode, with a notable drawback being the high energy consumption along with the emission of greenhouse gases such as CO_2, CO and CF_6 and also rigorous reaction conditions that need to be applied [40]. Recently, various methods for the preparation of aluminium nanoparticles have been reported [41–43]. For example, Pomfret *et al.* [44] have reported on the synthesis of Al nanorods with a length of 2–4 μm and a width of 10–30 nm by the electrochemical

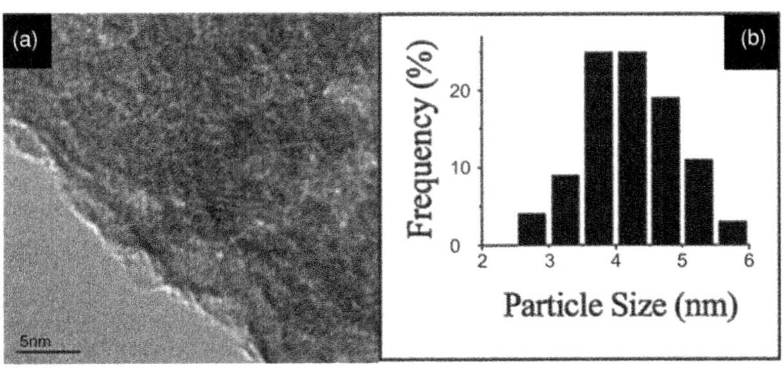

Figure 9.5 HR-TEM image of magnesium nanoparticles. (a) Magnesium nanoparticles produced sonoelectrochemically from BuMgCl in a 1.45 M DBDG solution, without $AlCl_3$. (b) A histogram of the size of individual magnesium nanoparticles (Reprinted with permission from [26] Copyright (2008) Royal Society of Chemistry).

template deposition method while Al particles 10–50 nm in size have also been prepared using the laser ablation method [45]. Abedin *et al.* [46] have reported the synthesis of nano- and micro-crystalline aluminium particles in the size range of 20–35 nm by using room temperature ionic liquids (RTILs). These methods do, however, suffer from numerous drawbacks [40]: (i) it is difficult to remove the template; (ii) a complicated experimental setup is required; (iii) expensive water-stable RTILs must be used; and (iv) the sizes of the aluminium particles prepared by the discussed methods are very large. The preparation of aluminium nanoparticles with sizes in the range of 10–20 nm by using a simple preparation method, at room temperature, is still a challenge to the scientific community [40].

Using the sonoelectrochemical approach, a cathodic current density of 100 mA cm^{-2} was applied to a solution of 4 g of $AlCl_3$ dissolved in 80 mL of a 1 M solution of $LiAlH_4$ in THF for 4 hours [40]. The duration of the current pulse was 600 s, and the t_{off} of the current phase was 60 s. with the duration of the ultrasonic pulse being 240 ms at a ultrasonic power of 76 W [40]. Applying this methodology enabled the production of nanoparticles in the size range of 25 $\pm$ 8 nm [40].

9.3.6 Lead and Cadmium

The production of highly dispersed, spherical, pure lead nanoparticles and dendritic structures in the presence of CTAB has been reported by Chen *et al.* [47]. The synthesis of cadmium selenide (CdSe) has also been reported up to sizes of 9 nm (diameter) using cadmium salts complexed with potassium nitrilotriacetate [1] which has been extended to PbSe [48] of ~12 nm particles using a similar approach.

9.3.7 Core Shell Nanoparticles

Recently Mancier and co-workers [49] have demonstrated Cu-Ag core shell nanoparticles are viably fabricated through the utilisation of sonoelectrochemistry. Bimetallic materials may have better or different chemical [1], magnetic [2, 3], optical [4, 5] and catalytic [50–53] properties compared to pure metals due to the nanoscale size of the layers and it has been reported that advantages of core shell structures are further improved: On the one hand, the replacement of an expensive reactive metal in the centre of the particles by another less expensive material makes it possible to achieve substantial economies (the only surface atoms of the nanoparticles are involved in the catalytic processes) while [54–55] on the other hand, when the core metal is magnetic, it may be possible to recover an expensive shell metal by simple magnetisation. The Cu-Ag system is of interest due to its the bactericidal properties of the external silver allowing potential interesting medical applications such as 'smart textiles' as wound dressings, contraceptive devices, surgical instruments and bone prostheses [56–58] with the lower cost compared with particles consisting entirely of silver. The approach is based on the sonoelectrochemical formation of copper, which is then extracted and added into a 'silver solution' in order to induce a displacement reaction which is as follows:

$$2Ag^+ + Cu \rightarrow 2Ag + Cu^{2+} \quad (9.2)$$

Using several displacement times from 0 to 476 minutes, copper core silver shell nanoparticles have been achieved.

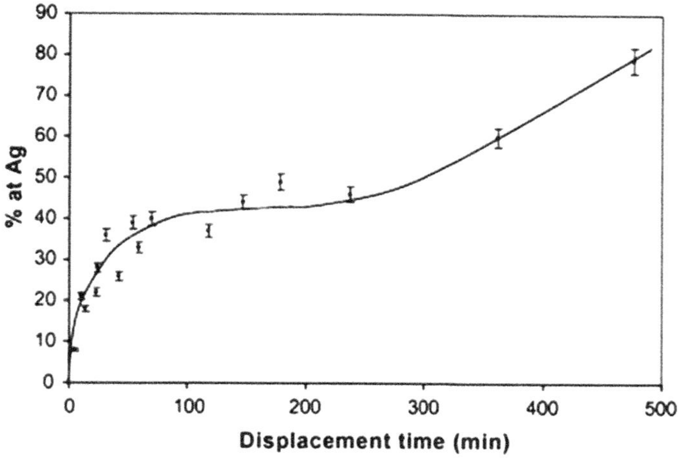

Figure 9.6 Evolution of the silver atomic percentage with displacement time (Reprinted with permission from [49] Copyright (2010) Elsevier Ltd).

Figure 9.6 demonstrates the effect of the applied displacement time as a function of % atomic silver as measured through XRD and complementary quartz crystal microbalance (QCM) measurements.

Figure 9.7(a) and (b) depict TEM images of Cu-Ag particles obtained after 476 minutes as an example revealing that a broad size range of the nanomaterials are produced. The nanoparticle diameters vary between ~10 and ~80 nm. Note that the diameters are larger

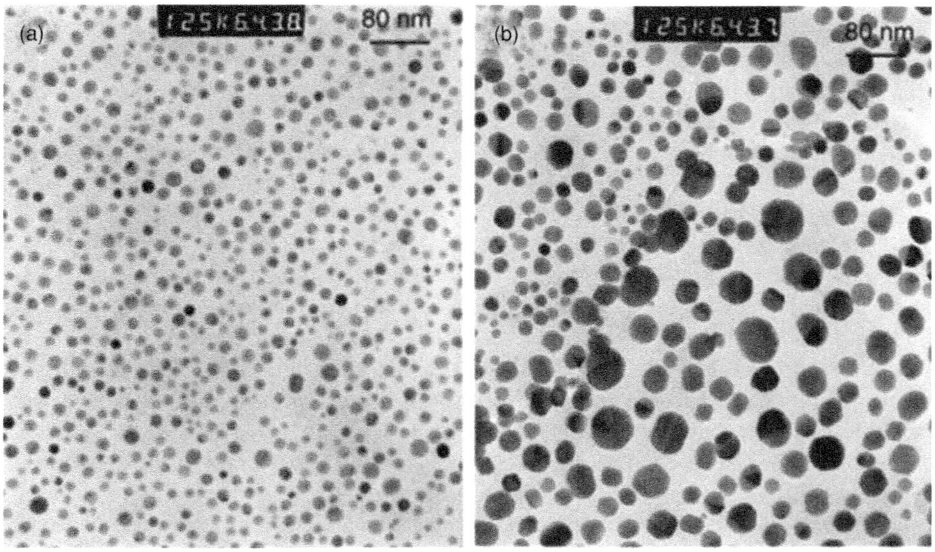

Figure 9.7 (a) and (b) show transmission electron micrographs of a Cu–Ag nanoparticles in sample (time = 476 minutes) (Reprinted with permission from [49] Copyright (2010) Elsevier Ltd).

than that of the initial copper powder because of the stoichiometric coefficients of the displacement reaction where one copper atom is 'replaced' by two silver atoms [49]. Furthermore, some Cu-Ag particles are joined and form small agglomerates [Figure 9.7(b)] which is consistent with the literature where nanoparticles tend to form agglomerates [52, 59–62].

9.3.8 Gold

Recently Shen and co-workers have reported on the sonoelectrochemical synthesis of stabiliser-'free' gold nanoparticles. Gold nanoparticle synthesis was first reported in 1857 by Faraday using an aqueous solution containing gold salt ($AuCl_4^-$) and phosphorus [63]. Since this pioneering work, a range of methods have been reported for fabricating gold nanoparticles such as metal evaporation [64], laser ablation [65], sonochemical [66] and electrochemical techniques, allowing size-selective gold nanoparticles to be readily fabricated by applying suitable deposition potentials. Another approach is the wet-chemical synthesis approach originally developed by Turkevich [67, 68] and later refined by Frens [69, 70], which employs an aqueous solution containing a gold salt and citrate as a reducing agent and also acts to ensure that the gold nanoparticles do not coagulate. Variants on this have been reported and recent fundamental studies have shown that formation goes through nanowires as an intermediate before forming nanoparticles [71]. For the reason of coagulation, many researchers are using dispersants [72–74] to synthesise nanoparticles [49]. For example, investigations involving PVP and a PEO disulfide polymer synthesised by anionic polymerisation, with the former allowing 10–50 nm diameters ranges while the latter stabilised particles up to 35 nm [72], have been reported.

However, such stabilisers and reducing agents are reported not to be compatible. Hence the move towards gold nanoparticles, which do not have this limitation but instead remain colloidal rather than agglomerating [75]. Using the sonoelectrochemical approach, with a pulse width of applied current of 0.5 s followed by a resting period of 0.5 s and t_{us} of 0.5 s and employing a current intensity of $\sim$16.3 mA cm^{-2}, an ultrasonic power of approximately 20 W, and a reaction time of 1.5 hours, spherical gold nanoparticles of low nanometre range were produced [75]. The reduction is shown to be pH dependent with low pH values of around 2 failing to enable production of gold nanoparticles; the effect of pH on nanoparticle formation is depicted in Figure 9.8.

The pH value was found to have an important effect on the size and dispersion of the produced gold nanoparticles. Figure 9.8 shows the TEM images of gold nanoparticles prepared in aqueous solutions at various pH values. When the pH was 4, the agglomerated gold nanoparticles with a diameter in the range 20–100 nm were obtained [Figure 9.8(a)]. As the pH was increased to 7, the produced gold nanoparticles were also aggregated [Figure 9.8(b)], and the size was in the range of 10–50 nm. In solution of pH 10, uniform gold nanoparticles with a diameter of $\sim$20 nm were formed [Figure 9.8(c)]. As the pH value was continuously increased to 12, the gold nanoparticles with a polyhedral shape were obtained [Figure 9.8(d)]. As discussed, a reduction in pH results in *no* gold nanoparticles being formed, which is likely to be due to the Au^{3+} not being electrochemically reduced at the pH value, because Cl^- anions are co-ordinating with Au^{3+}. inhibiting electro-reduction; additionally H^+ ions are more easily electrochemically reduced at a lower pH value, and the production of gold nanoparticles is severely inhibited [75].

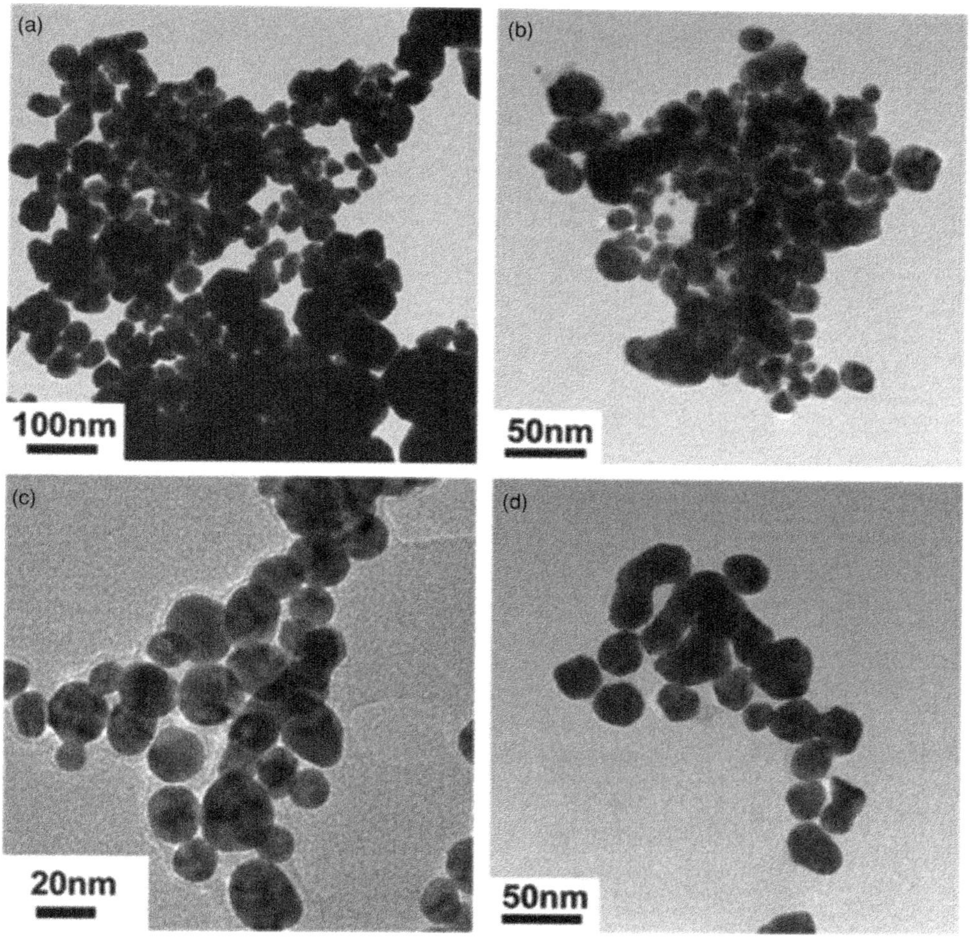

Figure 9.8 *TEM images of thegold nanoparticles synthesised by sonoelectrochemical method at different pH values: (a) pH 4; (b) pH 7; (c) pH 10; (d) pH 12 (Reprinted with permission from [75] Copyright (2010) Elsevier Ltd).*

Diligently, the authors explored the various experimental parameters used in the above work. In the first instance, current was explored over the range of 5–40 mA and was found have little effect on the morphology of the gold nanoparticles. The effect of pulse time of the current and insonation was also explored and found to be significant on the geometry and magnitude of the gold nanoparticles formed.

Figure 9.9 shows the TEM images of the gold nanoparticles at several pulse times ($t_{current}/t_{sono}$ = 0.3 s/0.3 s, 0.4 s/0.4 s, 0.5 s/0.5 s and 1.0 s/1.0 s) at pH 10. From Figure 9.9(a), it can be observed that, at the pulse time of $t_{current}/t_{sono}$ = 0.3 s/0.3 s, the products are gold nanoparticles with a diameter of ~10 nm with a few larger gold nanoparticles (~30 nm). The reason is as the pulse time is shorter, the gold nanoparticles

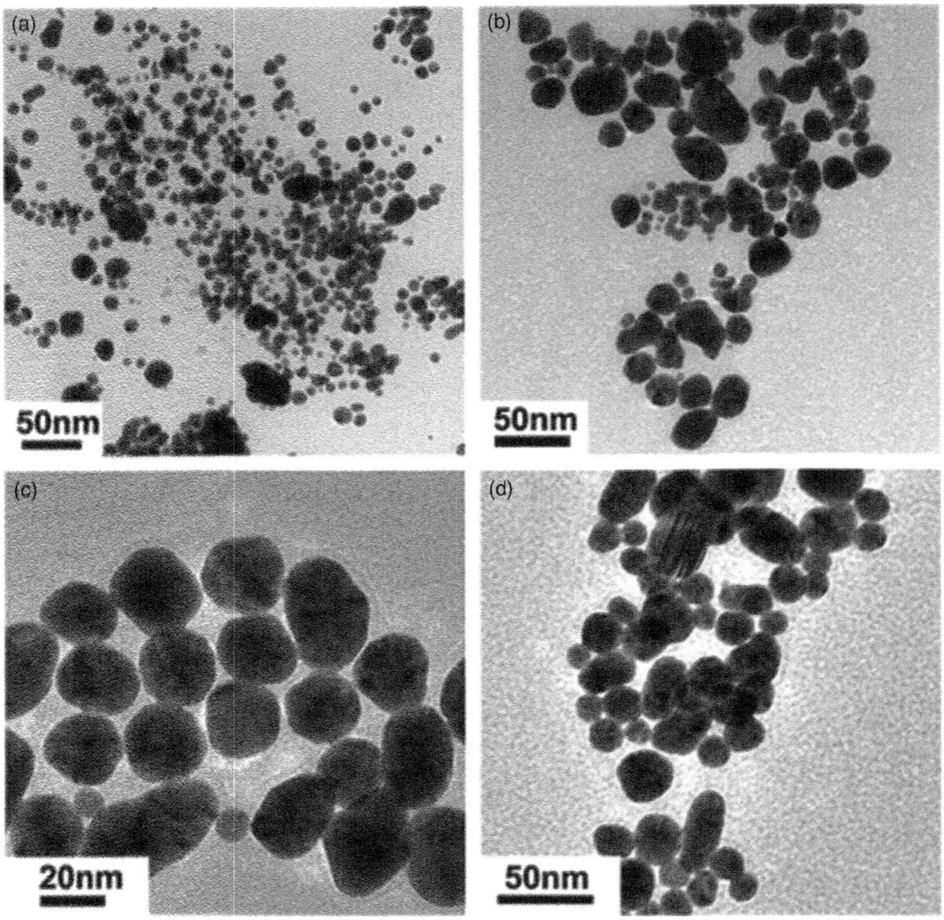

Figure 9.9 *TEM images of thegold nanoparticles synthesised by sonoelectrochemical method at different pulse times ($t_{current}/t_{sono}$): (a) 0.3 s/0.3 s; (b) 0.4 s/0.4 s; (c) 0.5 s/0.5 s; (d) 1.0 s/1.0 s (Reprinted with permission from [75] Copyright (2010) Elsevier Ltd).*

formed on the sonoelectrode is smaller. Smaller gold nanoparticles are attached on the electrode tightly and grow up until they are removed by ultrasound. When the pulse time is increased to $t_{current}/t_{sono} = 0.4$ s/0.4 s, the nanoparticle size is increased to ~20 nm [Figure 9.9(b)]. This means that the size of gold nanoparticles is increased with increases in the pulse time of current. From Figure 9.9(c), it can be seen that at the pulse time of $t_{current}/t_{sono} = 0.5$ s/0.5 s, uniform gold nanoparticles with a diameter of ~20 nm are obtained. As the pulse time is increased further to $t_{current}/t_{sono} = 1.0$ s/1.0 s, the particles increase up to ~20–50 nm and are aggregated together.

The reaction time was also explored; especially at 50 minutes, gold nanoparticles with a diameter of ~30 nm were obtained. At a reaction time of 70 minutes, the average diameter of the gold nanoparticles was decreased to 25 nm. When the reaction time was prolonged to 90 minutes, the uniform spherical gold nanoparticles with a diameter of ~20 nm were produced [75]. This work was explored towards the direct electrochemistry of horse radish peroxidase (HRP) forming a biosensor towards H_2O_2; such an approach clearly indicates that biocompatible –OH stabilised gold nanoparticles can be fabricated [75].

9.3.9 Tungsten

Lei and co-workers have reported [76] the sonoelectrochemical synthesis of tungsten nanoparticles. It is well known that reports on the synthesis of tungsten nanomaterials are limited, with thermal decomposition being one of the few successful approaches, yet it still suffers from a wide distribution with micron-sized aggregates being formed. Lei *et al.* have reported tungsten nanoparticles with ~30 nm diameters but containing some iron within the tungsten nanoparticles acting as impurities originating from the electrolyte [76].

9.4 Alloy Nanoparticles

Delplancke *et al.* have explored the binary alloys of Co-Ni, Fe-Ni and Fe-Co, which were found to have mean diameters of ~100 nm, using a current density of ~8000 A m^{-2} with: t_{on} = 300 ms; t_{off} = 300 ms and t_{us} of 200 ms [16]. Such an approach was extended to Fe-Co-Ni ternary alloys with their magnetic properties evaluated, where it was used to tailor the % composition of the alloys in both A-B and A-B-C type alloys. In other work, Guzman *et al.* [59] produced nano-powdered alloys of Pd(100)-xFe$_x$, where $x = 4, 6, 8$ and 12, which could be formed in the ~5 nm (radius) range leading to interesting magnetic properties [49, 59]. Mancier *et al.* [60] have extended this study to $Fe_{75}Co_{25}$ with radii of ~2–4 nm, forming agglomerates of larger sizes and have explored the magnetic and Mössbauer spectral properties. More recent work has reported $Co_{65}Fe_{35}$ alloys from Aotani solutions [77].

Fe-Cr alloys have implications in many technological areas due to their good mechanical, anti-corrosive and wear-resistant properties [78, 79]. Additionally, Fe-Cr alloys have been proposed as a more cost-effective alternative for $CaCrO_3$ in solid oxide fuel cells (SOFCs) [78] (see Chapter 10). More recently, Zin and Dabalà [78] have reported $Fe_{80}Cr_{20}$ alloys and separate Fe and Cr nanoparticles, with the aim of investigating the influence of Fe and Cr amounts on the final nanopowder morphologies and structures. The sonoelectrochemical synthesis of $Fe_{50}Cr_{50}$ nanoparticles were carried out with a current density of 120 mA cm^{-2}, t_{on} = 0.5 s, t_{us} = 0.3 s, and a t_p (rest time) = 0.2 s. The effect of temperature on the synthesis of $Fe_{50}Cr_{50}$ was explored such that the efficiency of ~21% at 10 °C was reduced dramatically to ~2% at 60 °C, which was attributed to the contribution of increased metal salt solubility and mobility of ions as well as the decrease of solution viscosity and variations in pH as the temperature was increased [78]. Changing the pH has also appeared to affect the % Fe content in the produced $Fe_{50}Cr_{50}$ alloy nanoparticles from 49% at 10 °C increasing to 59% at 25 °C and 38% at 60 °C [78].

Last but not least, PbTe nanorods have been prepared via sonoelectrochemistry. One-dimensional nanomaterials are of importance in the fabrication of high performance thermoelectric devices. Qiu and co-workers have reported a facile approach for 7 nm (diameter) nanorods with aspect ratios of 7. Note that the efficiency of the thermoelectric devices

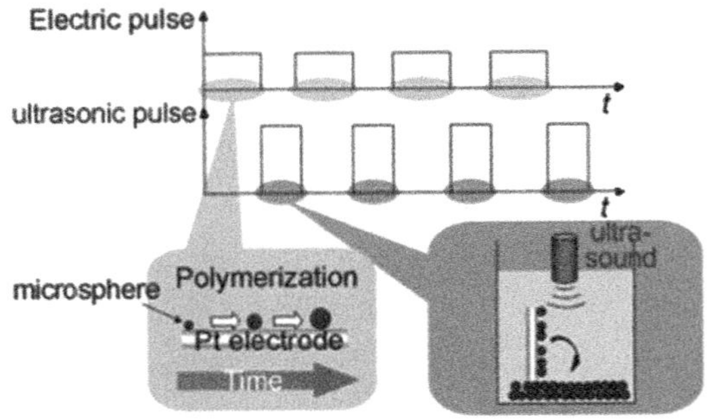

Figure 9.10 *Pulsed sonoelectrochemical method for the production and recovery of conducting-polymer microspheres (Reprinted with permission from [81] Copyright (2009) Wiley).*

improve as the diameter of the 1-D structures approach those of a few nanomaterials, with templates methods limited to well above 10 nm [80]. In this study, the authors found that the ratio of metal ion/ligand can be used to control the purity of the PbTe nanostructures.

9.5 Polymer Nanoparticles

Other notable work extending away from metallic nanoparticles is the fabrication of conducting polymer microspheres [81]. Figure 9.10 shows the methodology whereby poly(*N*-methylaniline (PNMA) microspheres are readily produced.

Through careful choices of the applied potential pulse, the average nanoparticle size can be tailored around the low micron size [81]. It was found that polyaniline nanoparticles are produced in the range of 20–40 nm.

9.6 Conclusions

The field of sonoelectrochemistry applied to forming micro- and nano-particles has been overviewed, highlighting the ever-growing promise of its applications due to its simplicity and cost effectiveness, which is based on the combination of electrochemistry and power ultrasound. We expect this field to significantly expand and be applied to fabricating nanomaterials that are difficult and costly to make using other conventional and traditional methodologies. Clearly the breadth of nanomaterials that sonoelectrochemistry can fabricate is truly breath taking!

References

(1) Mastai, Y., Polsky, R., Koltypin, Y., Gedanken, A. and Hodes, G. (1999) Pulsed sonoelectrochemical synthesis of cadmium selenide nanoparticles. *J. Am. Chem. Soc.*, **121**, 10047.
(2) Favre, L., Stanescu, S., Dupuis, V., Bernstein, E., Epicier, T., Melinon, P. and Perez, A. (2004) Nanostructured thin films from mixed magnetic Co–Ag clusters. *Appl. Surf. Sci.*, **226**, 265.

(3) Zeng, H., Li, J., Wang, Z.L. and Liu, J.P. (2004) Tailoring magnetic properties of core/shell nanoparticles. *Appl. Phys. Lett.*, **85**, 792.
(4) Basu, S. and Chakravorty, D. (2006) Optical properties of nanocomposites with iron core–iron oxide shell structure. *J. Non-Cryst. Solids*, **352**, 380.
(5) Gaudry, M., Cottancin, E., Pallarom, M., Leme, J., Arnaud, L., Huntzinger, J.R., Vialle, J.L., Broyer, M., Rousset, J.L., Treilleux, M. and Melinon, P. (2003) Size and composition dependence in the optical properties of mixed (transition metal/noble metal) embedded clusters. *Phys. Rev. B: Condens. Matter*, **67**, 155409.
(6) www.cytimmune.com (accessed Dec 2010).
(7) www.unidym.com/products/transparent.html (accessed Dec 2010).
(8) www.aquanova.de/index/php?site=index.html&dir=&nav=53 (accessed Dec 2010).
(9) www.grid-shift.com (accessed Dec 2010).
(10) Gedanken, A. (2004) Using sonochemistry for the fabrication of nanomaterials. *Ultrason. Sonochem.*, **11**, 47.
(11) He, C.N., Zhao, N.Q., Shi, C.S. and Song, S.Z. (2009) Fabrication of carbon nanomaterials by chemical vapor deposition. *J. Alloys Compd.*, **484**, 6.
(12) Simm, A.O., Ji, X., Banks, C.E., Hyde, M.E. and Compton, R.G. (2006) AFM studies of metal deposition: instantaneous nucleation and the growth of cobalt nanoparticles on boron-doped diamond electrodes. *Chem. Phys. Chem.*, **7**, 704.
(13) Campbell, F.W., Belding, S.R., Baron, R., Xiao, L. and Compton, R.G. (2009) Hydrogen peroxide electroreduction at a silver-nanoparticle array: Investigating nanoparticle size and coverage effects. *J. Phys. Chem. C*, **113**, 9053.
(14) Belding, S.R. and Compton, R.G. (2010) Transient voltammetry at electrodes modified with a random array of spherical nanoparticles: Theory. *J. Phys. Chem. C*, **114**, 8309.
(15) Reisse, J., Francois, H., Vandercammen, J., Fabre, O. and Kirschdemesmeker, A. (1994) Sonoelectrochemistry in aqueous electrolyte: A new type of sonoelectroreactor. *Electrochim. Acta*, **39**, 37.
(16) Delpancke, J.-L., Dille, J., Reisse, J., Long, G.J., Mohan, A. and Grandjean, F. (2000) Magnetic nanopowders: ultrasound-assisted electrochemical preparation and properties. *Chem. Mater.*, **12**, 946.
(17) Maisonhaute, E., Prado, C., White, P.C. and Compton, R.G. (2002) Surface acoustic cavitation understood via nanosecond electrochemistry. Part III: shear stress in ultrasonic cleaning. *Ultrason. Sonochem.*, **9**, 297.
(18) Gedanken, A. (2004) Using sonochemistry for the fabrication of nanomaterials. *Ultrason. Sonochem.*, **11**, 47. and references cited there in.
(19) Zhu, J., Liu, S., Palchik, O., Koltypin, Y. and Gedanken, A. (2000) Shape-controlled synthesis of silver nanoparticles by pulse sonoelectrochemical methods. *Langmuir*, **16**, 6396.
(20) Liu, S., Huang, W., Chen, S., Avivi, S. and Gedanken, A. (2001) Synthesis of X-ray amorphous silver nanoparticles by the pulse sonoelectrochemical method. *J. Non-Cryst. Solids*, **283**, 231.
(21) Socol, Y., Abramson, O., Gedanken, A., Meshorer, Y., Berenstein, L. and Zaban, A. (2002) Suspensive electrode formation in pulsed sonoelectrochemical synthesis of silver nanoparticles. *Langmuir*, **18**, 4736.
(22) Tang, S., Meng, X., Lu, H. and Zhu, S. (2009) PVP-assisted sonoelectrochemical growth of silver nanostructures with various shapes. *Mater. Chem. Phys.*, **116**, 464.
(23) Haas, I., Shanmugam, S. and Gedanken, A. (2006) Pulsed sonoelectrochemical synthesis of size-controlled copper nanoparticles stabilized by poly(*N*-vinylpyrrolidone). *J. Phys. Chem. B*, **110**, 16947.
(24) Haas, I., Shanmugam, S. and Gedanken, A. (2008) Synthesis of copper dendrites nanostructure by the sonoelectrochemical method. *Chem. Eur.*, **14**, 4696.

(25) Mancier, V., Daltin, A.L. and Leclercq, D. (2008) Synthesis and characterization of copper oxide (I) nanoparticles produced by pulsed sonoelectrochemistry. *Ultrason. Sonochem.*, **15**, 157.
(26) Haas, I. and Gedanken, A. (2008) Synthesis of metallic magnesium nanoparticles by sonoelectrochemistry. *Chem. Commun.*, **15**, 1795.
(27) Akiba, E. (1999) Hydrogen-absorbing alloys. *Curr. Opin. Solid State Mater. Sci.*, **4**, 267.
(28) Sandrock, G. (1999) A panoramic overview of hydrogen storage alloys from a gas reaction point of view. *J. Alloys Compd.*, **293**, 877.
(29) Schlapbach, L. and Züttel, A. (2001) Hydrogen-storage materials for mobile applications. *Nature*, **414**, 353.
(30) Grochala, W. and Edwards, P. (2004) Thermal decomposition of the non-interstitial hydrides for the storage and production of hydrogen. *Chem. Rev.*, **104**, 1283.
(31) Schüth, F., Bogdanovie, B. and Felderhoff, M. (2004) Light metal hydrides and complex hydrides for hydrogen storage. *Chem. Commun.*, 2249.
(32) Dillon, A.C., Jones, K.M., Bekkedahl, T.A., Kiang, C.H., Bethune, D.S. and Heben, M.J. (1997) Storage of hydrogen in single-walled carbon nanotubes. *Nature*, **386**, 377.
(33) Chen, P., Wu, X., Lin, J. and Tan, K.L. (1999) High H_2 uptake by alkali-doped carbon nanotubes under ambient pressure and moderate temperatures. *Science*, **285**, 91.
(34) Sergeev, G.B. (2003) Cryochemistry of metal nanoparticles. *J. Nanopart. Res.*, **5**, 529.
(35) Kooi, B.J., Palasantzas, G. and De Hosson, J.Th.M. (2006) Gas-phase synthesis of magnesium nanoparticles: A high-resolution transmission electron microscopy study. *Appl. Phys. Lett.*, **89**, 161914.
(36) Connor, J.H., Reid, W.E. and Wood, G.B. (1957) Electrodeposition of metals from organic solutions. *J. Electrochem. Soc.*, **104**, 38.
(37) Liebenow, C. (1997) Reversibility of electrochemical magnesium deposition from Grignard solutions. *J. Appl. Electrochem.*, **27**, 221.
(38) Aurbach, D., Lu, Z., Schechter, A., Gofer, Y., Gizbar, H., Turgeman, R., Cohen, Y., Moshkovich, M. and Levi, E. (2000) Prototype systems for rechargeable magnesium batteries. *Nature*, **407**, 724.
(39) Kisza, J.K.A., Borressen, B., Haarberg, G.M. and Tunold, R.J. (1995) Kinetics and mechanism of the magnesium electrode reaction in molten magnesium chloride. *Appl. Electrochem.*, **25**, 940.
(40) Mahendiran, C., Ganesan, R. and Gedanken, A. (2009) Sonoelectrochemical synthesis of metallic aluminum nanoparticles. *Eur. J. Inorg. Chem. A*, 2050.
(41) Xu, C., Dobatkin, S.V., Horita, Z. and Langdon, T.G. (2009) Superplastic flow in a nanostructured aluminum alloy produced using high-pressure torsion. *Mater. Sci. Eng. A*, **500**, 170.
(42) Gerlich, A.P. and Shibayanagi, T. (2009) Grain boundary sliding during friction stir spot welding of an aluminum alloy. *Scr. Mater.*, **60**, 236.
(43) Froustey, C. and Lataillade, J.L. (2009) Influence of the microstructure of aluminium alloys on their residual impact properties after a fatigue loading program. *Mater. Sci. Eng. A*, **500**, 155.
(44) Pomfret, M.B., Brown, D.J., Epshteyn, A., Purdy, A.P. and Owrutsky, J.C. (2008) Electrochemical template deposition of aluminum nanorods using ionic liquids. *Chem. Mater.*, **20**, 5945.
(45) www.aero.kyushu-u.ac.jp/am/laser/Al%20nano.pdf (accessed Dec 2010).
(46) Zein El Abedin, S., Moustafa, E.M., Hempelmann, R., Natter, H. and Endres, F. (2006) Electrodeposition of nano- and microcrystalline aluminium in three different air and water stable ionic liquids. *Chem. Phys. Chem.*, **7**, 1535.
(47) Qiu, X.-F., Xu, J.-Z., Zhu, J.-M., Zhu, J.-J., Zu, S. and Chen, H.-Y. (2003) Controllable synthesis of palladium nanoparticles via a simple sonoelectrochemical method. *J. Mater. Res.*, **18**, 1399.

(48) Zhu, J., Aruna, S.T., Koltypin, Y. and Gedanken, A. (2000) A novel method for the preparation of lead selenide: pulse sonoelectrochemical synthesis of lead selenide nanoparticles. *Chem. Mater.*, **12**, 143.
(49) Mancier, V., Rousse-Bertrand, C., Michel, J. and Fricoteaux, P. (2010) Sono and electrochemical synthesis and characterization of copper core–silver shell nanoparticles. *Ultrason. Sonochem.*, **17**, 690.
(50) Rolison, D.R. (2003) Catalytic nanoarchitectures – the importance of nothing and the unimportance of periodicity. *Science*, **299**, 1698.
(51) Kristian, N. and Wang, X. (2008) Pt shell–Au core/C electrocatalyst with a controlled shell thickness and improved Pt utilization for fuel cell reactions. *Electrochem. Commun.*, **10**, 12.
(52) Toshima, M.H.N., Yamazaki, Y. and Asakura, A. (1992) Catalytic activity and structural analysis of polymer-protected gold-palladium bimetallic clusters prepared by the simultaneous reduction of hydrogen tetrachloroaurate and palladium dichloride. *J. Phys. Chem.*, **96**, 9927.
(53) Wang, S., Kristian, N., Jiang, S. and Wang, X. (2009) Controlled synthesis of dendritic Au/Pt core–shell nanomaterials for use as an effective fuel cell electrocatalyst. *Nanotechnology*, **20**, 25605.
(54) Zhang, J., Lima, F.H.B., Shao, M.H., Sasaki, K., Wang, J.X., Hanson, J. and Adzic, R.R. (2005) Platinum monolayer on nonnoble metal–noble metal core–shell nanoparticle electrocatalysts for O_2 reduction. *J. Phys. Chem. B*, **109**, 2701.
(55) Zhang, J., Mo, Y., Vukmirovic, M.B., Klie, R., Sasaki, K. and Adzic, R.R. (2004) Platinum monolayer electrocatalysts for O_2 reduction: Pt monolayer on Pd(111) and on Carbon-supported Pd nanoparticles. *J. Phys. Chem. B*, **108**, 10955.
(56) Lee, H.J. and Jeong, S.H. (2004) Bacteriostasis of nanosized colloidal silver on polyester nonwovens. *Text. Res. J.*, **74**, 442.
(57) Lee, H.Y., Parkm, H.K., Lee, Y.M., Kim, K. and Park, S.B. (2007) A practical procedure for producing silver nanocoated fabric and its antibacterial evaluation for biomedical applications. *Chem. Commun.*, **28**, 2959.
(58) Vigneshwaran, N., Kathe, A.A., Varadarajan, P.V., Nachane, R.P. and Balasubramaya, R.H.J. (2007) Functional finishing of cotton fabrics using silver nanoparticles. *J. Nanosci. Nanotechnol.*, **7**, 1893.
(59) Guzman, M., Delplancke, J.-L., Long, G.J., Delwiche, J., Hubin-Franskin, M.J. and Grandjean, F. (2002) Morphologic and magnetic properties of Pd100–xFex nanoparticles prepared by ultrasound assisted electrochemistry. *J. Appl. Phys.*, **92**, 2634.
(60) Mancier, V., Delplancke, J.L., Delwiche, J., Hubin-Franskin, M.J., Piquer, C., Rebbouh, L. and Grandjean, F. (2004) Morphologic, magnetic, and Mössbauer spectral properties of $Fe_{75}Co_{25}$ nanoparticles prepared by ultrasound-assisted electrochemistry. *J. Magn. Magn. Mater.*, **281**, 27.
(61) Mancier, V., Daltin, A.L. and Leclercq, D. (2008) Synthesis and characterization of copper oxide (I) nanoparticles produced by pulsed sonoelectrochemistry. *Ultrason. Sonochem.*, **15**, 157.
(62) Jiang, J., Oberdorster, G. and Biswas, P. (2009) Characterization of size, surface charge, and agglomeration state of nanoparticle dispersions for toxicological studies. *J. Nanopart. Res.*, **11**, 77.
(63) He, X., Zhao, X., Chen, Y. and Feng, J. (2008) The evidence for synthesis of truncated triangular silver nanoplates in the presence of CTAB. *Mater. Charact.*, **59**, 380.
(64) Karadeniz, A.E.H., Caliskan, A., Pereira, C.M., Pereira, E.M. and Riheiro, J.A. (2007) Electrochemical sensing of silver tags labelled DNA immobilized onto disposable graphite electrodes. *Electrochem. Commun.*, **9**, 2167.
(65) Wang, D., An, J., Luo, Q., Li, X. and Li, M. (2008) A convenient approach to synthesize stable silver nanoparticles and silver/polystyrene nanocomposite particles. *J. Appl. Polym. Sci.*, **110**, 3038.

(66) Ng, C.H.B., Yang, J. and Fan, W.Y. (2008) Synthesis and self-assembly of one-dimensional sub-10 nm Ag nanoparticles with cyclodextrin. *J. Phys. Chem. C*, **112**, 4141.
(67) Turkevich, J., Stevenson, P.C. and Hiller, J. (1951) A study of the nucleation and growth processes in the synthesis of colloidal gold. *Faraday. Discuss. Chem. Soc.*, **11**, 55.
(68) Kimling, J., Maier, M., Okenve, B., Kotaidis, V., Ballot, H. and Plech, A. (2006) Turkevich method for gold nanoparticle synthesis revisited. *J. Phys. Chem. B*, **110**, 15700.
(69) Frens, G. (1972) Particle size and sol stability in metal colloids. *Colloid. Polym. Sci.*, **250**, 736.
(70) Frens, G. (1973) Controlled nucleation for the regulation of the particle size in monodisperse gold suspensions. *Nature*, **241**, 20.
(71) Pong, B.K., Elim, H.I., Chong, J.X., Ji, W., Trout, B.L. and Lee, J.Y. (2007) New insights on the nanoparticle growth mechanism in the citrate reduction of gold(III) salt: Formation of the Au nanowire intermediate and its nonlinear optical properties. *J. Phys. Chem. C*, **111**, 6281.
(72) Aqil, A., Serwas, H., Delplancke, J.L., Jerome, R., Jerome, C. and Canet, L. (2008) Preparation of stable suspensions of gold nanoparticles in water by sonoelectrochemistry. *Ultrason. Sonochem.*, **15**, 1055.
(73) Jiang, L.P., Wang, A.N., Zhao, Y., Zhang, J.R. and Zhu, J.J. (2004) A novel route for the preparation of monodisperse silver nanoparticles via a pulsed sonoelectrochemical technique. *Inorg. Chem. Commun.*, **7**, 506.
(74) Yu, Y.Y., Chang, S.S., Lee, C.L. and Wang, C.R.C. (1997) Gold nanorods: Electrochemical synthesis and optical properties. *J. Phys. Chem. B*, **101**, 6661.
(75) Shen, Q., Min, Q., Shi, J., Jiang, L., Hou, W. and Zhu, J.J. (2011) Synthesis of stabilizer-free gold nanoparticles by pulse sonoelectrochemical method. *Ultrason. Sonochem.*, **18**, 231.
(76) Lei, H., Tang, Y.-J., Wei, J.-J., Li, J., Li, X.-B. and Shi, H.-L. (2007) Synthesis of tungsten nanoparticles by sonoelectrochemistry. *Ultrason. Sonochem.*, **14**, 81.
(77) Dabala, M., Pollet, B.G., Zin, V. and Campadello, E. (2008) Sonoelectrochemical (20 kHz) production of $Co_{65}Fe_{35}$ alloy nanoparticles from Aotani solutions. *J. Appl. Electrochem.*, **38**, 395.
(78) Zin, V. and Dabala, M. (2010) Iron–chromium alloy nanoparticles produced by pulsed sonoelectrochemistry: Synthesis and characterization. *Acta Mater.*, **58**, 311.
(79) Costa, B.F.O., Le Caer, G., Amado, M.M., Sousa, J.B. and Ayres De Campos, N. (2000) Magnetic properties of coarse-grained and nanocrystalline Fe–Cr–Sn alloys. *J. Alloys Compd.*, **308**, 49.
(80) Qiu, Y.L.X., Samia, A.C.S., Devadoss, A., Burgess, J.D., Dayal, S. and Bursa, C. (2005) PbTe nanorods by sonoelectrochemistry. *Angew. Chem., Int. Ed. Engl.*, **44**, 5855.
(81) Atobe, M., Ishikawa, K., Asami, R. and Fuchigami, T. (2009) Size-controlled synthesis of conducting-polymer microspheres by pulsed sonoelectrochemical polymerization. *Angew. Chem., Int. Ed. Engl.*, **48**, 6069.

10

Sonochemistry and Sonoelectrochemistry in Hydrogen and Fuel Cell Technologies

Bruno G. Pollet

10.1 Introduction

Carbon dioxide (CO_2), greenhouse gas (GHG) and particulate emissions are the result of extracting, processing and utilising fossil fuels for our everyday energy needs. Attempts to reduce these emissions are a challenge and are continually on-going through processes such as carbon capture and storage (CCS) and amine scrubbing. However, these processes, are not sustainable solutions and are seen as temporary solutions to reducing GHG emissions.

The present energy paradigm whereby fossil fuels are exploited as energy sources and hydrocarbon fuels are employed as energy vectors is both environmentally and economically unsustainable. For example, oil production is expected to peak within 10 years, and there is uncertainty as to how long reserves will remain economical to exploit, together with significant increases in oil prices in the near future. The search for sustainable and renewable energy sources and energy vectors is currently a major worldwide R&D focus. The ideal energy system (from primary production to end use) should be carbon neutral, in other words with no GHG emissions throughout its lifecycle, therefore having no adverse effect on the planet and mankind.

Hydrogen is currently seen as a potential alternative fuel, especially for transportation applications [1–6]. Compared to gasoline, hydrogen has a significantly higher gravimetric energy density (gasoline = 48.6 MJ kg^{-1}; hydrogen = 140.4 MJ kg^{-1}). It is important to note that hydrogen is not a naturally occurring gas, even though it is abundant in the

Power Ultrasound in Electrochemistry: From Versatile Laboratory Tool to Engineering Solution, First Edition. Edited by Bruno G. Pollet.

Universe. It is usually found attached to other atoms in compounds such as water (H_2O) and methane (CH_4). Hydrogen is therefore known as a secondary energy source or as an energy 'vector'. Currently 50 million tonnes of hydrogen are produced globally every year. It is predicted that its production will rise by 10% year-on-year over the next decade. Steam methane reforming (SMR) is the most widely used method to produce hydrogen and currently accounts for 95–96% of global hydrogen production, whereas electrolysis only contributes 4–5% [4].

Today, fuel cells are widely considered to be efficient and non-polluting power sources offering much higher energy densities and energy conversion efficiencies compared to any other current energy storage devices. Fuel cells are therefore considered to be promising energy storage devices for the transport, mobile and stationary sectors [1–6]. A fuel cell is an 'electrochemical' device operating at various temperatures (up to 1000 °C) that transforms the chemical energy of a fuel (hydrogen, methanol, natural gas etc.) and an oxidant (air or pure oxygen) in the presence of a catalyst into water, heat and electricity. Furthermore, the power generated by a fuel cell depends largely upon catalytic electrodes and the materials used [1–6].

There are currently six main groupings of fuel cell available namely [1–6]:

(i) proton exchange membrane fuel cell (PEMFC) including direct methanol fuel cell (DMFC),
(ii) alkaline fuel cell (AFC),
(iii) phosphoric acid fuel cell (PAFC),
(iv) molten carbonate fuel cell (MCFC),
(v) solid oxide fuel cell (SOFC) and
(vi) microbial fuel cell (MFC).

PEMFC, AFC, PAFC and MFC operate at low temperatures in the range of 50–200 °C and MCFC and SOFC at high operating temperatures in the range of 650–1000 °C.

The heart of a fuel cell consists of a non-conductive electrolyte material sandwiched between two electrodes – the anode and cathode [6]. The fuel and the oxidant are fed continuously to the anode and the cathode sides respectively. At the anode side the fuel is decomposed into ions and electrons. The insulator electrolyte material allows only ions to flow from both the anode and cathode sides. The free electrons generated at the anode flow to the cathode side through an external electrical circuit. The recombination of the ions with oxidant occurs at the cathode to form pure water. Being a galvanic cell, as opposed to an electrolytic cell, the polarity of a fuel cell on the anode is negative and on the cathode is positive [6].

As all fuel cells are normally distinguished by the materials used, for example, the electrolyte and catalyst materials, although the manufacturing of the fuel cell electrodes is also different in each case, the main objective in fuel cell technologies is to develop low-cost, high-performance and durable materials. Currently, fuel cell systems are too expensive and not durable [4]. However, there are several ways to reduce the cost and increase the performance of a fuel cell by [1–6]:

(i) decreasing the catalyst loading in fuel cell electrodes,
(ii) decreasing the catalyst nanoparticle size,
(iii) developing Pt-alloys (binary and ternary) and Pt-free electrocatalysts,

(iv) using novel fabrication methods to synthesise catalysts and producing better catalyst dispersion on fuel cell electrodes,
(v) developing fuel cell electrode fabrication methods enabling better catalyst dispersion and utilisation, and
(vi) using new techniques to increase mass transport at the fuel cell electrode surface.

One of the most promising methods and techniques for producing performing and efficient fuel cell catalysts, electrodes and electrolyte materials is to employ ultrasound. For the last 20 years, there have been a few reports on the use of ultrasound either for fabricating noble metal and catalysts, preparing fuel cell materials or enhancing mass transport of electroactive species from the bulk solution to the fuel cell electrode surface. This chapter highlights the main uses of sonochemistry and sonoelectrochemistry in hydrogen and fuel cell technologies.

10.2 Sonoelectrochemical Production of Hydrogen

Cataldo [7] was the first researcher to study the effect of ultrasound (30 kHz and 1–2 W cm^{-2}) on the yield of gases released at platinum electrodes during electrolysis of aqueous saturated sodium chloride (NaCl, 6.0 M) and hydrochloric acid (HCl, 6.0 M) solutions. Cataldo found that ultrasound dramatically increases the yield of chlorine and hydrogen gases produced at the anode and cathode respectively in the electrolysis of NaCl and HCl [7]. Cataldo [7] stated that the most important phenomena under sonication is the strong degassing effect at the electrode surfaces due to bubble coalescence induced by cavitation. He also observed that the subsequent mechanical removal of bubbles from the electrode surface improved the yield of hydrogen and chlorine gases produced [7]. It is thought that depolarisation under ultrasound is due to cavitation and ultrasonic waves, which are able not only to stir efficiently the solution in bulk and near the electrodes, eliminating or decreasing the contribution to the overvoltage due to concentration gradients, but are also able to favour the passage of the ions across the electrode double layer and probably to favour their discharge. The adsorbed atomic species at the electrode may then be encouraged by ultrasonic waves to dimerise in molecules and to desorb from the electrode surface. Each step described means the elimination or at least the attenuation of the contribution of each single step to the total overvoltage (η, also known as overpotential), as well as the degassing effect which limits the gas bubble accumulation at the electrodes and in the solution, which otherwise is another contribution to the resistance of the cell.

Cataldo concluded that during sonication the bubbles are forced to coalesce into large bubbles which offer a lower total gas/liquid surface. Furthermore, the contact time between each bubble and the liquid phase is very small due to the gas bubbles being ‘pushed out’ at high speed from the solution, therefore the solution remains virtually ‘free’ from dissolved gases. It is known that the gases generated during electrolysis and dispersed into the liquid decrease its electrical conductivity (bubble effect); there is therefore a direct relationship between the concentration of the gas dispersed in the liquid and the drop in conductivity of an electrolyte.

Walton *et al.* [8] investigated the effect of ultrasound (28 kHz) upon hydrogen evolution on platinum electrodes in aqueous solutions, such as 1.0 M sulfuric acid (H_2SO_4), 1.0 M sodium chloride (NaCl) and 1.0 M hydrochloric acid (HCl). They found that ultrasound has a positive effect on the electrolytic system by increasing rates of hydrogen and chlorine gas evolution at the electrodes (cathode and anode respectively). It was also observed that whilst electrolysing 1.0 M H_2SO_4 solution, the oxygen evolution rate at the anode was not enhanced by sonication, whereas for the electrolyses of NaCl and HCl solutions, the formation of chlorine gas at the electrode was enhanced by insonation. They concluded that the use of ultrasound in this particular system is ideal as chlorine gas (Cl_2) is the preferred anodic product rather than the oxygen gas (O_2) [8].

McMurray *et al.* [9] studied the effect of ultrasound by using a titanium sonotrode and graphite electrodes dipped in 0.7 mol dm^{-3} Na_2SO_4. Their findings showed that ultrasound (20 kHz and 26 W cm^{-2}) significantly increases the rate of oxygen and hydrogen evolution. Furthermore, the oxygen reduction reaction (ORR) rate was increased due to enhanced mass transport under sonication. A more significant rate increase was observed for the hydrogen evolution reaction (HER) as a result of lowered activation overpotential (η_a) for hydrogen formation on graphite electrodes. McMurray *et al.* [9] concluded that the activation overpotential for hydrogen evolution can be reduced by sonication via the use of a sonotrode.

Budischak *et al.* [10] was one of the first researchers to study the sonoelectrochemical effects of the HER from aqueous potassium hydroxide (KOH) solutions. This was performed by linear sweep voltammetry (LSV) and chronoamperometry (CA) to analyse the influence of ultrasound (42 kHz) on the electrochemical reactions. It was shown that ultrasound improves the electrolytic efficiency, especially at intermediate current densities. Even after factoring in the ultrasonic power required from the sonicator, Budischak *et al.* [10] demonstrated that ultrasonic irradiation improves the overall efficiency of the system (ratio of input electrical current : hydrogen produced from the system – see later). Budischak *et al.* [10] also stated that photoelectrochemical cells could benefit from this effect.

Sasikala *et al.* [11] studied the decomposition of water in the presence of ultrasound (40 kHz and 200 W) by suspending solid particulates in the solution. This resulted in an increase in the number of cavitation bubbles created by sonication. It was found that a suitable balance of suspended particles and methanol (CH_3OH) added to the solution increased the yield of hydrogen gas produced. It was found that even with decreasing size of the particles in suspension, the yield of hydrogen gas produced also increased. It was understood that the particles had a larger surface area, and therefore more active sites to create cavitation bubbles, leading to an increase in hydrogen yields.

Recently, Li *et al.* [12] discussed the advantages of applying an ultrasonic field (60 kHz frequency and 50 W) to water electrolysis for the production of hydrogen. In this study, they showed that by lowering the cell voltage (V_{cell}), an increase in efficiency and a decrease in energy consumption for hydrogen gas were observed. In this work, the cell voltage was reduced by increasing the current density and by lowering the electrolyte concentration. This resulted in an increase in hydrogen production, and a decrease in the overall energy consumption. Li *et al.* [12] also showed that varying the current density and concentration of the electrolyte had little effect on the rate of O_2 production from the sonoelectrolytic cell.

Whereas in previous studies it was shown that ultrasound enhances mass transfer of electroactive species from the bulk solution to the electrode surface in any electrolytic systems, many investigations aim to quantify the change in hydrogen gas evolution at the cathode. Previous work has shown that ultrasound mechanically removes gas bubbles on the electrode surface, thus lowering the so-called 'bubble overpotential'. Therefore the use of ultrasound in an electrolytic solution would be expected to increase significantly the rate of hydrogen production.

10.3 Sonochemical Production of Noble Metals and Fuel Cell Electrocatalysts

Fuel cells strongly depend on the use of effective catalysts, either mono-metallic or as bi-metallic, especially for the methanol oxidation reaction (MOR) and hydrogen oxidation reaction (HOR) at the anode and the oxygen reduction reaction (ORR) at the cathode. One of the most important mechanisms that controls the metallic cluster reactivity is nucleation and growth, which strongly depend on the synthetic routes under which clusters are formed.

A number of methods have been proposed to prepare mono- and bi-metallic nanoparticles, including alcohol reduction, citrate reduction, polyol reduction, borohydride reduction, photolytic reduction, radiolytic reduction, laser ablation, metal evaporation-condensation [13] and, more recently, sonochemical reduction, that is, the use of ultrasound to reduce metallic ions in aqueous solutions by using special ultrasonic devices (probes and baths) and the experimental setup is as shown in Figures 10.1(a) and (b).

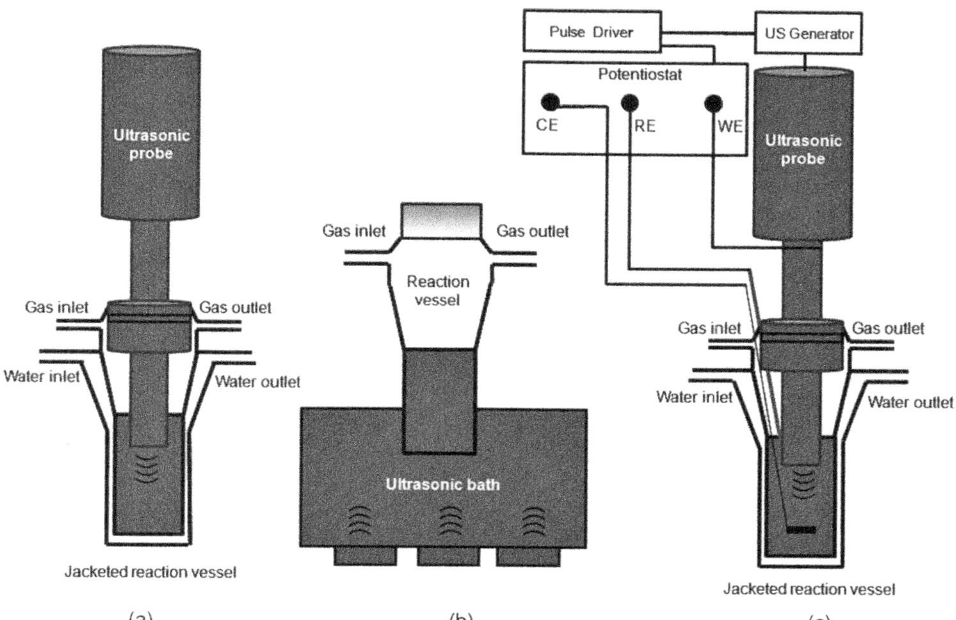

Figure 10.1 Sonochemical (a) – ultrasonic probe system; (b) ultrasonic bath system and sonoelectrochemical (c) – ultrasonic probe used as the cathode to prepare nano-size metals and alloys (Reprinted with permission from International Journal of Hydrogen Energy, The use of ultrasound for the fabrication of fuel cell materials by Bruno G. Pollet, 11986–12004, Copyright (2010) Elsevier).

10.3.1 Sonochemical Mono-Metallic Syntheses

The first report on the use of ultrasound for the fabrication of noble metals was in 1987 by the work of Gutierrez *et al.* [14], followed by the work of Nagata *et al.* [15] in 1992, who studied the reduction of silver ions in solution. In their study, stable colloidal dispersions of silver were prepared by sonication (200 kHz; 200 W) of aqueous $AgClO_4$ and $AgNO_3$ solutions in the presence of surfactants [sodium dodecylsulfate (SDS), polyoxyethylene(40)-monostearate (PEG40-MS) and polyoxyethylene(20)sorbitan monolaurate (Tween20)] and propan-2-ol. They proposed a mechanism whereby: (i) silver ions are reduced to zero-valent silver in several steps triggered by water sonolysis and radical formation, and (ii) the surfactant type plays an important role in the final silver colloidal size.

10.3.1.1 Effect of Surfactants

Okitsu *et al.* [16] studied the role of the surfactant on the size of the produced Pd nanoparticles and the sonochemical reduction mechanisms for Pd^{II}. They were the first to produce Pd nanoparticles by the sonochemical reduction (200 kHz; 6.0 W cm^{-2}) of tetrachloropalladate ions (as $PdCl_2$) in the presence of surfactants, for example, SDS, Tween20, PEG40-MS and poly(vinylpyrrolidone) (PVP) in aqueous solutions. They found that the Pd nanoparticle size (between 6 and 110 nm) depends on the type of surfactants (some of them more prone to sonolysis to yield radicals) and starting concentrations of Pd^{II} used. Recently, Nemamcha *et al.* [17] synthesised Pd nanoparticles of smaller sizes (3–6 nm) than those obtained by Okitsu *et al.* [16] by sonicating (50 kHz) palladium(II) nitrate solution in ethylene glycol (EG) and PVP for 180 min. They found that: (i) the PVP protects the Pd nanoparticles by its adsorption on its surface via the coordination of the C=O group with the Pd atoms to form un-aggregated nanoparticles, and (ii) an increase of the Pd^{II}/PVP molar ratio decreases the number of Pd nanoparticles and increases their size from 3 to 6 nm for low Pd^{II} content.

Okitsu *et al.* [18] also worked on the preparation of nano-size Ag, Au, Pt and Rh in either non-ionic surfactants (Tween20 and PEG40-MS) or anionic surfactants (SDS and PVP) to stabilise the metallic nanoparticles (circa 5 nm) in the presence of ultrasound (200 kHz; 6.0 W cm^{-2}) and under argon at 293 K. They found that four pathways are possible for the sonochemical reduction of Ag^{I}, Au^{III}, Pt^{II} and Rh^{III}.

From Okitsu's early findings on platinum, Mizukoshi *et al.* [19] studied the preparation of platinum nanoparticles by the sonochemical (200 kHz; 6.0 W cm^{-2}) reduction of Pt^{II} in the presence of SDS, where they found that the average diameters of the stable Pt nanoparticles formed were 2.6 ± 0.9 nm, nearly spherical in shape and well dispersed (Figure 10.2).

They attributed the formation of Pt nanoparticles to the reducing species (radicals) generated by water sonolysis induced by cavitation, which also followed the mechanisms described by Nagata *et al.* [15] and Okitsu *et al.* [16,18]. Mizukoshi *et al.* [20] also investigated the effect of surfactant such as SDS and sodium dodecylbenzenesulfonate (DBS) as anionic surfactants, PEG40-MS as non-ionic, dodecyltrimethylammonium chloride (DTAC) and bromide (DTAB) as cationic surfactants on the sonochemical (200 kHz; 200 W) reduction of Pt^{IV} ions. They showed that two types of reducing radicals induced by sonolysis and thermolysis are formed in some of the surfactants used which are directly contributing to a two-step mechanism where Pt nanoparticles (1–3 nm) are formed: step (1) – Pt^{IV} ions to Pt^{II} ions and step (2) – Pt^{II} ions to Pt^{0}.

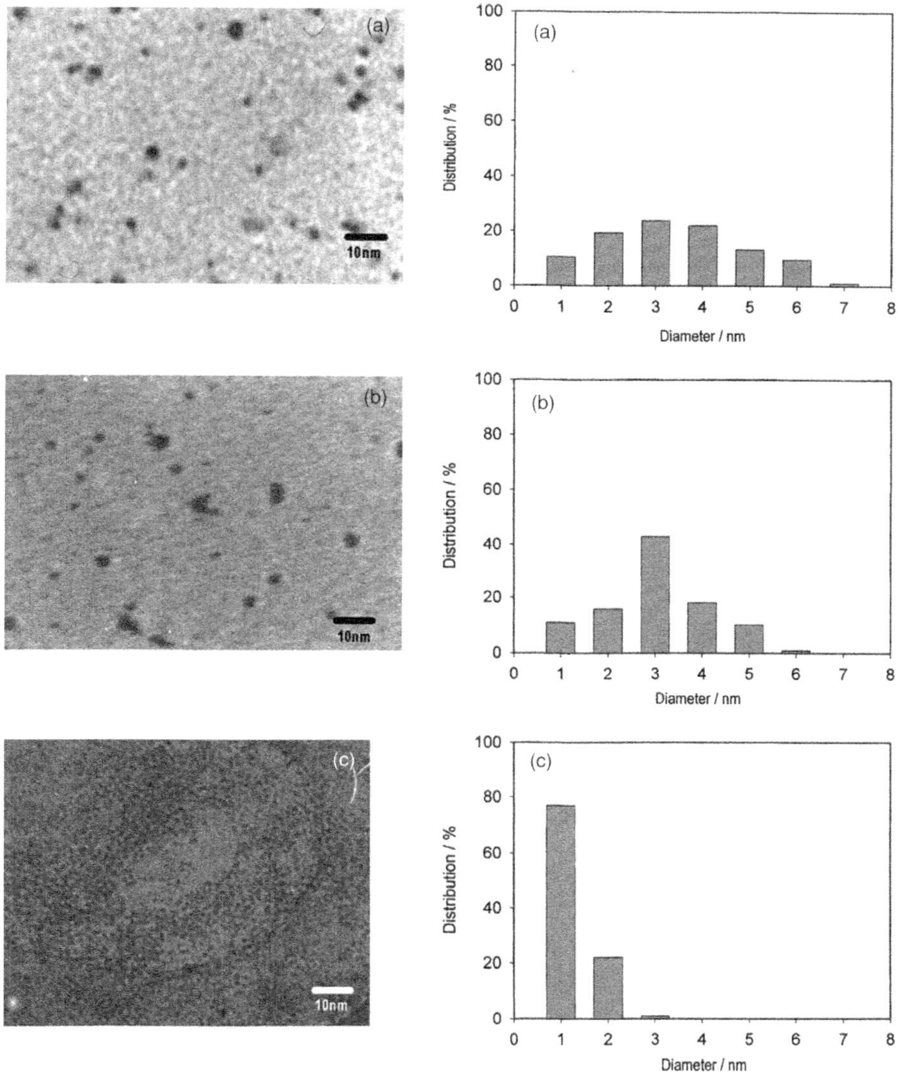

Figure 10.2 *TEM image and the diameter distribution of platinum particles sonochemically prepared in the presence of (a) 8 mM of SDS, (b) 3 mM of DBS and (c) 0.4 mM of PEG-MS. Initial concentration of H_2PtCl_6 = 1 mM [20] (Reprinted with permission from [20] Copyright (2001) Elsevier Ltd).*

10.3.1.2 Effect of Alcohols

Caruso *et al.* [21] also investigated the sonochemical (20 kHz; 10 W cm^{-2}) reduction of Pt^{IV} ions, but in the presence of various alcohols (ethanol, propan-1-ol and pentan-1-ol), to form Pt nanoparticles (≈ 2.6 nm). They attributed the Pt nanoparticle synthesis to the water sonolysis process induced by cavitation where the primary H• and OH• radicals formed are scavenged by interfacial adsorbed alcohol molecules which subsequently diffuse away from the cavitation bubble and react with Pt^{IV} in the bulk solution. They concluded that alcohol enhances the sonochemical reduction of the Pt^{IV} and two possible pathways may occur for the reduction of Pt^{IV}, that is, the reduction by radicals or via a thermal process occurring in the 'hot shell' surrounding the collapsing cavitation bubble.

Yeung *et al.* [22] showed the first study of the formation of gold nanoparticles (less than 10 nm) from $HAuCl_4$ solutions in the presence of ultrasound (20 kHz; 23–47 W cm^{-3}) and concluded that the particle size and efficiency of the formed nanoparticles strongly depends on the presence, type and concentration of the alcohol (methanol, propan-2-ol and *n*-pentanol) used. For example, they found that greater alcohol hydrophobicity leads to smaller nanoparticle mean diameters. They ascribed the order of particle size to the stabilisation of primary 'seed' nuclei by adsorbed alcohol and argued that the hydrophobicity of the alcohol influences its adsorption onto the particles. Caruso *et al.* [23] studied the sonochemical (20 kHz; 9 W cm^{-2}) reduction of Au^{IV} ions to nano-size Au (9–25 nm) and showed that the gold nanoparticle size depends on the presence and concentration of aliphatic alcohols and SDS in aqueous solutions, in other words on the gas-solution interfacial activity of the solutes used. They found that the Au particle sizes decrease with increasing alcohol concentration and alkyl chain length, that is, hydrophobicity of the alcohol. They suggested that in the chemical formation of Au nanoparticles in the presence of surfactants, particle growth is inhibited by surfactant adsorption onto the particles.

10.3.1.3 Effect of Atmospheric Gases

Fujimoto *et al.* [24] showed that the use of atmospheric gas (Ar, N_2 and Xe) affects the sonochemical (20 kHz; 100 W) reduction of Pt(IV) and Pd(IV) ions and hence the size of Pt and Pd nanoparticles. They found that Pd and Pt nanoparticles were formed under N_2 and Xe atmospheres respectively, which was thought to be attributed to the heat and thermal conductivities of the dissolved gases and the 'hot-spot' temperature caused by sonication. Salkar *et al.* [25] also showed that amorphous silver nanoparticles of circa 20 nm size can be prepared by the sonochemical (20 kHz; 100 W cm^{-2}) reduction of an aqueous silver nitrate solution in an atmosphere of argon-hydrogen where the reduction is triggered through the generation of hydrogen radicals during the sonication.

10.3.1.4 Effect of Sonication Time

Recently, Radziuk *et al.* [26] showed for the first time the effect of ultrasonic treatment (20 kHz, 22 and 27 W cm^{-2}) on the crystal structure of preformed platinum nanoparticles ($<$ 1 nm) in aqueous citrate, PVP and EG solutions. They successfully demonstrated that: (i) the time of sonication treatment of preformed Pt nanoparticles is important, for example, after 20 min of sonication, fewer crystalline or amorphous Pt nanoparticles were produced whereas after 60 min of ultrasonic treatment led to new Pt nanoparticles with higher crystallinity, (ii) the Pt nanoparticle crystallinity in aqueous solutions is reduced by sonication strongly up to its complete disappearance, while PVP and EG prevent the destruction of Pt nanoparticles, and (iii) Pt nanoparticles with higher crystallinity formed after 60 min of ultrasonic treatment in PVP showed better catalytic activity compared to Pt nanoparticles of low crystallinity formed after 20 min of sonication in EG. Wu *et al.* [27] also showed that sonication time affects the particle size when they described a new route for preparing of Au and Pt nanoparticles (10 and 6 nm respectively) by sonochemical (42 kHz) metal displacement reduction. The nanoparticles were produced by the displacement reduction of the precursor salts of $HAuCl_4.3H_2O$ and $H_2PtCl_6.6H_2O$ by Cu and Fe foils, respectively, where ultrasound was used to dislodge the metallic sites on the foils.

10.3.1.5 Effect of Ultrasonic Frequency

Okitsu *et al.* [28] investigated the effect of ultrasonic frequency (20, 213, 358, 647 and 1062 kHz) on the rate of the reduction of Au^{III} to produce Au nanoparticles in aqueous solutions containing 1-propanol. They found that the size and distribution of the Au nanoparticles produced depends on the rate of Au^{III} reduction, which in turn is influenced by the applied ultrasonic frequency. They concluded that the rate of Au^{III} reduction induced by sonication was the highest at 213 kHz due to the maximum amount of reducing radicals generated at that frequency. He *et al.* [29] also showed that ultrasonic frequencies (20, 213, 355, 647 and 1056 kHz) affect the rate of production of Ru nanoparticles (10–20 nm) from ruthenium chloride solutions. They found that ultrasound causes sequential reduction, proceeding from Ru^{III} to Ru^{II} to Ru^0 due to cavitation in solution. This is mainly caused by water sonolysis producing high reactive reducing radicals. In their study, they used a mixture of stabilisers such as SDS and propanol with the latter influencing the reduction rate of Ru^{III}. They also confirmed that at the ultrasonic frequencies of 213 and 255 kHz, higher metal ion reduction rates were observed due to higher primary radicals generated at these two frequencies compared to any other frequencies.

10.3.2 Sonochemical Bi-Metallic Syntheses

Bi-metallic catalysts are known to exhibit specific reactivities and their catalytic efficiencies strongly depend on their size. The production of core-shell bi-metallic nanoparticles is of great interest in fuel cells due to the fact that one of the metals (the shell) determines the surface properties of the nanoparticles while the other (the core) is responsible for specific functional catalytic properties. A few reports have shown that it is possible to produce core-shell bi-metallic nanoparticles by the sonochemical reduction of metallic ions in solutions.

10.3.2.1 Effect of Surfactants and Alcohols

Basnayake *et al.* [30] showed that a solution of tetrahydrofuran (THF) containing Pt^{IV} and Ru^{III} in a fixed mole ratio of either 1:10 or 1:1 subjected to ultrasound (20 kHz; 17 W cm^{-2}) gives Pt-Ru nanoparticles (around 2–3 nm). They also showed that sonication produces a Pt-Ru alloy phase with a nanometre composition close to the mole fraction of Pt^{IV} and Ru^{III} in solution. This observation was explained to be due to high mixing of Pt^{IV} and Ru^{III} in solution during synthesis induced by cavitation and acoustic streaming. Vinodgopal *et al.* [31] also prepared larger bi-metallic Pt-Ru nanoparticles (5–10 nm) by the sonochemical (213 kHz; 110–125 mW cm^{-3}) reduction of Pt^{II} and Ru^{III} in the presence of SDS and PVP as stabilisers under argon atmosphere. They showed that sequential reduction of the Pt^{II} followed by the Ru^{III} produces nanoparticles with a core shell (Pt-Ru) morphology.

Kim *et al.* [32] produced bi-metallic nanoparticles of palladium-tin (Pd–Sn) under ultrasonic irradiation (20 kHz; 42 W cm^{-2}) and showed that their electrocatalytic activity for ORR in 0.5 M KOH was greater than Pt or Pd nanoparticles alone in the same alkaline media. The Pd–Sn nanoparticles were prepared by sonicating a solution of $Pd(NH_4)_2Cl_4$ and $SnCl_2$ in an aqueous solution containing ethanol and citric acid. The resulting nanoparticles had an average size of about 3–5 nm with Pd in Pd–Sn nanoparticles found to be mostly in the metallic form. They also observed that the concentrations of $Pd(NH_4)_2Cl_4$, $SnCl_2$, ethanol, citric acid and corresponding pH values under sonication were found to affect the size distribution of the Pd–Sn nanoparticles mainly caused by radicals during sonolysis. They proposed a possible mechanism (similar to the one proposed by Okitsu *et al.* [18]) for the reduction of Pd^{2+} and Sn^{2+} and proposed a mechanism under insonation as follows:

$$H_2O \xrightarrow{)))} \bullet H + \bullet OH \tag{10.1}$$

$$RH + \bullet OH\,(\bullet H) \xrightarrow{)))} R\bullet + H_2O\,(H_2) \tag{10.2}$$

$$mM^{n+} + nmH\bullet \xrightarrow{)))} (M^0)m + nmH^+ \tag{10.3}$$

$$mM^{n+} + nmR\bullet \xrightarrow{)))} (M^0)m + nmR' + nmH^+ \tag{10.4}$$

where RH denotes the radical's scavenger, acting as protective agent (surfactant or alcohol), R• is the secondary radical and R′ is the deprotonated form. During sonolysis, it is known that water molecules are dissociated into hydrogen radicals (H•) and hydroxyl radicals (OH•), according to Equation (10.1). R• species are formed by hydrogen abstraction from RH with OH• and H• [Eq. (10.2)] and react with Pd and Sn cations to form Pd–Sn metal particles, according to Equations (10.3) and (10.4). Okitsu *et al.* [18] showed that citric acid acts as reducing agent in the formation of noble metals nanoparticles in addition to serving as a stabiliser.

Takatani *et al.* [33] reported that the sonochemical synthesis of bi-metallic nanoparticles of Au-Pt and Au-Pd depends greatly on the presence and nature of the surfactants. This was also observed by Nakanishi *et al.* [34] who prepared a colloidal Au(core)-Pt(shell) bi-metallic by sonicating (200 kHz; 4.2 W cm^{-2}) aqueous solutions containing Au^{3+} and Pt^{4+} ions and surfactants (SDS and PEG40-MS) at various sonication times. They showed that the structures of the Au-Pt nanoparticles (<10 nm) varied depending on the surfactants used and their sizes increased with increasing starting concentration of the metals. Anandan *et al.* [35] also showed that the Au-Ag nanoparticle structure and sizes depend on the surfactant. They produced Au(core)-Ag(shell) bi-metallic nanoparticles (20 nm) by the sonochemical (20 kHz; 23–47 W cm^{-2}) co-reduction of Au^{III} and Ag^{I} ions in PEG and EG for 30 mins where they observed that hydrogen atoms and alcohol radicals reduce gold and silver ions to generate gold-silver bi-metallic nanoparticles. They showed that in the presence of the surfactant as the stabiliser, it is also possible that the surfactant radicals, generated by thermal decomposition at the cavitation bubble/solution interface, reduce the metal ions to generate metal nanoparticles. They concluded that formation of core-shell morphology is most likely to be due to the difference in the reduction rates of the individual metal ions.

Kumar *et al.* [36] synthesised gold–ruthenium (Au–Ru) bi-metallic nanoparticles by the co-reduction of Au^{III} and Ru^{III} ions in aqueous solutions containing varying mole fractions (1:1, 1:3 and 1:5) of the metal ions using PEG as a stabiliser in the presence of ultrasound (355 kHz). TEM experiments revealed that bi-metallic nanoparticles were formed with higher ruthenium content in the shell and higher gold content in the core. UV-VIS spectra showed changes during the sonochemical co-reduction process indicating preferential reduction of metal ions (controlled by their redox potentials), that is, gold is first reduced followed by ruthenium. Mizukoshi *et al.* [37, 38] also observed similar findings when they synthesised bi-metallic nanoparticles of gold and palladium (Au-Pd) with a narrow size distribution by sonochemical (200 kHz; 200 W) reduction of an aqueous solution containing both Au^{III} and Pd^{II} ions. They reported core-shell morphology for the resulting nanoparticles with a Pd shell over the Au core. Bi-metallic nanoparticles with Au core and Pd shell have also been synthesised by Kan *et al.* [39] using successive sonochemical (40 kHz; 100 W) reduction of the precursors $Pd(NO_3)_2$ and $HAuCl_4$ in ethylene glycol.

10.3.3 Sonochemical Perovskite Oxides Syntheses

In the area of SOFC, perovskite oxides have been used as important materials, for example, $LaMnO_3$-based oxide is used as cathode and $LaCrO_3$ as the interconnects. Recently, Das *et al.* [40] reported the sonochemical preparation of $LaMnO_3$ nano-sized powder (19–55 nm) by using lanthanum nitrate and manganese acetate as starting materials in the presence of SDS as surfactant, in view of preventing agglomeration. They observed that the phase formation temperature of $LaMnO_3$ was 700 °C, which is much lower than any other conventional processes and the particle sizes of the powders were easily controlled by calcination conditions. Pang *et al.* [41] studied the production of $La_{1-x}Sr_xMnO_3$ ($x = 0.3$) [LSM, also used as SOFC cathode material] nanoparticles by sonication (20 kHz, 100 W cm^{-2}). They obtained fully crystalline LSM with an average size of 24 nm after the mixture was annealed at 900 °C for 2 hours. Michel *et al.* [42] produced single-phase $GdCo_{1-x}Cu_xO_{3-\delta}$ ($x = 0, 0.15, 0.30$) perovskites (another cathode material for SOFC) spherical-hollow nanoparticles with an average particle size of 49 nm by ultrasonic spray pyrolysis (USP), at a reactor temperature of 950 °C with starting aqueous solutions containing stoichiometric amounts of gadolinium, cobalt and copper nitrates, which were atomised using an ultrasonic transducer, operating at 1.7 MHz. In this set up the aerosol droplets were transported by O_2 inside a preheated tube furnace, where solvent evaporation and chemical reaction occurred. Table 10.1 shows a summary of the sonochemical production of noble mono- and bi-metallics at various ultrasonic frequencies and powers in several surfactants and alcohols.

10.4 Sonoelectrochemical Production of Noble Metals and Fuel Cell Electrocatalysts

The sonoelectrochemical method is the use of power ultrasound applied to various electrochemical processes leading to both chemical and physical effects, such as increase in mass transport, improved surface cleaning, increases in deposit hardness, deposit thickness, deposition rates and efficiencies as well as improved deposit adhesion and radical formation [43].

There are a few studies reporting the use of the sonoelectrochemical method to produce noble metals [44–47] and fuel cell electrocatalysts. Here, the sonoelectrochemical technique involves either depositing a metal under continuous electrical current and ultrasound or producing nano-size metals at various currents and ultrasonic pulses (a few ms) at a vibrating electrode [Figure 10.2(c)].

Table 10.1 *Sonochemical production of nano-size noble mono- and bi-metallics at various ultrasonic frequencies and powers in several surfactants and alcohols.*

Noble metals	Ultrasonic frequency/ (kHz)	Ultrasonic power	Surfactant	Solvent	Particle size/(nm)	Authors
Ag	200	200 W	SDS, PEG40-MS, Tween20	Propan-2-ol	13 ± 3	Gutierrez *et al.* [14] Nagata *et al.* [15]
Pd	200	6.0 W cm^{-2}	SDS, Tween20, PEG40-MS, PVP	—	6–110	Okitsu *et al.* [16]
Pd	50	—	PVP, EG	—	3–6	Nemamcha *et al.* [17]
Ag, Au, Pt, Rh	200	6.0 W cm^{-2}	Tween20, PEG-MS, SDS, PVP	—	5	Okitsu *et al.* [18]
Pt	200	6.0 W cm^{-2}	SDS	—	2.6 ± 0.9	Mizukoshi *et al.* [19]
Pt	200	200 W	SDS, DBS, PEG-MS, DTAC, DTAB	—	1 (PEG-MS) 3 (SDS) 3 (DBS)	Mizukoshi *et al.* [20]
Pt	20	10.0 W cm^{-2}	—	Ethanol, propan-1-ol, pentan-1-ol	2.6	Caruso *et al.* [21]

Pt, Pd	20	100 W	PVP	Propan-2-ol	2 (N_2) – Pd 3.6 (Ar) – Pd 3.2 (N_2) – Pt 2.9 (Ar) – Pt 1.5(Xe) – Pt	Fujimoto *et al.* [24]
Ag	20	100 W cm^{-2}	—	Ethanol	20 (Ar-H_2)	Salkar *et al.* [25]
Pt	20	22 & 27 W cm^{-2}	PVP, EG	Citrate	< 1	Radziuk *et al.* [26]
Pt, Au	42	—	PVP	—	6 (Pt) 10 (Au)	Wu *et al.* [27]
Au	20	23–47 W.cm^{-3}		Methanol, Propan-2-ol, *n*-pentanol	7 ± 2 (methanol) 6 ± 4 (propan-2-ol) 5 ± 3 (*n*-pentanol)	Yeung *et al.* [22]
Au	20	9 W.cm^{-2}	SDS	Methanol, 1-Propanol, Butanol	9–25 (depending on alacohol concentations)	Caruso *et al.* [23]
Au	20, 213, 358, 647 & 1062	0.1 ± 0.01 W.cm^{-3}	—	1-Propanol	15–30	Okitsu *et al.* [28]
Ru	20, 213, 355, 647 & 1056	—	SDS	1-Propanol	10–20	He *et al.* [29]

(continued)

Table 10.1 (Continued)

Noble metals	Ultrasonic frequency/ (kHz)	Ultrasonic power	Surfactant	Solvent	Particle size/(nm)	Authors
Pt-Ru	20	17 $W.cm^{-2}$	—	Methanol, THF	2.0 ± 0.1	Basnayake *et al.* [26]
Pt-Ru	213	110–125 $mW.cm^{-3}$	SDS, PVP	—	5–10	Vinodgopal *et al.* [31]
Pd-Sn	20	42 $W.cm^{-2}$	—	Citric acid, ethanol	3–5	Kim *et al.* [32]
Au-Ru	355	—	PEG	—	15	Kumar *et al.* [36]
Au(core)-Pd(shell)	200	200 W	SDS	—	8 [6 nm Au core & 1 nm Pd shell]	Mizukoshi *et al.* [37,38]
Au(core)-Pd(shell)	40	100 W	PVP, EG	—	8 [4 nm Au core & 4 nm Pd shell]	Kan *et. al.* [39]
Au(core)-Pt(shell)	200	4.2 $W.cm^{-2}$	SDS, PEG	—	6.5 ± 1.5 to 10.1 ± 3.6	Takatani *et al.* [33]
Au(core)-Ag(shell)	20	23–47 $W.cm^{-2}$	PEG, EG	—	20	Anandan *et al.* [35]

10.4.1 Effect of Surfactants and Polymers

Shen *et al.* [48] produced uniform spherical three-dimensional dendritic Pt nanostructures (DPNs) with an average dimension of 2.5 ± 0.5 nm, at room temperature sonoelectrochemically (20 kHz; 20 W; pulse width of the current = 1.0 s, resting time of the current pulse = 0.5 s; duration of the ultrasonic pulse = 0.3 s) by using mixtures of hexachloroplatinum(IV) acid hydrate in various surfactants [PVP, poly(ethylene glycol)$_{20}$-poly(propylene glycol)$_{20}$-poly(ethyleneglycol)$_{20}$ (P123), SDS and poly-diallyl-dimethyl ammonium chloride (PDDA)] at current densities up to 40 mA cm^{-2}. They showed that at low current densities the reduction rate of Pt is slower, the number of nuclei is small and the rate of growth is faster than that of nucleation, which lead to large nanoparticles (5–7 nm). As the current density increases the reduction rate of Pt^{4+} increases and the nucleation rate is faster than that of growth, leading to the generation of more nuclei and the formation of smaller primary Pt nanoparticles. They also showed that stabilised Pt nanoparticles do not aggregate in PVP due to stronger bonds between the Pt precursors with the C=O groups of the PVP, in other words PVP adheres to the nanoparticles through a charge-transfer interaction between the pyrrolidone rings and Pt atoms. The DPNs showed improved electrocatalytic activity towards MOR, due to monodisperse Pt nanoparticles, and improved porosity structure, leading to large effective surface area. They proposed a mechanism (Figure 10.3) whereby Pt ions are reduced by the electrical current and formed Pt primary nanoparticles which are subsequently dislodged by the vibrating electrode. The primary nanoparticles then in solution spontaneously assembled together and formed small spherical DPNs. They showed that ultrasound leads to small primary nanoparticles in favour of the crystallite reorganisation and growth of a stable near-single crystal.

Shen *et al.* [49] showed that by using the sonoelectrochemical (20 kHz; 20 W; pulse-on time of the current = 0.5 s; pulse-off time of the current = 0.5 s, duration of the ultrasonic pulse = 0.3 s) method, it is possible to realise the morphology-controlled synthesis of palladium nanostructures [spherical (SNP), multitwinned (MTP) and sponge-like (SSP)] at room temperature in the presence of various surfactants and polymers [cethyltrimethylammonium bromide (CTAB), PVP and PDDA] (Figure 10.4).

They showed that the size and shape of the Pd nanostructures may be controlled by varying the current density and the precursor solution pH value. Furthermore, the

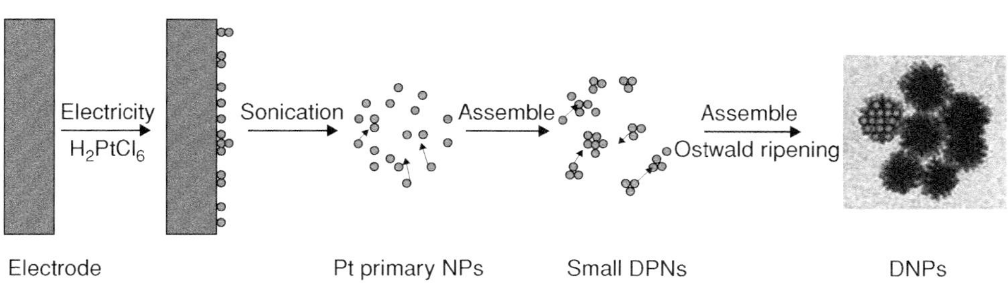

Figure 10.3 Schematic illustration of the formation of DPNs (Reprinted with permission from [48] Copyright (2008) Elsevier Ltd).

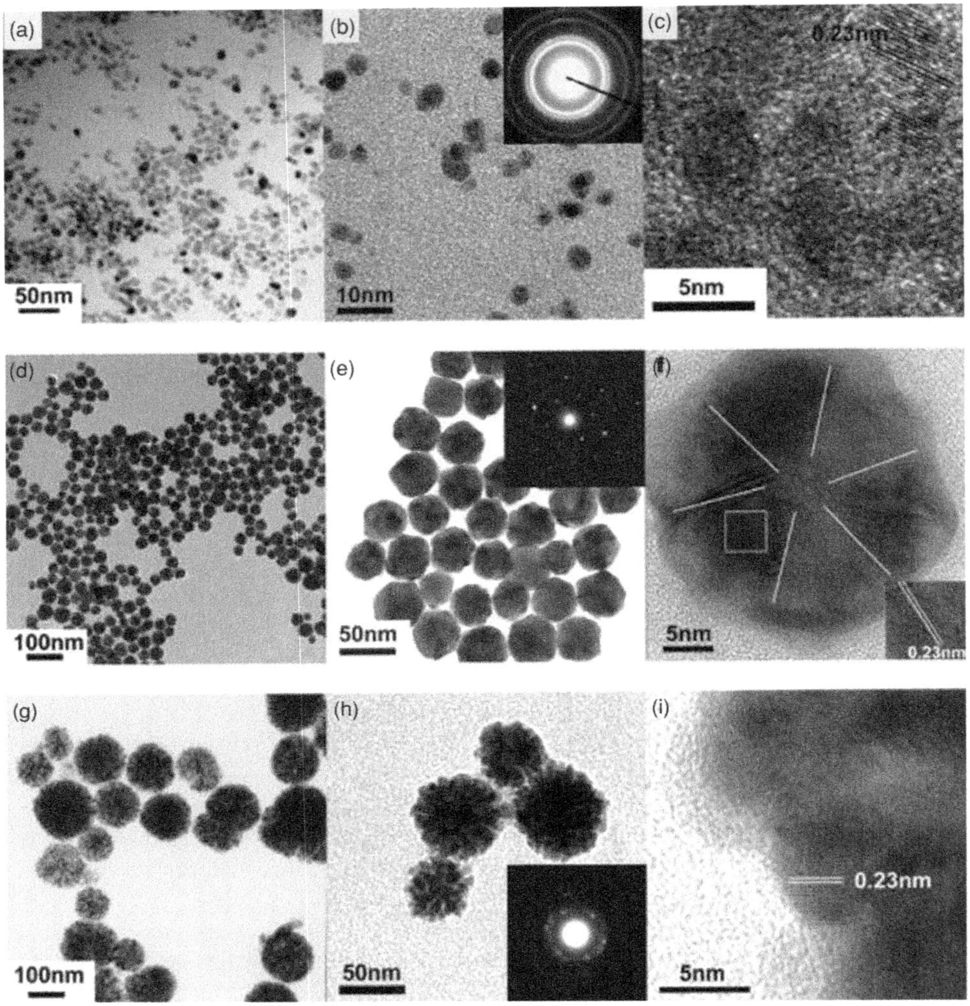

Figure 10.4 *TEM and HR-TEM images of the Pd nanostructures prepared in different solutions. (a–c) Spherical nanoparticles (SNPs) prepared in CTAB solution. (d–f) Multi-twinned particles (MTPs) prepared in PVP solution. (g–i) Spherical sponge-like particles (SSPs) prepared in PDDA solution. The insets in parts b, e, and h are SAED patterns of the corresponding Pd nanostructures. The inset in part f is the enlarged image of the square zone area. The white lines in part f indicate the twin boundaries separating the different adjacent facets (Reprinted with permission from [49] Copyright (2009) Elsevier Ltd).*

electrocatalytic activities of the produced sponge-like Pd nanostructures for direct alcohol oxidation in alkaline media showed higher electrochemical active surface than other Pd nanostructures. Qui *et al.* [50] also synthesised highly dispersed spherical Pd nanoparticles of a dendritic superstructure in the presence of CTAB by the pulse sonoelectrochemical (20 kHz) method at room temperature and a reaction time above 2.5 hours. They explained that the dendritic-structured Pd nanoparticles had a tree-like structure and agreed with the diffusion-limited aggregation (DLA) model, involving cluster formation by the adhesion of particles to a selected seed on contact and allowing the particle to diffuse and stick to the growing structure. They stipulated that it is possible that Pd particles reach the anode and grow into a dendritic structure. They concluded that the shape and size of spherical nanocrystalline Pd may be controlled by varying the current density (8–13 mA cm^{-2}), the interval between continuous ultrasonic pulses, ultrasonic intensity (20–120 W cm^{-2}) and the CTAB concentration. For example, they observed that the shape of the nanoparticles appeared irregular and agglomerated below 20 and 120 W cm^{-2}.

10.4.2 Effect of Aqueous Solutions

Recently, Zin *et al.* [51] produced platinum nanoparticles from aqueous chloroplatinic solutions by the pulse sonoelectrochemical (20 kHz, up to 118 W cm^{-2}) method on titanium alloy electrodes, in the absence of any surfactants, alcohols and polymers, by producing short applied current pulses triggered and followed immediately by ultrasonic pulses at the working electrode (in this case the cathode). The time management sequence employed was as follows:

1. A short current pulse of $|i| = 50$ mA cm^{-2} was applied to the sonoelectrode, whereby the titanium horn acted as an electrode only (t_{ON}); the time of this phase typically varied between 0.3 and 0.5 s.
2. Immediately after the electrochemical pulse was turned off, an ultrasonic pulse was sent to the sonoelectrode and here it acted only as a vibrating ultrasonic horn (t_{US}); this second phase lasted no more than 0.5 s.
3. A rest time, t_p, followed the two previous phases (this was useful to restore the initial electrolyte conditions close to the sonoelectrode).

They showed that a Pt mean grain size ranging from 11 to 15 nm was produced and globular clusters had a mean size ranging between 100 and 200 nm which in turn aggregated and built complex structures (Figure 10.5).

For all the sonoelectrochemical production of nano-size metals, a mechanism has been proposed where metallic ions are reduced by a short current pulse to produce metallic nanoparticles on the sonoelectrode surface, which are then dislodged by the ultrasonic pulse. The metallic nanoparticles in solution tend to spontaneously assemble together whereby under insonation, the Ostwald ripening process is accelerated, leading to smaller primary nanoparticles [50, 51]. Table 10.2 shows a summary of the sonoelectrochemical production of noble mono-metallics at various ultrasonic frequencies and powers in several surfactants.

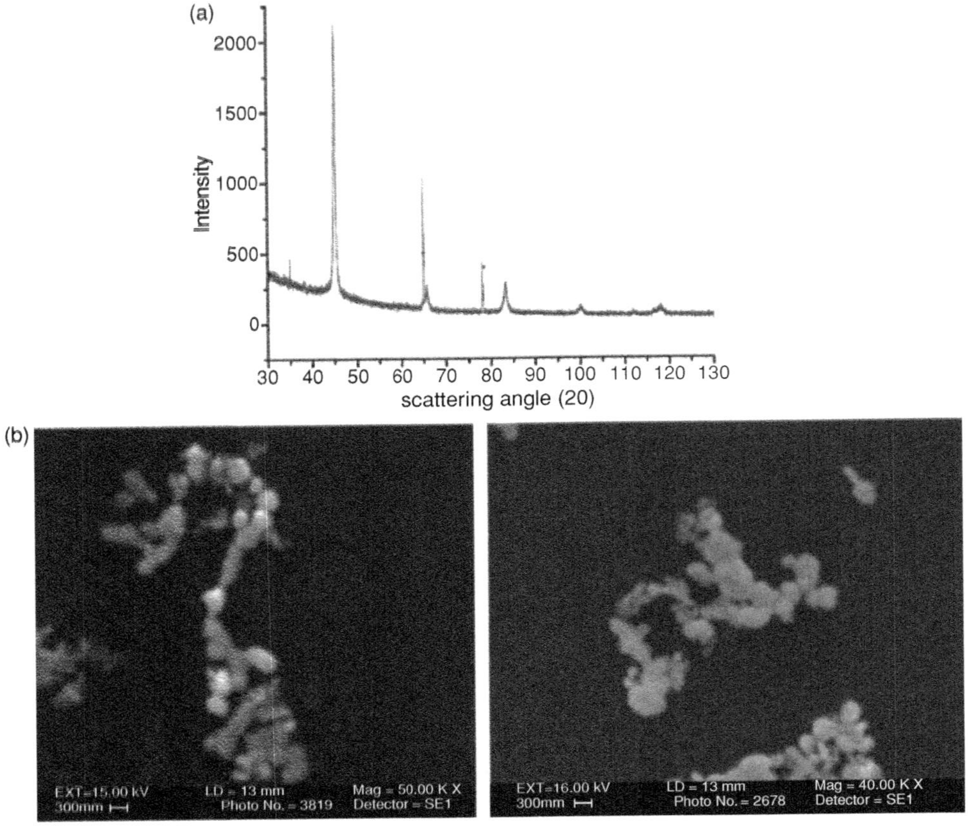

Figure 10.5 *(a) XRD pattern of Pt nanopowders and (b) SEM micrograph of Pt nanoparticles (Reprinted with permission from [51] Copyright (2009) Elsevier Ltd).*

10.5 Sonochemical and Sonoelectrochemical Preparation of Fuel Cell Electrodes

There are a few papers dealing with the use of sonochemical and sonoelectrochemical techniques to produce fuel cell electrodes; however, this section is added in this chapter as it merits some attention.

There is a strong interest in the electrodeposition of Co/CeO_2 composites for protection of cathodes in MCFC. Recently, Argirusis *et al.* [52] electrodeposited Co/CeO_2 and Ni/CeO_2 composites from electrolytes containing suspended nanoparticles of gadolinia-doped ceria in the presence of ultrasound (20 kHz; 29 W cm^{-2}). They showed that sonication improves the co-deposition of ceria nanoparticles with Ni and Co, but no MCFC tests were carried out to investigate the performance of these new materials.

Currently there are different methods for mixing and milling the carbon support, the carbon support electrocatalyst and the ionomer or PEM either by intense magnetic stirring,

Table 10.2 Sonoelectrochemical production of nano-size noble mono-metallics at various ultrasonic frequencies and powers in several surfactants.

Noble metals	Ultrasonic frequency/ (kHz)	Ultrasonic power	Surfactant	Solvent	Particle size/ (nm)	Authors
Pt	20	20 W	PVP, P123, SDS, PDDA	—	2.5 ± 0.5	Shen *et al.* [48]
Pt	20	up to 118 W cm^{-2}	—	—	<200	Zin *et al.* [51]
Pd	20	—	PVP, PDDA, CTAB	—	<200	Shen *et al.* [49]
Pd	20	20–120 W cm^{-2}	CTAB	—	5–10 (depending on current densities & CTAB concentrations)	Qui *et al.* [50]

ultrasonic agitation or ball-milling to produce a PEMFC and DMFC catalyst 'ink'. This ink is either applied to (i) the gas diffusion layer (GDL) to form a gas diffusion electrode (GDE) or a catalyst-coated substrate (CCS) or (ii) the PEM for form a catalyst-coated membrane (CCM). The CCSs are usually prepared by spreading, spraying, deposition, ionomer impregnation, electrodeposition (continuous and pulsed) and sputtering, whereas the CCMs are fabricated by impregnation reduction, evaporation deposition, sputtering, dry spraying and decaling. For further details about the above methods, the reader is invited to refer to the excellent reviews on analysis of PEM fuel cell design and manufacturing by Mehta and Cooper [53] and Litster and McLean on PEM fuel cell electrodes [54].

For the first time, Pollet [55] showed that the sonoelectrochemical method can be used for preparing PEMFC electrodes whereby platinum loaded on Nafion-bonded carbon anodes were prepared in K_2PtCl_4 aqueous solutions by galvanostatic pulse electrodeposition in the absence and presence of power ultrasound (20 kHz). Pollet found that PEMFC electrodes prepared sonoelectrochemically showed better performance compared to those prepared by (i) galvanostatic pulse method only (i.e. silent conditions) and (ii) conventional methods (Figure 10.6).

10.6 Industrial Applications of the Use of Ultrasound for the Fabrication of Fuel Cell Materials

There are few industrial ultrasonic applications for the fabrication of fuel cell materials; however, this section merits some attention. Recently, a US-based company, Sono-Tek Corporation, has developed and patented an innovative thin-film ultrasonic coating technology for the fabrication of PEMFC, DMFC and SOFC electrodes (Figure 10.6) [56]. They showed that fuel cell performances are improved compared to any other industrial methods of fabrication such as screen-printing.

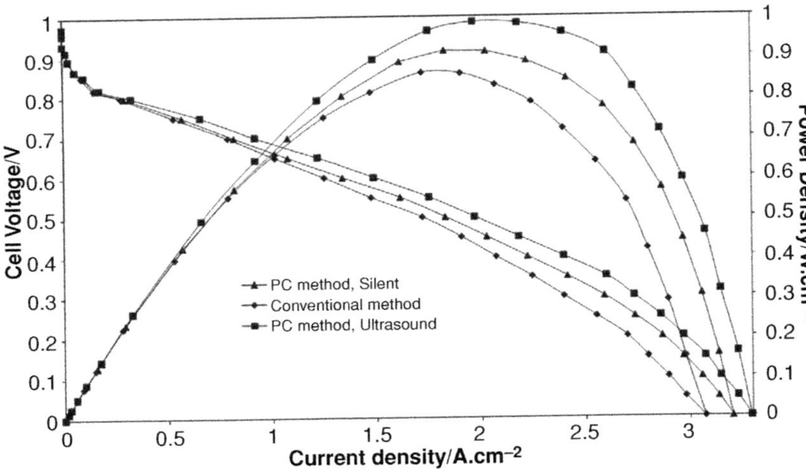

Figure 10.6 *Comparison of MEA performance between anodes prepared by: (a) the galvanostatic pulse method in the absence of ultrasound [▲], (b) the sono-galvanostatic pulse method (20 kHz, 20 W cm^{-2}) [■] and (c) conventional method (0.30 mg Pt cm^{-2} electrodes) [♦]. The fuel cell testing parameters were H_2/O_2 (1.5/2 stoics), 70 °C and 1 atm (Reprinted with permission from [55] Copyright (2009) Elsevier Ltd).*

The Sono-Tek systems incorporate one or several ultrasonic atomising nozzles, vibrating at high frequency ultrasound (up to 200 kHz) created by piezoelectric transducers inside the nozzles's titanium housing [Figure 10.7(b)]. The liquid (either a catalyst ink or a ceramic slurry) is pumped through the nozzle and is atomised into a fine mist at the nozzle tip [Figure 10.7(c)] to produce highly repeatable thin films of micron-sized droplets, with coating thicknesses from 200 nm to 50 μm [Figure 10.7(d)]. The main advantages of using the ultrasonic spray system are as follows: (i) minimal bounce-back and overspray of the liquid onto the substrate, (ii) non-clogging issues at the nozzle tip, (iii) de-agglomeration of the nanoparticles in solution due to the nozzle's high frequency vibration [Figure 10.7(e)]. Other benefits are high transfer efficiency, the ability to precisely control thickness of deposition and drop size (by varying the nozzle ultrasonic frequency), and tight drop distribution that results in highly uniform deposition. The soft and low velocity ultrasonic spray provides an excellent adherence to substrates of any geometry.

10.7 Conclusions

The ultrasonic, sonochemical and sonoelectrochemical methods for the preparation of efficient mono- and bi-metallic nanoparticles, carbon-supported electrocatalysts, fuel cell electrodes and membranes have an advantage over many other methods due to the unusual experimental conditions caused by cavitation, water sonolysis and enhanced mass transport phenomenon. It is shown in this chapter that ultrasound can be used as a 'reducing agent' to form mono- and bi-metallic catalysts of nanometre size with better catalytic properties for ORR, especially when highly dispersed (due to Van der Waal's force between nanoparticles)

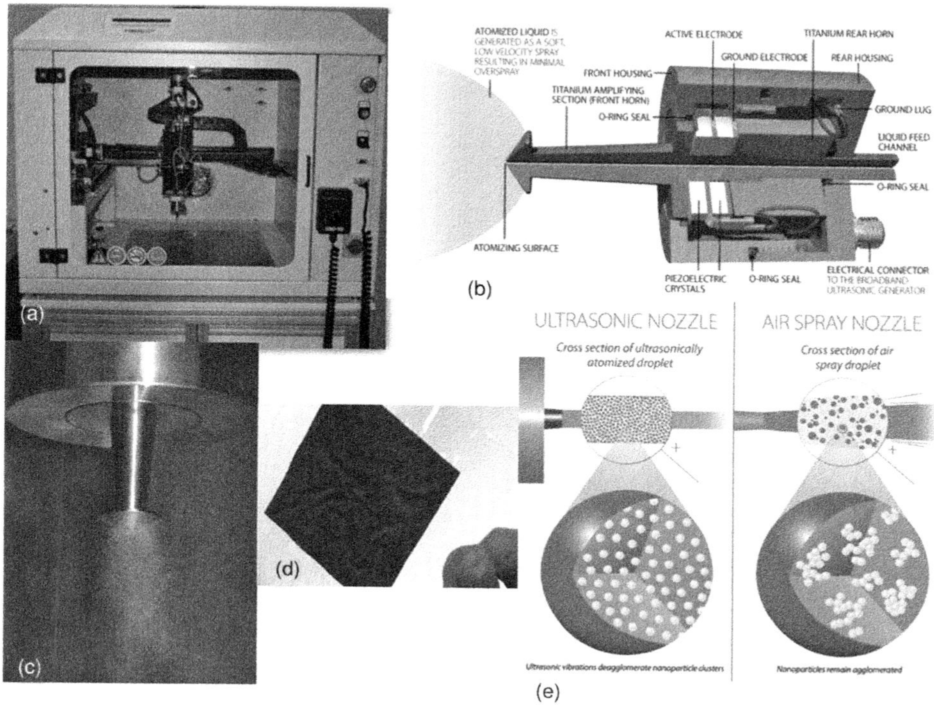

Figure 10.7 (a) – Sono-Tek Ultrasonic Spray system – 'ExactaCoat'; (b) Schematic of the vibrating nozzle cross-section; (c) Picture showing mist formation of the liquid; (d) Catalyst coated membrane (CCM) for PEMFC and DMFC fabricated by the Sono-Tek method; (e) Schematic of deagglomeration of nanoparticles by the ultrasonic spray method compared with the air spray method (Reprinted with permission from [56] Copyright (2011) Elsevier Ltd).

on carbon supports such as CNTs. Fuel cell electrodes for MCFCs and PEMFCs can also be produced by the ultrasonic and sonoelectrochemical methods where better fuel cell performances are observed. Overall, the ultrasonic method is expected to be promising in fuel cell technologies as it is cost effective, easy to use and less time consuming than any other conventional methods.

Acknowledgement

The author would like to thank Advantage West Midlands (AWM) under the Birmingham Science City Initiative and EPSRC (Contract No: EP/E034888/1).

List of Abbreviations

AEM	Anion Exchange Membrane
AFC	Alkaline Fuel Cell

AMDMFC	Alkaline Membrane Direct Methanol Fuel Cell
CCM	Catalyst Coated Membrane
CCS	Catalyst Coated Substrate
DMFC	Direct Methanol Fuel Cell
DPNs	Dendritic Platinum Nanostructures
EC	Electrocatalyst
ECSA	Electrochemical Surface Area
GDE	Gas Diffusion Electrode
GDL	Gas Diffusion Layer
HER	Hydrogen Evolution Reaction
HOR	Hydrogen Oxidation Reaction
MCFC	Molten Carbonate Fuel Cell
MEA	Membrane Electrode Assembly
MFC	Microbial Fuel Cell
MOR	Methanol Oxidation Reaction
PAFC	Phosphoric Acid Fuel Cell
ORR	Oxygen Reduction Reaction
PEM	Proton Exchange Membrane
PEMFC	Proton Exchange Membrane Fuel Cell
SOFC	Solid Oxide Fuel Cell
YSZ	Yttria-stabilised Zirconia

References

(1) Sammes, N. (ed.) (2006) *Fuel Cell Technology: Reaching Towards Commercialization*, Springer, London.
(2) Blomen, L.J.M.J. and Mugerwa, M.N. (1993) *Fuel Cell Systems*, Plenum Press, New York.
(3) Mench, M.M. (2008) *Fuel Cell Engines*, John Wiley and Sons Ltd, London.
(4) O'Hayre, R., Colella, W., Cha, S.-W. and Prinz, F.B. (2009) *Fuel Cell Fundamentals*, John Wiley and Sons Ltd, London.
(5) Harper, G.D.J. (2008) *Fuel Cell Projects for the Evil Genius*, McGraw-Hill/TAB Electronics, London.
(6) Srinivasan, S. (2006) *Fuel Cells: From Fundamentals to Applications*, Springer, New York.
(7) Cataldo, F. (1992) Effects of ultrasound on the yield of hydrogen and chlorine during electrolysis of aqueous solutions of NaCl or HCl. *J. Electroanal. Chem.*, **332**, 325–331.
(8) Walton, D.J., Burke, L.D. and Murphy, M.M. (1996) Sonoelectrochemistry: Chlorine, hydrogen and oxygen evolution at platinised platinum. *Electrochim. Acta*, **41**, 2747–2751.
(9) McMurray, H.N., Worsley, D.A. and Wilson, B.P. (1998) Hydrogen evolution and oxygen reduction at titanium sonotrode. *Chem. Commun.*, 887–888.
(10) Burdischak, C., Honsberg, C. and Opila, R.L. (2008) *Electroanalytical Effects of Ultrasound on Hydrogen Evolution Reaction in KOH*, University of Delaware.
(11) Sasikala, R., Jayakumar, O.D. and Kulshreshtha, S.K. (2007) Enhanced hydrogen generation by particles during sonochemical decomposition of water. *Ultrason. Sonochem.*, **14** (2), 153–156.
(12) Li, S.-D., Wang, C.-C. and Chen, C.-Y. (2009) Water electrolysis in the presence of an ultrasonic field. *Electrochim. Acta*, **54** (15), 3877–3883.
(13) Feldheim, D.L. and Foss, C.A. Jr (2002) *Metal Nanoparticles: Synthesis Characterization and Applications*, Marcel Dekker (ed.), New York.

(14) Gutierrez, M., Henglein, A. and Dohrmann, J.K. (1987) H atom reactions in the sonolysis of aqueous solutions. *J. Phys. Chem.*, **91**, 6687–6690.
(15) Nagata, Y., Watanabe, Y., Fujita, S., Dohmaru, T. and Taniguchi, S.J. (1992) Formation of colloidal silver in water by ultrasonic irradiation. *Chem. Soc., Chem. Commun.*, 1620–1622.
(16) Okitsu, K., Bandow, H. and Maeda, Y. (1996) Sonochemical preparation of ultrafine palladium particles. *Chem. Mater.*, **8**, 315–317.
(17) Nemancha, A., Rehspringer, J.-L. and Khatmi, D. (2006) Synthesis of palladium nanoparticles by sonochemical reduction of palladium(II) nitrate in aqueous solution. *J. Phys. Chem. B*, **110**, 383–387.
(18) Okitsu, K., Mizukoshi, Y., Bandow, H., Maeda, Y., Yamamoto, T. and Nagata Y. (1996) Formation of noble metal particles by ultrasonic irradiation. *Ultrason. Sonochem.*, **3** (3), S249–S251.
(19) Mizukoshi, Y., Oshima, R., Maeda, Y. and Nagata, Y. (1999) Preparation of platinum nanoparticles by sonochemical reduction of the Pt(II) ion. *Langmuir*, **15** (8), 2733–2737.
(20) Mizukoshi, Y., Takagi, E., Okuno, H., Oshima, R., Maeda, Y. and Nagata, Y. (2001) Preparation of platinum nanoparticles by sonochemical reduction of the Pt(IV) ions: role of surfactants. *Ultrason. Chem.*, **8**, 1–6.
(21) Caruso, R.A., Ashokkumar, M. and Grieser, F. (2000) Sonochemical formation of colloidal platinum. *Colloids Surfaces A: Physicochem. Eng. Aspects*, **169**, 219–225.
(22) Yeung, S.A., Hobson, R., Biggs, S. and Grieser, F. (1993) Formation of gold sols using ultrasound. *Chem. Soc., Chem. Commun.*, 378–379.
(23) Caruso, R.A., Ashokkumar, M. and Grieser, F. (2002) Sonochemical formation of gold sols. *Langmuir*, **18**, 7831–7836.
(24) Fujimoto, T., Terauchi, S.-Y., Umehara, H., Kojima, I. and Henderson, W. (2001) Sonochemical preparation of single-dispersion metal nanoparticles from metal salts. *Chem. Mater.*, **13**, 1057–1060.
(25) Salkar, R.A., Jeevanandam, P., Aruna, S.T., Koltypin, Y. and Gedanken, A. (1999) The sonochemical preparation of amorphous silver nanoparticles. *J. Mater. Chem.*, **9**, 1333–1335.
(26) Radziuk, D., Mohwald, H. and Shchukin, D. (2008) Ultrasonic activation of platinum catalysts. *J. Phys. Chem. C*, **112**, 19257–19262.
(27) Wu, C., Mosher, B.P. and Zeng, T. (2006) Rapid synthesis of gold and platinum nanoparticles using metal displacement reduction with sonomechanical assistance. *Chem. Mater.*, **18**, 2925–2928.
(28) Okitsu, K., Ashokkumar, M. and Grieser, F. (2005) Sonochemical synthesis of gold nanoparticles: effects of ultrasound frequency. *J. Phys. Chem. B*, **109**, 20673–20675.
(29) He, Y., Vinodgopal, K., Ashokkumar, M. and Griese, F. (2006) Sonochemical synthesis of ruthenium nanoparticles. *Res. Chem. Int.*, **32** (8), 709–715.
(30) Basnayake, R., Li, Z., Katar, S., Zhou, W., Rivera, H., Smotkin, E.S., Casadonte, D.J. and Korzeniewski, C. (2006) PtRu nanoparticle electrocatalyst with bulk alloy properties prepared through a sonochemical method. *Langmuir*, **22** (25), 10446–10450.
(31) Vinodgopal, K., He, Y., Ashokkumar, M. and Grieser, F. (2006) Sonochemically prepared platinum-ruthenium bimetallic nanoparticles. *J. Phys. Chem. B*, **110** (9), 3849–3852.
(32) Kim, J., Park, J.-E., Momma, T. and Osaka, T. (2009) Synthesis of Pd–Sn nanoparticles by ultrasonic irradiation and their electrocatalytic activity for oxygen reduction. *Electrochim. Acta*, **54** (12), 3412–3418.
(33) Takatani, H., Kago, H., Nakanishi, M., Kobayashi, Y., Hori, F. and Oshima, R. (2003) Characterisation of noble metal alloy nanoparticles prepared by ultrasound irradiation. *Rev. Adv. Mater. Sci.*, **5**, 232–238.
(34) Nakanishi, M., Takatani, H., Kobayashi, Y., Hori, F., Taniguchi, R., Iwase, A. and Oshima, R. (2005) Characterization of binary gold/platinum nanoparticles prepared by sonochemistry technique. *App. Surf. Sci.*, **241**, 209–212.

(35) Anandan, S., Grieser, F. and Ashokkumar, M. (2008) Sonochemical synthesis of au-ag core-shell bimetallic nanoparticles. *J. Phys. Chem. C*, **112** (39), 15102–15105.

(36) Kumar, P.S.S., Manivel, A., Anandan, S., Zhou, M., Grieser, F. and Ashokkumar, M. (2010) Sonochemical synthesis and characterization of gold–ruthenium bimetallic nanoparticles. *Colloids and Surfaces A: Physicochem. Eng. Aspects*, **356** (1–3), 140–144.

(37) Mizukoshi, Y., Okitsu, K., Maeda, Y., Yamamoto, T.A., Oshima, R. and Nagata, Y. (1997) Sonochemical preparation of bimetallic nanoparticles of gold/palladium in aqueous solution. *J. Phys. Chem. B*, **101** (36), 7033–7037.

(38) Mizukoshi, Y., Fujimoto, T., Nagata, Y., Oshima, R. and Maeda, Y. (2000) Characterization and catalytic activity of core-shell structured gold/palladium bimetallic nanoparticles synthesized by the sonochemical method. *J. Phys. Chem. B*, **104**, 6028–6032.

(39) Kan, C., Cai1, W., Li, C., Zhang, L. and Hofmeister, H. (2003) Ultrasonic synthesis and optical properties of Au/Pd bimetallic nanoparticles in ethylene glycol. *J. Phys. D: Appl. Phys.*, **36**, 1609–1614.

(40) Das, N., Bhattacharya, D., Sen, A. and Maiti, H.S. (2009) Sonochemical synthesis of $LaMnO_3$ nano-powder. *Ceramics Int.*, **35**, 21–24.

(41) Pang, G., Xu, X., Markovich, V., Avivi, S., Palchik, O., Koltypin, Y., Gorodetsky, G., Yeshurun, Y., Buchkremer, H.P. and Gedanken, A. (2003) Preparaion of $La_{1-x}Sr_xMnO_3$ nanoparticles by sonication-assisted co-precipitation. *Mater. Res. Bull.*, **38**, 11–16.

(42) Michel, C.R., Lopez, E.R. and Zea, H.R. (2006) Synthesis of $GdCo_{1-x}Cu_xO_{3-\delta}$ ($x = 0$, 0.15, 0.30) perovskites by ultrasonic spray pyrolysis. *Mater. Res. Bull.*, **41**, 209–216.

(43) Pollet, B.G. and Hihn, J.-Y. (2011) Sonoelectrochemistry – from theory to applications, Chapter 24, in *Handbook on Applications of Ultrasound – Sonochemistry for Sustainability* (eds D. Chen, S.K. Sharma and A. Mudhoo), CRC Press, pp. 623–657.

(44) Jiang, L.-P., Wang, A.-N., Zhao, Y., Zhang, J.-R. and Zhu, J.-J. (2004) Novel route for the preparation of monodisperse silver nanoparticles via a pulsed sonoelectrochemical technique, *Inorg. Chem. Commum.*, **7**, 506–509.

(45) Aqil, A., Serwas, H., Delplancke, J.L., Jérôme, R., Jérôme, C. and Canet, L. (2008) Preparation of stable suspensions of gold nanoparticles in water by sonoelectrochemistry. *Ultrason. Chem.*, **15**, 1055–1061.

(46) Vu, L.V., Long, N.N., Doanh, S.C. and Trung, B.Q. (2009) Preparation of silver nanoparticles by puls sonoelectrochemical method and studying their characteristics. *Journal of Physics: Conference Series*, **187**, 012077.

(47) Tang, S., Meng, X., Lu, H. and Zhu, S. (2009) PVP-assisted sonoelectrochemical growth of silver nanostructures with various shapes. *Mater. Chem. Phys.*, **116**, 464–468.

(48) Shen, Q., Jiang, L., Zhang, H., Min, Q., Hou, W. and Zhu, J.-J. (2008) Three-dimensional dendritic Pt nanostructures: Sonoelectrochemical synthesis and electrochemical applications. *J. Phys. Chem. C*, **112**, 16385–16392.

(49) Shen, Q., Min, Q., Shi, J., Jiang, L., Zhang, J.-R., Hou, W. and Zhu, J.-J. (2009) Morphology-controlled synthesis of palladium nanostructures by sonoelectrochemical method and their application in direct alcohol oxidation. *J. Phys. Chem. C*, (113), 1267–1273.

(50) Qiu, X.-F., Xu, J.-Z., Zhu, J.-M., Zhu, J.-J., Xu, S. and Chen, H.Y. (2003) Controllable synthesis of palladium nanoparticles via a simple sonoelectrochemical method. *J. Mater. Res.*, **18** (6), 1399–1404.

(51) Zin, V., Pollet, B.G. and Dabalà, M. (2009) Sonoelectrochemical (20 kHz) production of platinum nanoparticles from aqueous solutions. *Electrochim. Acta*, **54**, 7201–7206.

(52) Argirusis, Chr., Matic, S. and Schneider, O. (2008) An EQCM study of ultrasonically assisted electrodeposition of Co/CeO_2 and Ni/CeO_2 composites for fuel cell applications. *Phys. Stat. Sol. (A)*, **205** (10), 2400–2404.

(53) Mehta, V. and Cooper, J.S. (2003) Review and analysis of PEM fuel cell design and manufacturing. *J. Power Sources*, **114** (1), 32–53.
(54) Litster, S. and McLean, G. (2004) PEM fuel cell electrodes. *J. Power Sources*, **130**, 61–76.
(55) Pollet, B.G. (2009) A novel method for preparing PEMFC electrodes by the ultrasonic and sonoelectrochemical techniques. *Electrochem. Commun.*, **11**, 1445–1448.
(56) Millington, B.A., Whipple, V. and Pollet, B.G. (2011) A novel method for preparing proton exchange membrane fuel cell electrodes by the ultrasonic-spray technique. *J. Power Sources*, **196** (20), 8500–8508.

Appendix: Sonochemical Effects on Electrode Kinetics

Several workers have shown that ultrasound enhances mass-transport processes and electrode potentials [1]. Compton *et al.* [2–6] have observed sigmoidal shape voltammograms and named these new shapes 'sonovoltammograms' for quasi-reversible systems [see Figure 1.8, Chapter 1]. They concluded that sonovoltammetry is closely related to hydrodyamics methods, for example, polarography, dropping mercury electrode (DPE), rotating disc electrode (RDE), wall-jet electrode (WJE) and showed that Equation A.1 [2–6], which relates the limiting current (I_{lim}) to the diffusion layer thickness (δ), still applies in the presence of ultrasound.

$$I_{\text{lim}} = \frac{nF\text{A}D_{\text{o}}C^*}{\delta} \qquad \text{(A.1)}$$

where I_{lim} is the limiting current, n is the number of electrons transferred during the electrochemical process, F is the Faraday constant, A is the electrode area, D_{o} is the diffusion coefficient of the electroactive species, C^* is the bulk concentration of the electroactive species and δ is the diffusion layer thickness.

In contrast to any conventional hydrodynamics responses (e.g. S-shaped voltammograms by RDE), sonovoltammograms usually exhibit 'noisy' responses or a series of individual current pulses responses in the mass-transport region of the voltammograms (see Figure 1.8, Chapter 1) corresponding to an intense transient cavitation in the vicinity of the electrode surface caused by individual collapse of cavitation bubbles.

Compton *et al.* also demonstrated that under sonication, millielectrodes (i.e. electrodes with radii larger than 1 mm) behave as microelectrodes (i.e. radii smaller than 1 mm) [6]. They found that limiting currents obtained with a microelectrode in the absence of ultrasound were comparable with those obtained with a millielectrode in the presence of ultrasound. However, controversy exists in identifying if the steady-state voltammogram obtained in the presence of ultrasound is due only to cavitation effects or to micro-jetting and acoustic streaming or to a combination of the three.

An early report by Pollet and Hihn [7] showed that, for a 'face-on geometry' in the presence of ultrasound, the limiting current (I_{lim}) is inversely proportional to the horn–electrode

Power Ultrasound in Electrochemistry: From Versatile Laboratory Tool to Engineering Solution, First Edition. Edited by Bruno G. Pollet.

distance (d), the electrode geometry (r_e) and proportional to the ultrasonic intensity (Ψ) [at constant temperature, concentration, etc.] according to Equation A.2, also known as the **Pollet equation**:

$$I_{\mathrm{lim}} = 0.84nFAD_o^{2/3}\nu^{-1/6}r_e^{-0.5}d^{-0.5}\Psi^{0.5}C^* \quad \text{(A.2)}$$

where I_{lim} is the limiting current, n is the number of electrons transferred during the electrochemical process, F is the Faraday constant, A is the electrode area, D_o is the diffusion coefficient of the electroactive species, d is the horn–electrode distance, r_e is the electrode radius, ν is the kinematic viscosity and C^* is the bulk concentration of the electroactive species.

They showed that the electrode–ultrasonic horn positioning is an important factor in assigning hydrodynamics parameters such as the limiting current (I_{lim}), the mass transport coefficient (m_o), the limiting solution velocity (U_{lim}), the Sherwood (Sh) number and the Reynolds (Re) number.

Pollet *et al.* [8] also showed that the geometry of the sonoelectrochemical set up affects mass-transport phenomenon. For example, Pollet investigated the effect of changing the position of an ultrasonic horn tip (i.e. vertical and horizontal) and a cylinder electrode (CE) on limiting currents (for the reduction of Ag^I in sodium thiosulfate), in an attempt to find the optimum position required for maximum sonoelectrochemical effect. The importance of the ultrasonic intensity, the electrode–horn distance and positioning (angle) in assigning limiting currents was also investigated. For the cylinder electrode (CE) placed at an angle of 45° with respect to the ultrasonic horn, it was suggested that the 50% increase in limiting current for the 'face-on' geometry is caused by an approximately 50% decrease in diffusion layer thickness for the 'face-on' geometry compared to the 'angular' geometry due to the difference in the sonicated areas for both geometries (see Figure A.1).

Figure 1(c) shows the limiting currents plotted against the square root of the emitted ultrasonic intensity ($\Psi^{1/2}$) for both the 'face-on' and 'angular' geometry. This figure shows linear relationships between the limiting current and $\Psi^{1/2}$ for the two electrode configurations. This observation suggests, as for the disc electrode (DE), that limiting currents obtained on a sonicated (CE) are also related to $\Psi^{1/2}$. In other words, whatever the positioning of the electrode, a square root dependence of ultrasonic intensity relationship is obtained for the two geometries employed.

The figure also shows that, for an identical ultrasonic intensity transmitted to the total volume, the limiting current obtained for the 'angular' geometry is much lower than that obtained for the 'face-on' geometry. For example, at 30 W cm^{-2} power output, the limiting current obtained for the 'face-on' geometry (I_{lim} = 150.4 mA) is 1.5 times higher than that obtained for the 'angular' geometry (I_{lim} = 101.5 mA). Furthermore, in our conditions, this would correspond to an equivalent 'effective rotation speed' of 15 000 rpm and 8500 rpm for the 'face-on' and 'angular' geometries respectively at maximum power (30 W cm^{-2}). These observations indicate that the positioning of the electrode with respect to the source of ultrasound is an important parameter in assigning ultrasonic hydrodynamics.

To account for the difference between the 'face-on' and 'angular' geometries, there are three possible explanations, which are detailed below. Please note that the following subscripts are used for the ultrasonic intensity (Ψ): x and y are spacial components,

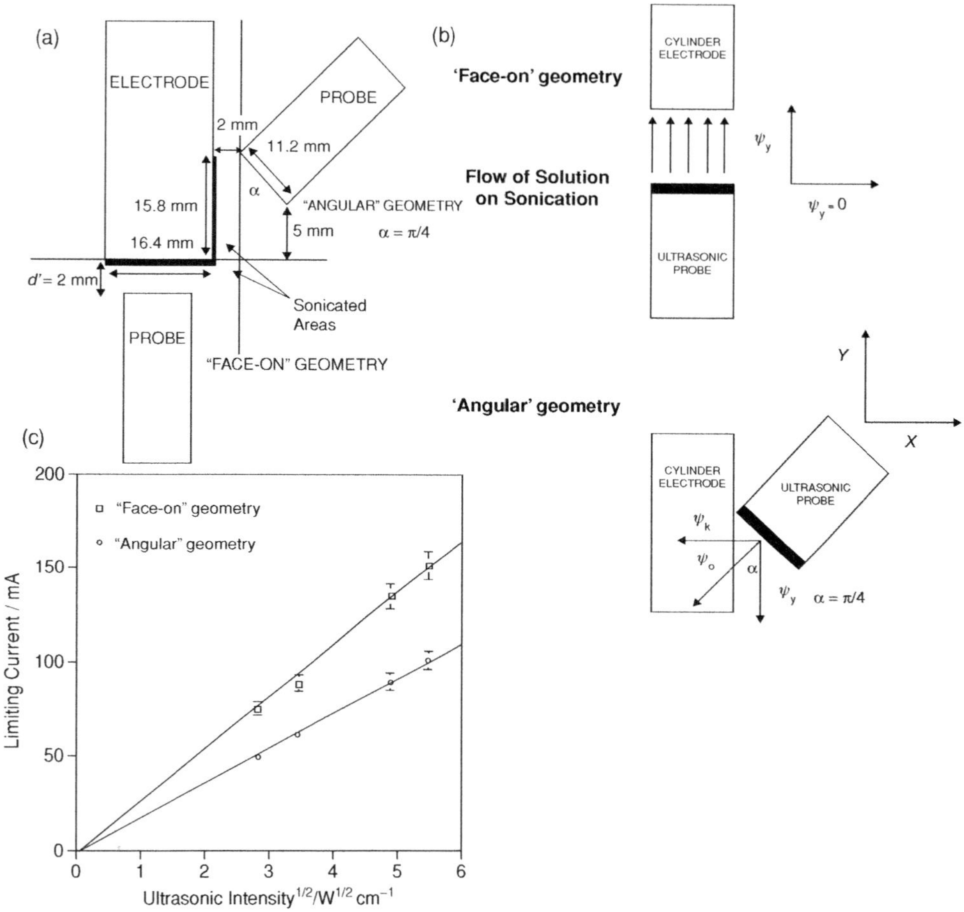

Figure A.1 (a) Two geometries employed during the study – the face-on' and 'angular' geometries. The figure shows the two sonicated areas for both geometries. (b) Co-ordinate system used to describe the 'face-on' geometry and 'angular' geometry in terms of ultrasonic intensity. (c) Limiting currents plotted against the square root of ultrasonic intensity (20 kHz) for the reduction of silver (4 g dm^{-3}) on a sonicated cylinder electrode at 298 (±1) K for the face-on' and 'angular' geometries.

0 indicates the ultrasonic intensity at $x = 0$ or/and $y = 0$, f.o. the 'face-on' geometry and ang. the 'angular' geometry.

(i) *Influence of the incident flow* – The first possible explanation is to consider the two geometries and resolve the flows into the component parts [see Figure A.1(a,b)]. For example, it is assumed that the total ultrasonic intensity, Ψ, is given by Equation A.3:

$$\Psi = \Psi_x + \Psi_y \qquad \text{(A.3)}$$

It has been also demonstrated that the ultrasonic intensity within a fluid depends on distance d'' travelled thought that fluid with an absorption coefficient α, according to:

$$\Psi = \Psi_0 \exp(-2\alpha d'') \tag{A.4}$$

Using 8.6×10^{-8} cm^{-1} for the absorption coefficient of water at 298 K and at 20 kHz, Equation A.4 leads to $\Psi = \Psi_0$ (incident ultrasonic intensity) in the range 2 and 10 mm electrode–probe distances. Thus, if the acoustic absorption is neglected, Equation A.4 may be reduced to:

$$\Psi_0 = \Psi_{0,x} + \Psi_{0,y} \tag{A.5}$$

For the 'face-on' geometry (see Figure A.1), the ultrasonic intensity components are in the x and y directions, that is, $\Psi_{0,x,\mathrm{f.o.}}$ $(= 0)$ and $\Psi_{0,y,\mathrm{f.o.}}$ $(= \Psi_0)$ respectively.

For the 'angular' geometry [see Figure A.1(b)], the ultrasonic intensity components are in the x and y directions, that is, $\Psi_{0,x,\mathrm{ang.}} = \Psi_0 \sin(45°) = 0.707\ \Psi_0$ and $\Psi_{0,y,\mathrm{s.o.}} = \Psi_0 \cos(45°) = 0.707\ \Psi_0$ respectively.

The ratio of the ultrasonic intensity on the y-axis for the two geometries gives

$$\frac{\Psi_{\mathrm{o},y,\mathrm{f.o.}}}{\Psi_{\mathrm{o},y,\mathrm{ang.}}} = \frac{\Psi_0}{0.707\Psi_0} = 1.41 \tag{A.6}$$

Assuming that the limiting current (I_{lim}) is proportional to the square root of the ultrasonic intensity ($\Psi_0{}^{1/2}$) [see Figure A.1(c)], Equation A.6 may be reduced to:

$$\frac{I_{\mathrm{lim,f.o.}}}{I_{\mathrm{lim,ang.}}} = \sqrt{1.41} = 1.18 \tag{A.7}$$

Whilst the value of 1.18 can account in part for some of the observed 1.5 value for $I_{\mathrm{lim,f.o.}}/I_{\mathrm{lim,ang.}}$, it cannot account for it all and again suggests that the ultrasonic probe positioning is crucial for the determination of limiting currents.

For both cases, 'face-on' and 'angular' geometries, the current density is variable over the total electrode area and therefore simple mathematical treatments and equations have obvious limitations. It is interesting to note that the electrode area in both cases may be regarded as two areas: (i) a high current density area and (ii) a low current density area. Ideally, it would be useful to treat the resulting system using Equation A.6 containing two terms at high current density and low current density. However, low current density area is difficult to evaluate because of the complexity of the cylinder electrode and the 'angular' geometries.

(ii) *Influence of hydrodynamics on the diffusion layer thickness* – This difference in limiting current between the two geometries may be attributed to a greater thinning of the diffusion layer in the case of the 'face-on' geometry where electroactive species reaches all points of the electrode uniformly. In the 'angular' geometry, the diffusion layer is non-uniform and roughly resembles that of a channel electrode. Further, it is interesting to note that the sonicated areas for the two geometries are different [see Figure A.1(a)]. It was deduced that the sonicated area for the 'face-on' geometry is 2.13 cm^2 (CE base area) and for the 'angular' geometry is 2.65 cm^2 (ellipsoid area). In order to compare diffusion layer thicknesses for both 'face-on' and 'angular' geometries, a hydrodynamic study can be made as follows.

For the 'face-on' geometry, the limiting current is given by

$$I_{\mathrm{lim,f.o.}} = \frac{nFA_{\mathrm{f.o.}}DC^*}{\delta_{\mathrm{f.o.}}} \quad \text{(A.8)}$$

and for the 'angular' geometry, the limiting current is given by

$$I_{\mathrm{lim,ang.}} = \frac{nFA_{\mathrm{ang.}}DC^*}{\delta_{\mathrm{ang.}}} \quad \text{(A.9)}$$

provided that the non-sonicated areas of the electrode for both the 'face-on' and 'angular' geometries are not considered.

The ratio of the limiting currents for the two geometries gives

$$\frac{I_{\mathrm{lim,f.o.}}}{I_{\mathrm{ang.}}} = \frac{A_{\mathrm{f.o.}}\delta_{\mathrm{ang.}}}{A_{\mathrm{ang.}}\delta_{\mathrm{f.o.}}} \quad \text{(A.10)}$$

[Assuming that the diffusion coefficients (D) are similar for both geometries.]

Inserting experimental values of limiting current density and surface area for both geometries in Equation A.10, a ratio of diffusion layer thicknesses at 30 W $\mathrm{cm^{-2}}$ power output may be deduced:

$$\frac{\delta_{\mathrm{ang.}}}{\delta_{\mathrm{f.o.}}} = \frac{I_{\mathrm{lim,f.o.}}A_{\mathrm{ang.}}}{I_{\mathrm{lim,ang.}}A_{\mathrm{f.o.}}} = \frac{150.4 \times 2.65}{101.5 \times 2.13} = 1.84 \quad \text{(A.11)}$$

Thus, the overall 'average' diffusion layer thickness is 1.8 times thinner for a 'face-on' geometry than a 'angular' geometry under identical conditions.

(iii) *Influence of localised temperature variations on diffusion coefficients* – The observed difference in limiting current between the two geometries may also be attributed to a change in diffusion coefficients due to localised temperature variations. Assuming that the diffusion coefficient for the 'face-on' ($D_{0,\mathrm{f.o.}}$) and the 'angular' ($D_{0,\mathrm{ang.}}$) geometries are different and assuming that the diffusion layer for the 'face-on' geometry is 'hotter' than that for the 'angular' geometry [for example, 400 K for the 'face-on' geometry (value deduced in Lorimer *et al.* [9]) compared to 298 K for the 'angular' geometry], it is possible to deduce the ratio of the diffusion coefficients of silver for the two geometries at maximum power (30 W $\mathrm{cm^{-2}}$), that is

$$\frac{D_{0,\mathrm{f.o.}}}{D_{0,\mathrm{ang.}}} = \frac{I_{\mathrm{lim,f.o.}}}{I_{\mathrm{lim,ang.}}} = \frac{6.17 \times 10^{-5}}{1.64 \times 10^{-5}} = 3.76 \quad \text{(A.12)}$$

Since, in our conditions, the ratio of the limiting currents for the two geometries for 30 W $\mathrm{cm^{-2}}$ power output is

$$\frac{I_{\mathrm{lim,f.o.}}}{I_{\mathrm{lim,ang.}}} = 1.5 \quad \text{(A.13)}$$

this suggestion may be dismissed.

[The diffusion coefficient of silver at 400 K was deduced by assuming linearity of the plot of ln(D) versus $1/T$. The equation of the line was in the form of $\ln(D) = -1553/T - 5.8$. Values of D were obtained from the rotating disc experiments at 298, 310.5 and 323 K].

However, it is possible to deduce an 'apparent' temperature by deducing an 'apparent' diffusion coefficient for the 'face-on' geometry. Assuming that the ratio of the limiting currents for both geometries is true (Equation A.13), a value of the 'apparent' diffusion coefficient may be deduced employing Equation A.12:

$$D_{\text{f.o.}} = \frac{I_{\text{lim,f.o.}}}{I_{\text{lim,ang.}}} D_{\text{ang.}} = 1.5 \times 1.62 \times 10^{-5} = 2.43 \times 10^{-5} \text{cm}^2 \text{s}^{-1} \tag{A.14}$$

value of $D = 1.62 \times 10^{-5}$ cm^2 s^{-1} at 298 K.

Thus, by assuming linearity between ln(D) versus 1/T, the 'apparent' temperature was deduced to be approximately 350 K. This result suggests that if the 1.5-increase in limiting current for the 'face-on' geometry compared to the 'angular' geometry was solely due to an increase in diffusion coefficient for the 'face-on' geometry, this would lead to a bulk temperature of approximately 350 K, a value which was never achieved under sonication at 298 K. However, it is interesting to note that this result compares favourably with the temperature of the diffusion layer for the 'face-on' geometry is 400 K (value deduced in Lorimer *et al.* [9, 10]).

Apart from the remarkable effects of ultrasound on mass-transport processes, it has been shown that sonication alters the electrode potential. Several workers such as Morigushi [11] and Pollet *et al.* [12] showed that the overpotential of hydrogen evolution on gold and platinum electrodes decreases in a sonicated environment. This finding was later observed by the Compton group on other metals a few years ago [13]. Pollet *et al.* speculated that this decrease in overpotential was due to both a reduction in concentration gradients and in nucleation overpotential at the electrode surface and in the removal of adhered hydrogen bubbles and adsorbed materials. Even now, some controversy still surrounds the effect of ultrasound upon electrode kinetic parameters such as the formal (E^0) and half-wave ($E_{1/2}$) potentials and apparent heterogeneous rate constant (k_0), which are related to equilibrium potentials.

However, a detailed study by Pollet *et al.* in 1998 [14] showed that ultrasound affects heterogeneous electron-transfer kinetic parameters of a typical quasi-reversible system providing that the electrode receives a strict cleaning procedure (for surface reproducibility) prior to any sonoelectrochemical experiments. It was shown that the half-wave potential ($E_{1/2}$) shifts cathodically and the apparent heterogeneous rate constant (k_0) increases with increasing ultrasonic intensity. Also, it was shown that these parameters appear to be little affected by the frequency of simultaneous ultrasonic irradiation in the range 20–800 kHz, and is not influenced by choice of ultrasonic bath or probe as sonic source, provided measurements are made at constant ultrasonic intensity.

All these aspects at low and high ultrasonic frequencies have been reviewed in 2003 by Compton *et al.* [6] and recently by Pollet and Hihn [15]. The authors concluded that the electrochemical processes in the presence of high-frequency ultrasound are governed by processes (microjetting and micromixing) considerably different from those that are important at lower frequencies (acoustic streaming).

References

(1) González-García, J., Esclapez, M.D., Bonete, P., Hernández, Y.V., Garretón, L.G. and Sáez, V. 2010. Current topics on sonoelectrochemistry, *Ultrasonics*. **50**: 318–322.

(2) Compton, R.G., Eklund, J.C., Page, S.D., Sanders, G.H.W. and Booth, J. 1994. Voltammetry in the presence of ultrasound. Sonovoltammetry and surface effects. *Journal of Physical Chemistry*, **98**: 12410–12414.

(3) Compton, R.G., Eklund, J.C. and Page, S.D. 1995. Sonovoltammetry – heterogeneous electron transfer processes with coupled ultrasonically induced chemical reaction – The sono-EC reaction. *Journal of Physical Chemistry*, **99**: 4211–4214.
(4) Compton, R.G., Eklund, J.C. and Marken, F. 1997. Sonoelectrochemical processes: A review. *Electroanalysis*, **9**: 509–522.
(5) Compton, R.G., Eklund, J.C., Marken, F., Rebitt, T.O., Akkermans R.P. and Waller, D.N. 1997. Dual activation: coupling ultrasound to electrochemistry – An overview. *Electrochimica Acta*, **42**: 2919–2927.
(6) Compton, R.G., Hardcastle, J.L. and del Campo, J. 2003. Sonoelectrochemistry, Physical Aspects, in: Bard-Stratmann (Ed.), *Encyclopedia of Electrochemistry*, in: Pat Unwin (Ed.), Instrumentation and Electrochemical Chemistry, vol. 3: pp. 312–327.
(7) Pollet, B.G., Hihn, J.Y., Doche, M.L., Lorimer, J.P., Mandroyan A. and Mason, T.J. 2007. Transport limited current close to an ultrasonic horn: equivalent flow velocity determination. *Journal of Electrochemical Society*, **154**: E131–R138.
(8) Pollet, B.G., Lorimer, J.P., Phull, S.S., Mason, T.J. and Hihn, J.Y. 2003. A novel angular geometry for the sonochemical silver recovery process at cylinder electrodes. *Ultrasonic Sonochemistry*, **10**: 217–222.
(9) Lorimer, J.P., Pollet, B., Phull, S.S., Mason, T.J. and Walton, D.J. 1998. Disc and cylindrical electrode with ultrasound. *Electrochimica Acta*, **43**: 449–455.
(10) Lorimer, J.P., Pollet, B., Phull, S.S., Mason, T.J., Walton, D.J. and Geissler, U. 1996. The effect of ultrasonic frequency and intensity upon limiting currents at rotating disc and stationary electrodes. *Electrochimica Acta*, **41**: 2737–2741.
(11) Moriguchi, N. 1934, The influence of supersonic waves on chemical phenomena. III – the influence on the concentration polarisation. *Journal of Chemical Society Japan*, **55**: 349.
(12) Pollet, B.G., Hihn, J.Y., Lorimer, J.P., Phull, S.S., Mason, T.J. and Walton, D.J. 2002. The effect of ultrasound upon the oxidation of thiosulphate on stainless steel and platinum electrodes. *Ultrasonic Sonochemistry*, **9**: 267–274.
(13) Hyde, M.E., Compton, R.G. 2002. How ultrasound influences the electrodeposition of metals. *Journal of Electroanalytical Chemistry*, **531**: 19–24.
(14) Pollet, B.G. 1998. The effect of ultrasound upon electrochemical processes, PhD Thesis. Coventry University.
(15) Pollet, B.G. and Hihn, J.-Y. 2012. *Sonoelectrochemistry: from Theory to Applications*, in: Chen, D., Sharma, S.K. and Mudhoo, A. (Eds), Handbook on Applications of Ultrasound: Sonochemistry for Sustainability, CRC Press, USA.

Index

Power Ultrasound in Electrochemistry: From Versatile Laboratory Tool to Engineering Solution, First Edition. Edited by Bruno G. Pollet.

Printed and bound by CPI Group (UK) Ltd, Croydon, CR0 4YY
17/09/2026
14964179-0001